Thiol-X Chemistries in Polymer and Materials Science

**RSC Polymer Chemistry Series**



*Titles in the Series:*

1: Renewable Resources for Functional Polymers and Biomaterials
2: Molecular Design and Applications of Photofunctional Polymers and Materials
3: Functional Polymers for Nanomedicine
4: Fundamentals of Controlled/Living Radical Polymerization
5: Healable Polymer Systems
6: Thiol-X Chemistries in Polymer and Materials Science

*How to obtain future titles on publication:*
A standing order plan is available for this series. A standing order will bring delivery of each new volume immediately on publication.

*For further information please contact:*
Book Sales Department, Royal Society of Chemistry, Thomas Graham House, Science Park, Milton Road, Cambridge, CB4 0WF, UK
Telephone: +44 (0)1223 420066, Fax: +44 (0)1223 420247
Email: booksales@rsc.org
Visit our website at www.rsc.org/books

# *Thiol-X Chemistries in Polymer and Materials Science*

Edited by

**Andrew B. Lowe**
*University of New South Wales, Sydney, Australia*
*Email: a.lowe@unsw.edu.au*

**Christopher N. Bowman**
*University of Colorado, USA*
*Email: Christopher.bowman@colorado.edu*

RSC Publishing

RSC Polymer Chemistry Series No. 6

ISBN: 978-1-84973-660-2
ISSN: 2044-0790

A catalogue record for this book is available from the British Library



Published by The Royal Society of Chemistry,
Thomas Graham House, Science Park, Milton Road,
Cambridge CB4 0WF, UK

Registered Charity Number 207890

For further information see our web site at www.rsc.org

Printed in the United Kingdom by CPI Group (UK) Ltd, Croydon, CR0 4YY, UK

PREFACE

# *Thiol-X Reactions: Click Reaction Mechanisms and Applications*

## Introduction

In order for chemical reactions to gain traction in new processes for polymerizations, synthesis, and polymer modifications it is desirable that they be simple, efficient, highly selective, yield a single product, and occur under mild and preferably ambient conditions. Such 'Click' chemistries have an incredible potential to achieve significant utilization and advancement in all basic disciplines of science and engineering but particularly in materials and polymer science. Among the toolbox of 'Click' chemistries, the thiol-X reaction family's utility for synthesis, polymerizations and functionalization are currently a topic of immense interest. Since thiols react to high yields under benign conditions with a vast range of other chemical functional groups, their utility extends to a vast number of applications in the chemical, biological, physical, materials and engineering fields. While in general the thiol-X reaction chemistries are not new, many have only recently been recognized and exploited for their exceptional qualities in the polymer/materials field. As such, the thiol-click chemistry is rapidly becoming heavily utilized and has the potential for enabling numerous advancements in polymerizations, polymer synthesis, and polymer modification. In this book, a wide range of various thiol-X chemistries is highlighted over a range of topics for which they have been successfully used. Research spanning reaction chemistries such as the ubiquitous thiol-ene, thiol Michael addition, thiol-isocyanate and thiol-halogen reactions are all discussed along with applications of these materials in systems ranging from tissue

RSC Polymer Chemistry Series No. 6
Thiol-X Chemistries in Polymer and Materials Science
Edited by Andrew B. Lowe and Christopher N. Bowman

Published by the Royal Society of Chemistry, www.rsc.org

engineering matrices to surface modification and their implementation in reactions involving monomers derived from naturally occurring substrates.

## Book Summary

The book begins with Chapter 1 by Bowman and coworkers which begins the discussion of specific applications of thiol-X chemistries in polymer and materials science, particularly the radical-mediated thiol reactions involving both vinyl and alkyne functional groups. This chapter highlights the thiol-ene and thiol-yne chemistry, particularly the polymerization reaction, as it is used in ideal network synthesis. The chapter begins with a general discussion and applications of such network materials followed by a more detailed account of the mechanism and kinetics associated with the thiol-ene reaction including its functional group tolerance. A similar section follows that is focused on the thiol-yne chemistry. Subsequently, the material properties of thiol-ene/yne network materials are discussed followed by sections on applications and issues associated with odor, shelf-life, and oxidative and photostability.

Chapter 2 by Lowe *et al.* details the use of the suite of thiol-X chemistries for the preparation of end-functional (co)polymers prepared *via* reversible addition fragmentation chain transfer (RAFT) radical polymerization. The chapter begins with an introduction to the RAFT mechanism in radical polymerization (general features and mechanism) and includes methods for effecting removal of thiocarbonylthio groups from (co)polymers. This ability to reduce the typical RAFT end groups to thiols renders this mechanism of polymer formation ideal for coupling with thiol-X reactions. The introduction is followed by a short discussion on the possible thiol-based chemistries than can, and have, been employed with thiol-terminated RAFT (co)polymers and some of the issues associated with conducting such reactions with macromolecular thiols effectively. The remainder of the chapter discusses recent examples of thiol-X chemistry that have been employed to modify RAFT polymers with an emphasis on end-group functionalization. This discussion includes highlights of thiol-Michael, thiol-yne, thiol-isocyanate, thiol-epoxide, thiol-halogen as well as thiol-thiol coupling reactions. The chapter concludes with a summary and outlook for these reactions.

Shipp reviews the use of thiol-X chemistries in the synthesis of degradable polymers in Chapter 3. In particular, thiol-ene and thiol-Michael chemistries have been widely used because of their ability to be high yielding, orthogonal and functional group tolerant. They also exhibit the capacity to be conducted under physiological conditions, thus allowing them to be utilized *in vivo*. Furthermore, a wide range of thiol-containing systems, such as cysteine residues in peptides, can be incorporated into the polymeric material, thereby providing functionality that may be specifically tailored, *e.g.* for targeted delivery, degradation, adhesion or cell growth. Degradation of such polymers may be achieved through a number of bond scission processes, including hydrolysis (*e.g.* of ester linkages) but may also be catalyzed by enzymes. Continued interest in the development of new materials for tissue engineering and the delivery of therapeutics is bound to further the involvement of thiol-X chemistries, and should see them become mainstays in the production of biomedical devices in the near future.

Chapter 4 by Roth and Theato focuses on a specific functionalization method for (co)polymers employing simple thiol-thiol coupling reactions resulting in disulfide linkages with an emphasis on the reaction of thiols with thiosulfonates yielding asymmetric disulfides. After the brief introduction, the chapter highlights the reactivity and synthesis of small molecule as well as polymeric functional thiosulfonates. This introduction is followed by a discussion on the modification of RAFT-prepared (co)polymers that have been cleaved to the corresponding macromolecular thiols and is complementary to the discussion in Chapter 2 on the modification of such macromolecular thiols. Aside from the straightforward end-group functionalization, the authors also highlight tandem end- and side-group functional approaches employing a combination of thiosulfonate chemistry and nucleophilic acyl substitution of pentafluorophenyl (meth)acrylate (co)polymers as well the preparation of self-assembled monolayers prepared with methyl methanethiosulfonate.

Jones and Haddleton highlight the use of the thiol-Michael conjugate addition reaction in Chapter 5 as a method for the functionalization of polymers and as a route to polymer-biomolecule conjugates. Beginning with conjugation reactions involving cysteine, the chapter moves on to highlight the use of maleimide substrates with an emphasis on (co)polymers bearing such Michael acceptors as chain end functionality available for post-polymerization conjugation. This discussion is followed by a section highlighting alternative efficient Michael acceptors employed as bioconjugation handles on (co)polymers including the vinyl sulfone functional group and the ability to form disulphide bridges. Protein/peptide modification employing both radical and base-mediated thiol-ene reactions is then highlighted. The application of novel mono- and di-bromomaleimides, in the small molecule as well as polymeric forms (as end-groups), for the conjugation of cysteine species is then noted. The chapter concludes with two short sections on the application of thiol-ene chemistry in siRNA and aptamer conjugation as well as antibody drug conjugates.

Chapters 6 and 7 by Son and coworkers and Perrier and coworkers, respectively, focus on the application of thiol-ene and thiol-yne chemistries for the synthesis of non-linear (macro)molecules. Chapter 6 gives a broader overview on the use of thiol-X chemistries for the synthesis of branched/star and dendritic species beginning with examples employing the thiol-Michael reaction for coupling followed by approaches utilizing the radical thiol-ene, radical thiol-yne and even sequential thiol-ene/yne reactions. In contrast, Chapter 7 is more specific and focuses exclusively on the application of the radical thiol-yne reaction as a means of preparing dendritic (co)polymers.

Anseth and coworkers highlight the application of thiol-X chemistries in tissue engineering applications in Chapter 8. Important features associated with cytocompatible monomer design are discussed as well as subsequent encapsulation in hydrogel formulations. The authors subsequently review the major thiol-X chemistries that have been employed to prepare hydrogel cell scaffold systems, providing generalized methods for the synthesis of cell-laden hydrogels *via* thiol-X reactions. Distinct advantages and disadvantages of each of the approaches are discussed.

In Chapter 9, Du Prez and coworkers highlight thiolactone chemistry as a tool to fabricate tailor-made polymer architectures using the thiol-X approach. The use of thiolactone chemistry in polymer science has only been introduced very recently and in this chapter the potential virtues regarding versatility and synthetic scope are highlighted. Thiolactones are cyclic esters of mercapto-acids. Their use provides a remarkable versatility of modular synthesis and modification by opening the thiolactones with a wide variety of functional amines to release thiol groups that can then react *via* any number of thiol-click reactions. The structural features, synthesis and properties of homocysteine-γ-thiolactone and derivatives are highlighted. The use of thiolactones as functional handles along the backbone of a variety of linear polymers enables the use of a wide variety of amines for double, modular post-polymerization modification consisting of aminolysis and thiol-click reactions.

Chapter 10 by Boutevin and coworkers discusses a variety of different routes for utilizing thiol-ene coupling reactions to synthesize oligomers or polymers with controlled functionality and molecular weight. This includes thiol-ene radical coupling to synthesize both block and graft copolymer reactive precursors including both monofunctional and telechelic precursors. Mono-functional reactive precursors are achievable with either thiol or vinyl bearing compounds. The versatility of the thiol reactions enables utilization of a wide range of functional groups that may undergo additional condensation reactions. The functional groups also promote the ability to tailor for targeted precursor properties, such as hydrophilic, lipophobic, and adhesion. Telechelic precursors are achievable from the addition of a functional thiol to a vinyl group, addition of two functional thiols to non-functional dienes, and the addition of non-functional dithiols onto non-functional dienes. Thiol-ene radical coupling is also highlighted as a route to modify polymers containing vinyl groups on the backbone as a means to partially modify polymeric backbone properties. The synthesis of block copolymers and ionomers by thiol-ene radical coupling is also discussed.

In Chapter 11, Stenzel presents different types of thiol-X reactions and their use to create hybrids of synthetic polymers with nature's natural building blocks such as sugars, peptides and proteins. Through the functionalization of soluble polymer chains, the aim is to combine the best properties of both to create new materials that enable the design of polymers that can take on the properties of synthetic polymers while also being bioactive. Glycopolymers are a class of commonly synthesized biomaterials and their direct polymerization is well established. Post-polymerization modification of these polymers *via* thiol-X reactions allows the production of libraries of glycopolymers with the same macromolecular architecture. Thiol-X chemistry also generally provides a simpler and more benign synthesis procedure for these materials. Glyco-polymers are common biomaterials and though their synthesis is well established, the use of thiol-X reactions enables more facile synthesis routes and increased options for functionalization. Functionalization of glycopolymers with thiol-X reactions has been achieved *via* several routes including thio-halogen reactions, thio-para fluoro reactions, and thiol-ene and thiol-yne reactions. Polymer-peptide conjugates are another common biomaterial where enhanced bioactivity is envisaged through enhanced solubility that changes the mode of delivery as well as the aim to design new polymer hybrids. The synthesis of polymer-peptide conjugates through the use of thio-halogen reactions, thiol-ene reactions, thiol-vinylsulfone reactions, thiol-maleimide reactions, and thiol-pyridyl disulfide reactions is described and the various benefits of each are discussed. Utilizing similar conjugation chemistry to those employed for polymer-peptide conjugates, thiol-X reactions have also been used to synthesize polymer-protein conjugates, polymer-DNA conjugates, and polymer-monoclonal antibodies conjugates.

In Chapter 12 Hensarling and Patton highlight the application of the suite of thiol-X chemistries as a means of modifying surfaces. The authors highlight how these chemistries have been successfully employed to modify monolayers and ultrathin films, polymeric surfaces, microporous membranes, particle surfaces as well as biological surfaces. It is shown that these thiol-based chemistries represent an extremely powerful set of tools for the efficient modification and patterning of the above surfaces and have established themselves as some of the preferred methods for achieving facile functionalization.

In Chapter 13, Liang *et al.* introduce the implementation of thiol-ene/thiol-yne radical mediated reactions as a means to achieve robust and patternable surface modification and functionalization. Thiol-ene/yne reactions are compatible with a wide range of functional groups and exhibit good orthogonality with other common organic synthesis reactions. Hence, thiol-ene/yne reactions represent a versatile approach for tailoring solid surfaces with specific properties, immobilizing macromolecules such as proteins, surface engineering, and patterning. Thiol-ene reactions are discussed for surface immobilization of proteins for biochips, surface patterning functional organic monolayers, multilayer structures, polymer brushes, crosslinked polymer films, as well as bioorganic coatings. Tailoring of surface properties is also discussed *via* attachment of functionalized particles, functionalization of trialkoxysilanes, and biofunctionalization of nanoparticles. Similar to the thiol-ene reactions, thiol-yne reactions are also utilized for surface patterning and modification of biomaterials. Surface functionalization of silica-based stationary phases is utilized to achieve different selectivities in chromatographic stationary phases. Thiol-ene/yne reactions represent an ideal pathway to achieve a range of tailored surface functionalizations. In this chapter, the preparation of chiral (ionic exchange), reversed-phase, mixed-mode, and hydrophilic interaction stationary phases are highlighted.

Neil B. Cramer and Christopher N. Bowman,
*University of Colorado, USA*

Andrew B. Lowe,
*University of New South Wales, Australia*

# Contents



RSC Polymer Chemistry Series No. 6
Thiol-X Chemistries in Polymer and Materials Science
Edited by Andrew B. Lowe and Christopher N. Bowman

Published by the Royal Society of Chemistry, www.rsc.org

CHAPTER 1

# Thiol-ene and Thiol-yne Chemistry in Ideal Network Synthesis

NEIL B. CRAMER AND CHRISTOPHER N. BOWMAN*

Department of Chemical and Biological Engineering, University of Colorado, Boulder, CO 80309-0596, USA
*Email: Christopher.Bowman@Colorado.Edu

## 1.1 Introduction

Thiol-ene polymers are formed by the stoichiometric reaction of multifunctional enes with multifunctional thiols. While numerous other thiol-vinyl reactions are possible through a variety of mechanisms as presented throughout this book, the reaction considered in this chapter as the "thiol-ene" reaction proceeds *via* a radical-mediated step growth mechanism that can be initiated by light, peroxides, thermal initiators or any system whereby radicals are generated. The reaction of thiols with enes was first observed in 1905,[1] their use in forming polymers was discovered in 1926,[2] and in the 1930s, the thiol-ene reaction mechanism for radical polymerization was first proposed.[3] A systematic study of the reactions of dienes with dithiols was performed in the 1940s with the findings summarized in a patent.[4] The systematic examination of these reactions demonstrated that the reaction was general in nature, proceeded under mild reaction conditions, reacted rapidly to achieve high yields, gave a specific composition of products (the anti-Markovnikov addition product of thiols to enes), and exhibited few or no side products – whence this collection of

RSC Polymer Chemistry Series No. 6
Thiol-X Chemistries in Polymer and Materials Science
Edited by Andrew B. Lowe and Christopher N. Bowman

Published by the Royal Society of Chemistry, www.rsc.org

attributes more recently has led to the thiol-ene reaction meeting all of the basic requirements for being considered a click reaction.[5] This study and numerous others since have clearly demonstrated that thiol-ene radical reactions uniquely combine the benefits of click reactions with the spatial and temporal advantages that come from its ability to be photoinitiated.

It wasn't until the 1970s that thiol-ene polymerizations were developed for commercial applications, such as relief printing plates, conformal coatings on electronics, and floor coatings by W. R. Grace and Co.[6–8] However, these and other potential applications for thiol-enes gave way to acrylate and methacrylate systems, which became the commercial photopolymerization systems of choice. The most notable exception was for optical adhesives developed and sold by Norland since the 1970s. In the early 1990s there was again renewed interest in thiol-ene photopolymerizations, largely in concert with an increased interest in all non-acrylate photopolymers due to concerns over toxicological issues associated with acrylate monomers. Additionally, the rapid polymerization rate of thiol-enes, as well as their inherent ability to polymerize in the presence of oxygen, are highly desirable characteristics. These unique properties, along with many others, led to continued interest and development in thiol-ene polymerizations. Another review was published in 1993,[9] resulting from Loctite's interest in thiol-ene systems as coatings. The fundamental understanding, application potential, and overall interest in thiol-enes has continued to grow with several reviews since 2000,[10–14] indicating the resurgence of interest in thiol-ene chemistry, particularly as one of the click-reaction family.

Though thiol-ene chemistry and polymerization have been known for many decades, the use of thiol-ene photopolymerizations is still not widely prevalent commercially, with the photopolymerization of acrylate and methacrylate systems continuing to dominate the commercial radiation curing markets. Acrylate and methacrylate systems are widely utilized commercially for applications including coatings, adhesives, dental materials, contact lenses, and photolithographic processes.[15–17] There exists significant potential to expand application of photopolymerization technology to applications in nanotechnology, biomaterials, high-resolution lithography, polymer and surface functionalization, electro-optics, high-impact energy absorbing devices, optical switching arrays, and many others. Critical to achieving expanded application is an improvement in polymerization and material performance. The classical acrylate and methacrylate photopolymerizations encompass several significant drawbacks including inhibition of the polymerization by oxygen,[18–20] polymerization induced stress development,[21,22] and the formation of highly heterogeneous polymer networks.[23–25] Significant research is aimed at improving the properties and performance of photopolymerizations to achieve expanded application and the resurgence of interest and development of thiol-ene photopolymerizations has been one of the thrusts of this research. Thiol-enes exhibit numerous polymerization characteristics that directly address the drawbacks of the classical acrylate/methacrylate photopolymerization systems. Thiol-ene photopolymerizations exhibit rapid and simplified polymerization kinetics, are relatively uninhibited by oxygen, form low stress homogeneous polymer networks, and are able to achieve a wide range of material properties. Each of these unique thiol-ene properties will be discussed in detail.

## 1.2 Polymerization Mechanism and Kinetics

The thiol-ene reaction occurs as a result of the radical-mediated addition of thiol functional groups across carbon–carbon double bonds (enes) (Scheme 1.1). The thiyl radical attacks the least substituted side of the double bond giving the anti-Markovnikov addition product.[26] Traditional thiol-ene systems utilize ene monomers, such as allyl ethers and vinyl ethers, that do not homopolymerize. However, thiol monomers can react with a multitude of different vinyl moieties, including those that do homopolymerize such as acrylates, methacrylates, and acrylamides.

The propagation mechanism of the thiol-ene photopolymerization is an alternation of addition and chain transfer reactions, as outlined in Scheme 1.2. The thiyl radical adds across carbon–carbon double bonds *via* an addition reaction, forming a carbon radical. Carbon radicals abstract hydrogen from thiol functional groups, forming thiyl radicals. For systems where the ene monomers are utilized that do not homopolymerize, such as vinyl ethers, allyl ethers, and norbornenes, this addition/chain transfer process continues

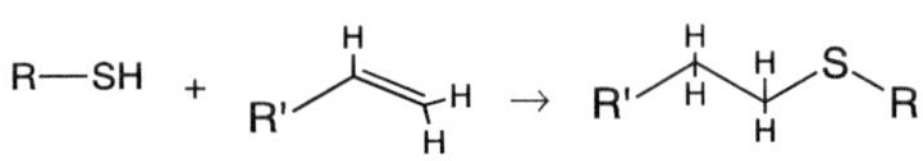

**Scheme 1.1** Thiol and ene monomers react to form the anti-Markovnikov thiol-ene addition product.

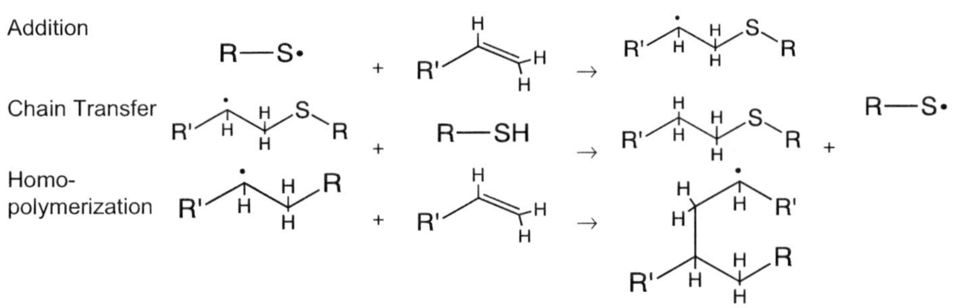


**Scheme 1.2** Propagation and chain transfer reactions in thiol-ene polymerizations. Addition: thiyl radicals add across an ene functional group *via* an addition reaction that generates a carbon radical. Chain transfer: carbon radicals abstract hydrogen from a thiol functional group forming the anti-Markovnikov addition product and a thiyl radical. Homopolymerization: carbon radicals add across an ene functional group via an addition reaction that generates another carbon-centered radical.

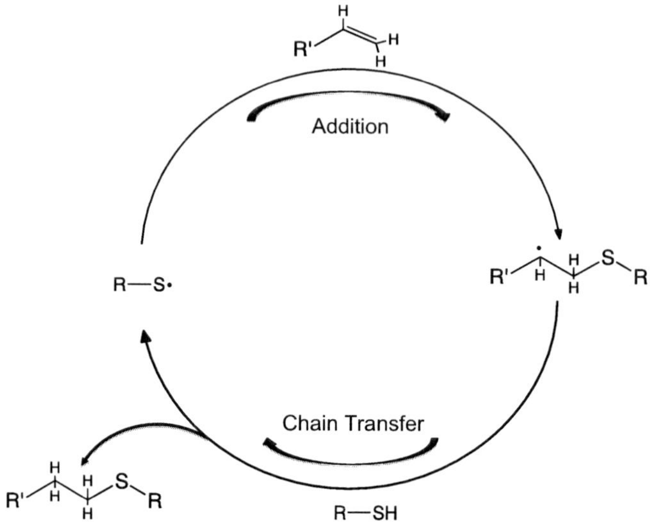


**Figure 1.1** Thiol-ene reaction mechanism with no ene homopolymerization. Thiyl radical addition to an ene functional group forms a carbon-centered radical, which then chain transfers to a thiol functional group to regenerate a thiyl radical and the thiol-ene addition product.

cyclically as seen in Figure 1.1, forming the basis of the thiol-ene step growth polymerization mechanism. The step growth mechanism results in a 1 : 1 stoichiometric consumption of thiol and ene functional groups resulting in near ideal molecular weight and network structure development as predicted for step growth polymer molecular weight evolution. The molecular weight of the system increases slowly, resultant from the chain transfer step following every addition step, building up dimers, trimers, *etc.* before higher molecular weight species and gelation are ultimately achieved. As such, the polymerization kinetics are also relatively simple over the majority of the polymerization, in contrast to conventional acrylic free radical chain growth polymerizations where high molecular weight species are rapidly formed resulting in rapid gelation and complex diffusion mediated kinetics. The use of multiple types of ene functional groups and/or enes that also homopolymerize (such as acrylates and methacrylates) leads to an additional homopolymerization propagation step (Scheme 1.2). When homopolymerizable enes are utilized, such as with acrylates and methacrylates, significant homopolymerization of the (meth)acrylate occurs, leading to a combination of both step and chain growth molecular weight evolution which can be characterized as dual cyclical reactions, as seen in Figure 1.2. These step-chain growth systems are kinetically more complicated, but also provide a unique platform enabling much greater control and greater range of achievable polymerization and polymer properties.

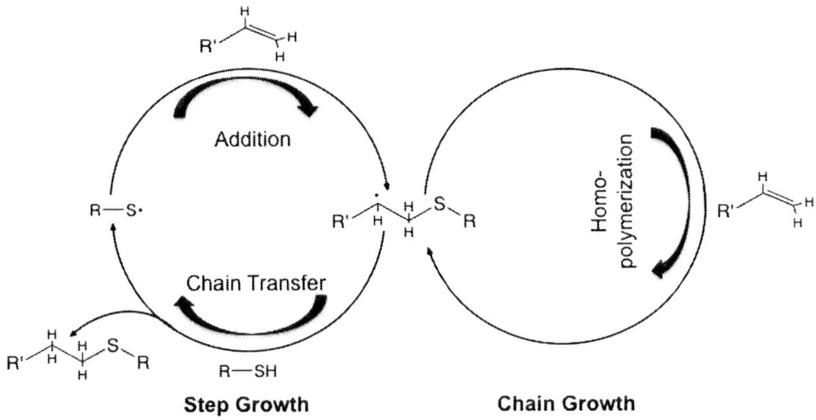


**Figure 1.2** Thiol-ene reaction mechanism with homopolymerization of ene functional groups. In the step growth cycle, thiyl radical addition to an ene functional group generates a carbon-centered radical and the carbon-centered radical reacts with a thiol functional group in a chain transfer reaction generating another thiyl radical and the thiol-ene step growth addition product. In the chain growth cycle, the carbon-centered radical homopolymerizes with another ene functional group generating the chain growth homopolymerization product and another carbon-centered radical.

## 1.2.1 Oxygen

Radicals are extremely efficient at catalyzing the reaction of thiols and enes. The reaction of radicals with oxygen is also an extremely efficient process with reactivities of radicals towards oxygen being around four orders of magnitude greater than the reactivity towards propagation.[27,28] In conventional radical photopolymerizations, the reaction with oxygen serves to inhibit the polymerization, as the ensuing peroxy radical that is formed is unreactive towards additional propagation. This inhibition of the polymerization occurs from oxygen that is dissolved in the monomer formulation as well as oxygen that is able to diffuse into the system during polymerization of non-laminated films. In both cases, oxygen inhibition has a huge impact on photopolymerizations resulting in tacky (uncured) surface layers and/or requiring the use of high irradiation intensities, high initiator concentrations, multiple initiator types, and inert/nitrogen blankets. Uniquely, the thiol-ene polymerization, due to its unique addition/chain transfer propagation mechanism, is not strongly impacted by oxygen inhibition. The propagating radicals are reactive with oxygen, forming peroxy radicals. However, the peroxy radicals, though not reactive towards additional propagation, readily abstract hydrogen from thiol groups (Scheme 1.3). The hydrogen abstraction generates an alkylhydroperoxide and a thiyl radical and the polymerization thus continues with only a single, rapid additional reaction step, rather than being inhibited or effectively terminated by a reaction with oxygen. Numerous studies have been performed

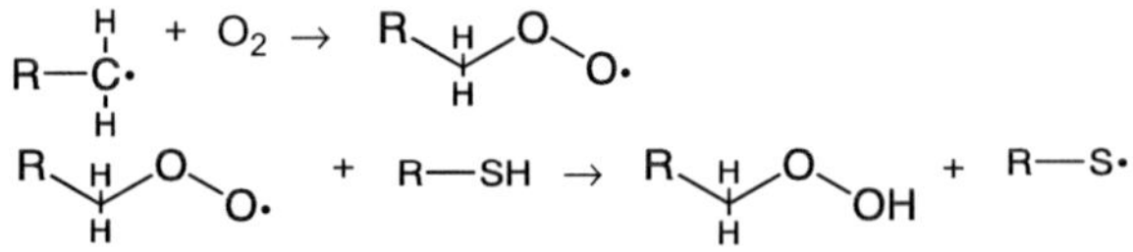


**Scheme 1.3** Oxygen incorporation in thiol-ene polymerizations. Molecular oxygen reacts with a carbon-centered radical to form a peroxy radical that is unreactive towards additional reaction. Subsequently, peroxy radicals abstract hydrogen from a thiol functional group, generating an alkyl-hydroperoxide and a thiyl radical.

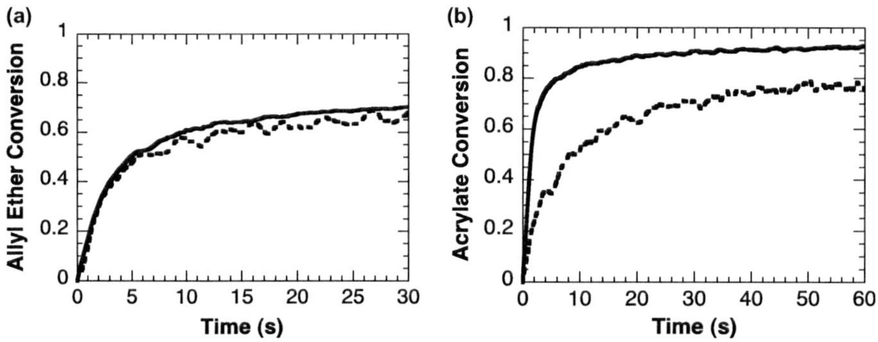


**Figure 1.3** Conversion *versus* time for 1 : 1 stoichiometric mixtures of pentaerythritol tetra(3-mercaptopropionate) and (a) triallyl triazine trione and (b) trimethylolpropane triacrylate for 6 μm thick samples that are exposed to ambient oxygen (---) as well as laminated (—). The samples contain 0.5 wt% 2,2-dimethoxy-2-phenylacetophenone and are irradiated at 5 mW cm$^{-2}$ UV light with principle output at 365 nm.

that have characterized polymerization rates in thiol-ene systems with and without the presence of ambient oxygen and in comparison to acrylic systems.[10,29–32] Figure 1.3 illustrates conversion *versus* time results for thiol-allyl ether and thiol-acrylate systems exposed to either ambient oxygen or as a laminate. In both cases the samples are spread to a thickness of 6 μm. In the thiol-allyl ether system the polymerization rate is almost unaffected by the ambient conditions. In the thiol-acrylate system there is a modest reduction in both polymerization rate and conversion, which is remarkable for an acrylate system with low initiator concentration, low irradiation intensity, and only 6 μm thick!

### 1.2.2 Initiation

Initation is the process whereby radicals are generated and subsequently react with either thiol or ene functional groups to generate thiyl or carbon radicals that start the polymerization process. Type II abstraction initiators (typically benzophenone) have long been utilized to initiate thiol-ene reactions.[9] In these systems, benzophenone or a similar compound absorbs ultraviolet irradiation

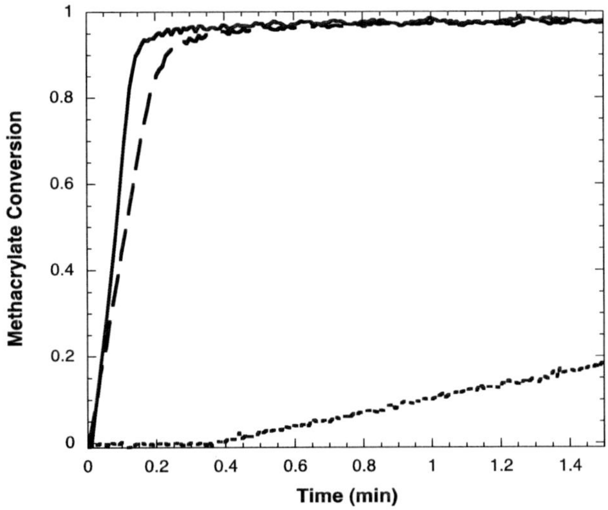


**Figure 1.4** Conversion versus time for a methacrylate/thiol-ene system (60 wt% ethoxylated bis-phenol A dimethacrylate and 40 wt% pentaerythritol tetra(3-mercaptopropionate):triallyl triazine trione with a 2 : 1 stoichiometric ratio) using 0.3 wt% 2, 2-dimethoxy-2-phenylacetophenone (——) irradiated at 77 mW cm$^{-2}$ using a 320–500 nm filter, and bis acyl phosphine oxide (— —), and camphorquinone/ethyl dimethyl amino benzoate (---) irradiated at 41 mW cm$^{-2}$ using a 400–500 nm filter.

and goes into the excited triplet state where it abstracts hydrogen from a thiol functional group generating a pinacol radical and a thiyl radical. The pinacol radical is relatively unreactive except towards termination. The thiyl radical enters the thiol-ene reaction cycle as outlined in Scheme 1.2. Type I cleavage initiators exhibit significantly better initiation efficiencies than abstraction initiators and are also effective at initiating thiol-ene polymerizations. Cleavage initiators, either ultraviolet, such as hydroxy cyclohexyl phenyl ketone and 2,2-dimethoxy-2-phenylacetophenone, or visible such as diphenyl (2,4,6-trimethylbenzoyl)- phosphine oxide (TPO) directly cleave into primary radicals that readily initiate thiol-ene reactions by either hydrogen abstraction from a thiol or addition across an ene. Interestingly, the camphorquinone/amine initiation system that is commonly utilized for many visible light initiated polymerizations is extremely inefficient in thiol-ene systems. Figure 1.4 illustrates the polymerization rate in a methacrylate/thiol-ene system with a UV photoinitiator (2,2-dimethoxy-2-phenylacetophenone), a TPO derivative (bis acyl phosphine oxide), and a camphorquinone/amine combination. Both of the cleavage initiators are extremely efficient, whereas the camphorquinone/amine system exhibits minimal polymerization. The improved efficiency of cleavage initiators relative to benzophenone is illustrated in Figure 1.5, where it is seen that a thiol-ene system proceeds much more rapidly when initiated with 2,2-dimethoxy-2-phenylacetophenone relative to a system initiated with benzophenone. Thermal initiators that degrade into initiating radicals upon

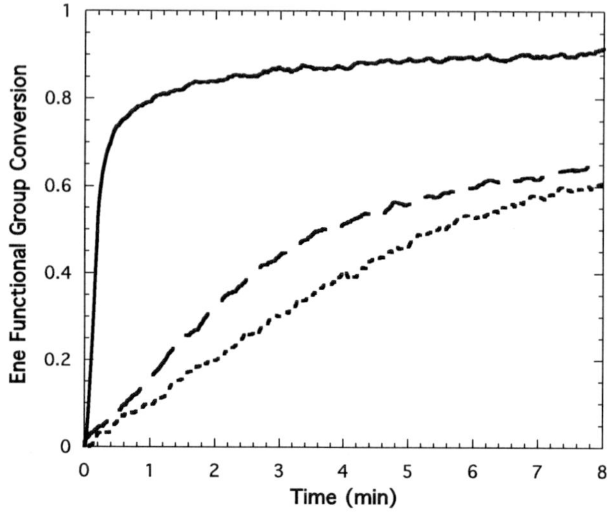


**Figure 1.5** Conversion *versus* time for pentaerythritol tetra(3-mercaptopropionate) and triallyl triazine trione initiated with 0.7 wt% Irgacure 651 (——), 1.0 wt% benzophenone (— —), and without added photoinitiator (---). Samples were irradiated with UV light with principle output at 365 nm with an irradiation intensity that was 0.25 mW cm$^{-2}$ for the Irgacure 651 and benzophenone systems and 3 mW cm$^{-2}$ for the system without added photoinitiator.
Figure adapted from ref. 35.

heating, such as the commonly utilized azoisobutylnitrile (AIBN), are also efficient at initiating thiol-ene reactions though they do have some drawbacks as well.[33,34] However, the use of thermal initiators in some cases, causes the thiol-ene monomer formulation to be unstable even at ambient conditions.

One of the unique aspects of thiol-ene polymerizations is that for ultraviolet irradiation, no added photoinitiator molecules are required to initiate the polymerizations.[10,30,35,36] In systems where ultraviolet light centered around 365 nm is utilized, polymerization depths of up to 25 inches have been achieved resultant from the low absorptivity of the monomers/polymers with the absence of added photoinitiator.[30] In these cases, and as illustrated in Figure 1.5, the initiation rate is dramatically reduced relative to systems containing added photoinitiator systems. When 254 nm light is utilized, initiation rates are extremely rapid, however polymerization depths are limited due to high absorptivity. The mechanism of initiation without added photoinitiators is not well understood. It has been postulated that a charge transfer complex between the thiol and ene functional groups results in a complex that can absorb photons and initiate polymerization.[37–40] The charge-transfer complex was substantially studied and demonstrated to initiate the polymerization *via* consumption of oxygen.[37–39] However, no absorption from this complex can be seen been spectroscopically and the charge transfer complex does not appear to be responsible for photoinitiation. The rate of photoinitiation without added photoinitiators and using 365 nm irradiation has been shown to be directly proportional to the concentration of ene functional groups,[36] however a specific mechanism has not yet been found that is consistent with the experimental data. The mechanism of initiation with 254 nm irradiation is ascribed to direct cleavage of thiol functional groups,[36] a process that is known to occur in gas phase thiols.[41,42]

### 1.2.3 Termination

Temination can occur by any number of reactions, but is often dominated by the various radical–radical recombinations of thiyl and carbon radicals. The termination mechanism(s) has a strong effect on polymerization kinetics. Given the thiol-ene polymerization mechanism, if one assumes that bimolecular radical–radical recombination is the dominant termination mechanism, the polymerization rate under pseudo-steady state conditions would be expected to scale by the initiation rate to the $\frac{1}{2}$ power ($R_p \propto R_i^{1/2}$). Indeed several investigations dating back to the 1970s have demonstrated that this classical polymerization rate scaling behavior[6,43] occurs in at least some of the thiol-ene reactions. In addition, the type of ene used as well as the presence of any dissolved oxygen or other impurities in the resin are critical to the termination mechanism, which was clearly identified in several systems not to be dominated by bimolecular termination reactions.[44]

In most (meth)acrylate chain growth polymerizations, radical propagation results in high molecular weight polymers and oligomers being formed almost immediately and diffusion limits termination reactions very early in the polymerization leading to the phenomenon of autoacceleration and complicated polymerization kinetics. By contrast, chain transfer is an integral component of the step growth thiol-ene mechanism, resulting in uniform molecular weight build-up with delayed molecular weight evolution. Thus, the termination reaction rates generally remain high throughout the polymerization, resulting in consistent polymerization kinetics throughout much of the polymerization. The reduced diffusion limitations also lead to relatively high overall functional group conversions.

## 1.3 Functional Group Chemistry

The thiol functional group is extremely versatile, participating in numerous chemical reactions. For radically mediated reactions, any non-sterically hindered terminal ene can react in the manner depicted in Scheme 1.1. As such there are an incredibly diverse number of different types of ene functional groups that can be utilized in thiol-ene reactions. Thiol functional groups have less diversity with the most common thiols including alkyl thiols, thiol propionates, thiol glycolates, and thiol phenols. Typical thiol and ene functional groups are depicted in Figure 1.6. The propagation reaction is exothermic with higher enthalpies for electron poor double bonds and lower enthalpies for electron rich double bonds. Reaction enthalpies range

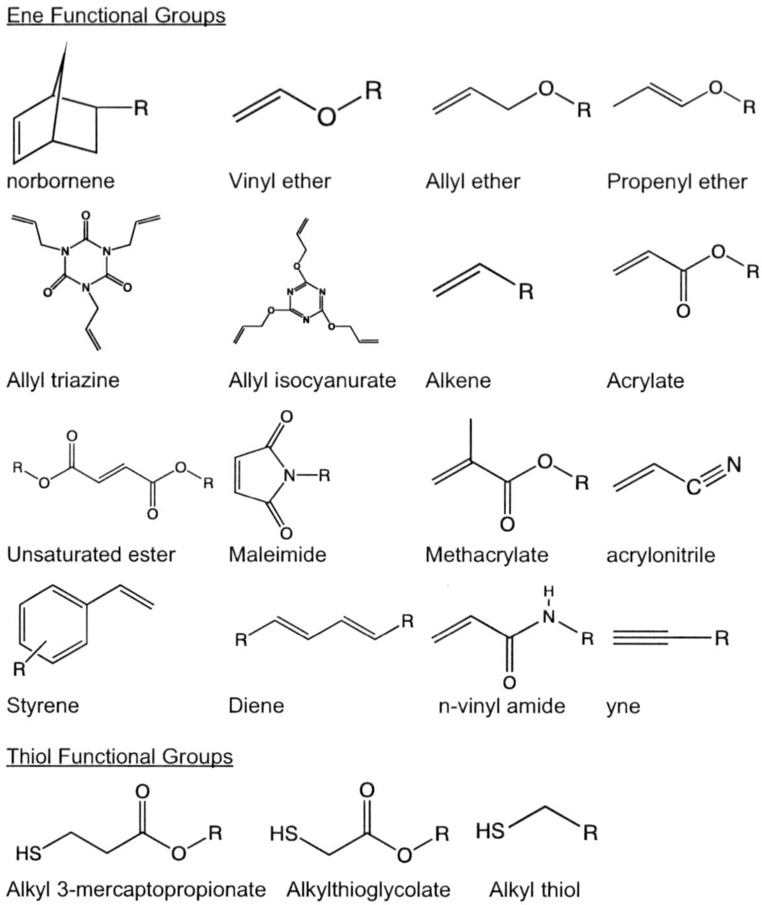


**Figure 1.6** General structures of typical thiol and ene functional groups.

from 10.5 kcal $mol^{-1}$ for an electron-rich vinyl ether to 22.6 kcal $mol^{-1}$ for an electron poor maleimide.[45]

Reaction rates vary dramatically with different types of enes with terminal enes being the most reactive and electron-rich (vinyl ethers and allyl ethers) and strained (norbornene) enes reacting more rapidly than electron-deficient enes; internal 1,2-substituted enes exhibit much lower overall reaction rates.[46] Numerous thiol and ene functional groups have been evaluated for their relative reactivities.[6,9,10,35] Figure 1.7 illustrates an example of a thiol monomer polymerized with a divinyl ether, a diallyl ether, and a diyne. It is readily observed that the vinyl ether reacts extremely rapidly followed by the allyl ether and with the yne group being the slowest. There are still questions about the specific order of reactivities, as numerous experiments have been performed over a wide range of experimental reaction conditions. The relative reactivities

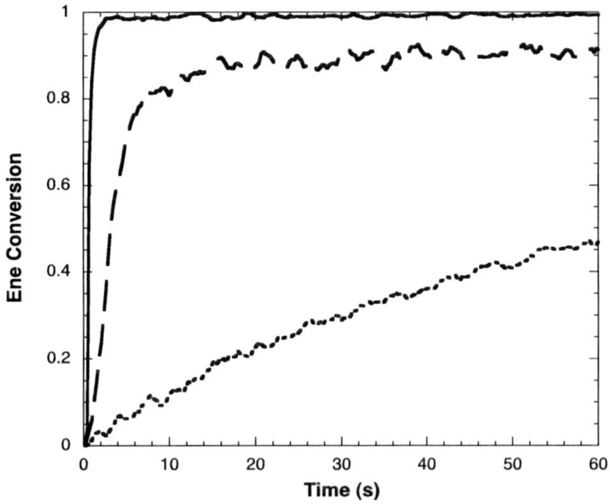


**Figure 1.7** Conversion versus irradiation time for a thiol monomer (pentaerythritol tetra(3-mercaptopropionate) polymerized with stoichiometric ratios of varying ene functional groups; triethylene glycol divinyl ether (——), trimethylolpropane diallyl ether (— —), and heptadiyne (---). The samples contain 0.5 wt% 2,2-dimethoxy-2-phenylacetophenone and are irradiated at 5 mW cm$^{-2}$ UV light with principle output at 365 nm.

loosely are as follows with norbornene > vinyl ether > propenyl > alkene ~ vinyl ester > *n*-vinyl amide > allyl ether ~ allyltriazine ~ allylisocyanurate > acrylate > unsaturated ester > maleimide > acrylonitrile ~ methacrylate > styrene > conjugated dienes.

### 1.3.1 Thiol-yne

One of the distinct features of thiol-enes is that each ene functional group reacts only once with a thiol group whereas in chain growth polymerizations, each ene group reacts twice. An advantage of the ene group only reacting once is that thiol-enes exhibit less volume shrinkage during polymerization than equivalent (meth)acrylate systems. One of the disadvantages of the single ene reaction is that thiol-enes also exhibit lower crosslink density, and hence often lower modulus and $T_g$, than equivalent (meth)acrylate systems. An exception to the single reaction paradigm of ene functional groups is the yne triple bond. As depicted in Scheme 1.4, an yne functional group reacts with thiols to form a vinyl sulfide and the vinyl sulfide can subsequently react with another thiol. Outside of the dual reactivity of the yne functional groups, all of the other characteristics of the reaction follow the ideal click reaction paradigm of thiol-ene systems, where the reaction is highly efficient, with the ensuing network being near ideal in homogeneity and

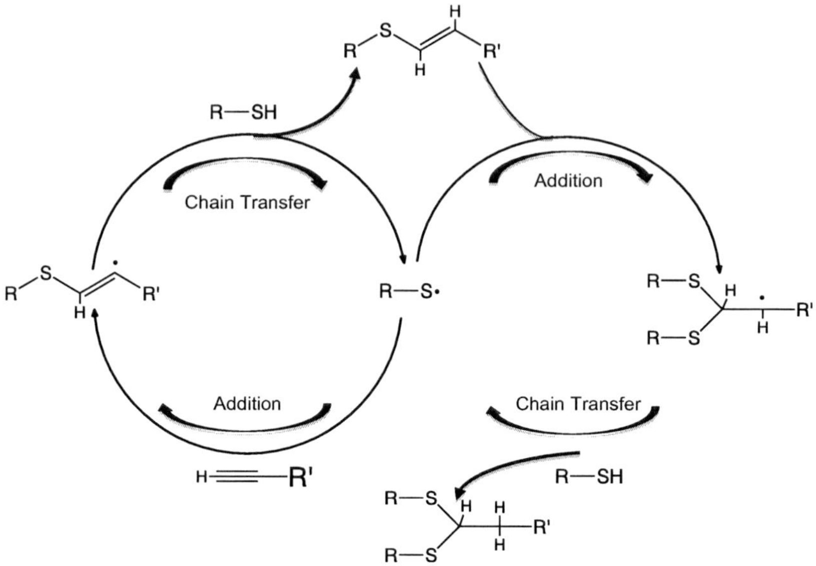


**Scheme 1.4** The thiol-yne polymerization mechanism. Thiyl radical addition to an yne functional forms a vinyl sulfide radical, which then chain transfers to a thiol functional group to regenerate a thiyl radical and form the vinyl sulfide addition product. Thiyl radical addition to a vinyl sulfide functional group forms a carbon-centered radical, which then chain transfers to another thiol functional group to regenerate a thiyl radical and form the thiol-vinyl sulfide addition product.

relatively unaffected by oxygen. The radical mediated thiol-yne reaction was initially observed and characterized in the 1940s–1960s.[47–51] More recent interest has followed from the overall resurgence in interest in the unique properties of thiol-ene polymerizations[52–58] with a review being published in 2010.[13]

From the standpoint of polymerization kinetics, the reaction of the thiol with the yne is slower than the subsequent reaction of thiol with vinyl sulfide.[52,56] The relative reactivities of a range of different yne moieties with an alkyl thiol has been studied with the relative reaction rates being 1-octyne > propargyl acetate > methyl propargyl ether > 2-octyne. In the case of cyclooctyne, methyl propargylamine, and ethyl propiolate the ynes react with thiols (rapidly for cyclooctyne and very slowly for methyl propargylamine and ethyl propiolate), but the subsequent addition of thiol with the vinyl sulfide was not observed to occur.[56] Resultant from the dual reactivity of the yne functional group, the crosslink density of thiol-yne based polymer networks is substantially increased relative to an equivalent thiol-ene polymer network. A comparison of properties of equivalent thiol-ene and thiol-yne systems in Table 1.1 illustrates this point, showing dramatic increases in crosslink density, modulus, and $T_g$.

**Table 1.1** Comparison of equivalent thiol-ene and thiol-yne systems.[52] The thiol monomer is pentaerythritol tetra(3-mercaptopropionate) (PETMP), the yne monomer is 1,9-dodecadiyne, and the analogous ene monomer is butanediol divinylether.

| *Thiol* | *Ene* | *Yne* |
|---|---|---|
| 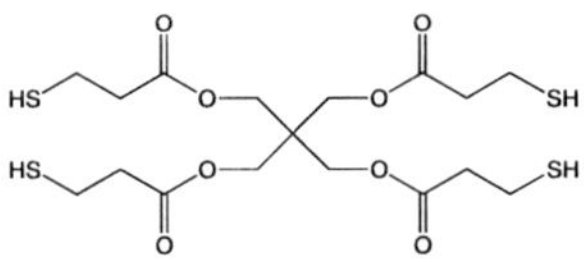 | 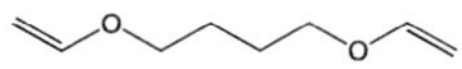 | 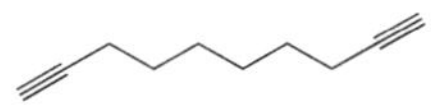 |

| | *Crosslink density (M)* | *Modulus (E′/MPa)* | $T_g$/°C |
|---|---|---|---|
| Thiol-yne | 8.4 | 69 | 40.7 |
| Thiol-ene | 1.5 | 13 | −22.3 |

# 1.4 Network Formation

The nature of the step growth polymerization reaction, as depicted in Figure 1.8, is such that if mono-thiol monomers are reacted with mono-ene monomers, only dimer products result. Reacting di-thiol and di-ene monomers in an ideal thiol-ene reaction results in the formation of linear polymers. However, it is worth noting that many starting materials are not 100% pure and contain fractions that are less than the specified functionality. These impurities significantly affect the molecular weight evolution, and especially in the case of di-thiols and di-enes high molecular weight linear polymers are therefore difficult to achieve. Only when multi-functional monomers (average functionality > 2) are utilized do crosslinked polymer networks results. In these cases, the thiol-ene polymerization proceeds *via* an ideal step growth polymerization with molecular weight evolution, gel point, and network structure similar to any other traditional step growth polymerization such as epoxy-amines, alcohol-isocyanates, *etc.* As such, the Flory–Stockmayer equation (eqn (1.1)) accurately predicts the gel point conversion ($\alpha$) in ideal thiol-ene systems

$$\alpha = \sqrt{\frac{1}{r(f_{\text{thiol}} - 1)(f_{\text{ene}} - 1)}} \tag{1.1}$$

where $r$ is the molar ratio of thiol to ene functional groups, $f_{\text{thiol}}$ is the thiol monomer functionality, and $f_{\text{ene}}$ is the ene monomer functionality. The Flory–Stockmayer equation and ideal step growth network evolution are only valid when negligible ene homopolymerization occurs.

To generate a more complete understanding of molecular weight evolution in thiol-ene systems, models have been developed based on combined kinetic and statistical approaches to describe molecular weight evolution, gel point conversion, and crosslink density of thiol-ene systems.[59–61] The models include

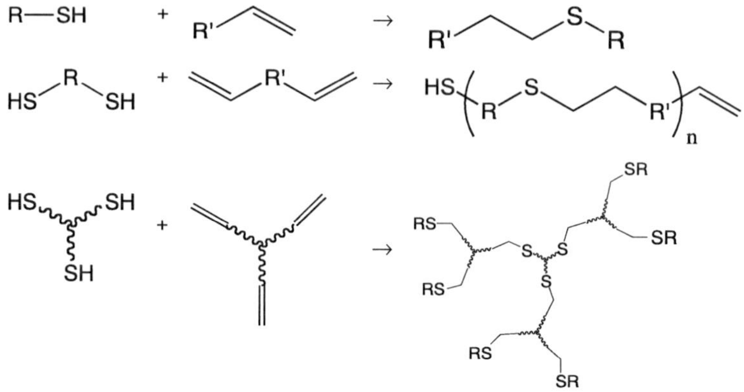


**Figure 1.8** Network formation for varying thiol-ene functionalities. Mono-thiol and mono-ene monomers react to form dimer products. Di-thiol and di-ene monomers react to form linear polymers. Multifunctional thiol and ene (functionality >2) react to form crosslinked polymer networks.

pure step growth and mixed mode step-chain growth systems. In traditional thiol-ene step growth systems, with negligible amounts of ene homopolymerization, molecular weight evolution is consistent with any other step growth polymerization process. Cyclization has a significant impact on gel point, particularly with lower functionality monomers, with increased levels of cyclization increasing gel point conversions. In cases where homopolymerization occurs, such as in thiol-acrylate systems, molecular weights (prior to gelation) are increased and the gel point is shifted towards lower conversions.[61] In these cases, the ratio of thiyl radical propagation to chain transfer ($k_p/k_{ct}$) impacts the gel point with increased ratios of propagation to chain transfer resulting in earlier gel points. Both thiol-acrylate and thiol-ene-acrylate systems demonstrate a readily controllable network evolution with the ability to manipulate network properties such as gel point conversion and crosslink density through changes in monomer functionality and the initial stoichiometric thiol to ene ratios. The ability to control network properties ultimately correlates with a concomitant ability to control polymerization and polymer properties in ways not available through either pure step growth or pure chain growth polymerizations.

## 1.5 Polymerization Kinetics and Modeling

Based on the initiation, propagation, and termination mechanisms of the thiol-ene polymerization, equations can be derived that describe the polymerization kinetics of ideal, binary thiol-ene systems.[62,63] The polymerization rates of thiol and ene functional groups are described by eqn (1.2) and (1.3), respectively. Eqn (1.3) describes the polymerization rate as it relates to thiol monomers ($R_{psh}$), where the rate of consumption of thiol functional groups is related to the chain transfer kinetic parameter ($k_{ct}$), the thiol functional group concentration, and the carbon radical concentration. Eqn (1.4) describes the polymerization rate as it relates to ene monomers ($R_{pcc}$), where the rate of consumption of ene functional groups is related to the thiyl radical propagation kinetic parameter ($k_{psc}$), the carbon radical propagation kinetic parameter ($k_{pcc}$), the concentration of ene functional groups, and the concentration of thiyl and carbon radicals.

$$R_{psh} = \frac{d[SH]}{dt} = -k_{ct}[SH][C^{\bullet}] \quad (1.2)$$

$$R_{pcc} = \frac{d[C{=}C]}{dt} = -k_{psc}[C{=}C][S^{\bullet}] - k_{pcc}[C{=}C][C^{\bullet}] \quad (1.3)$$

Utilizing these equations, along with equations describing rates of initiation and termination, models have accurately confirmed and predicted thiol-ene polymerization kinetics over a range of functional group chemistries, stoichiometric ratios, and polymerization conditions.[43,62–64] The modeling can also be expanded to include ternary systems.[64]

Given the cyclical nature of the reaction, and for systems where carbon radical propagation is negligible ($k_{pcc} \ll k_{ct}$), *i.e.* traditional binary thiol-ene step growth systems, thiol and ene functional groups are equivalently consumed at the same rate and eqn (1.3) and (1.4) represent equivalent reactions. Several limiting cases have been identified for these classical thiol-ene systems.[63] For cases where propagation and chain transfer are equivalent, $k_{CT} \approx k_p$, the polymerization rate is $\frac{1}{2}$ order in both thiol and ene functional group concentration ($R_p \propto [SH]^{1/2}[C{=}C]^{1/2}$). For cases where chain transfer is the rate limiting step, $k_{CT} \ll k_p$, the polymerization rate is first order in thiol functional group concentration ($R_p \propto [SH]^1$). For cases where thiyl radical addition is the rate limiting step, $k_{CT} \gg k_p$, the polymerization rate is first order in ene functional group concentration ($R_p \propto [C{=}C]^1$). In all cases, the reaction is overall first order in functional group concentration, with the particular dependencies being dictated by the nature and reactivitiy of the functional groups and radical species. The polymerization characteristics are more strongly dependent on the chemical nature of the ene functional groups. For allyl ether functional groups, chain transfer is the limiting step and the reaction is first order dependent on thiol concentration. For ring strained enes, such as norbornenes, and vinyl ethers, propagation and chain transfer are near equivalent and the reaction is $\frac{1}{2}$ order in both ene and thiol concentration. For vinyl silazanes, chain transfer is the limiting step and the reaction is first order dependent on ene concentration. Figure 1.9 illustrates an example of a thiol-allyl ether polymerization where chain transfer is the rate limiting step and the polymerization rate is first order in thiol functional group concentration and zero order in allyl ether functional group concentration.

The kinetic equations describing thiol-ene polymerization kinetics can also be used in combination with a modified rotating sector technique to experimentally quantify the kinetic parameters.[43,64] To date thiol-allyl ether, thiol-norbornene and thiol-vinyl ether systems have all been characterized *via* this technique and their propagation, chain transfer, and termination kinetic

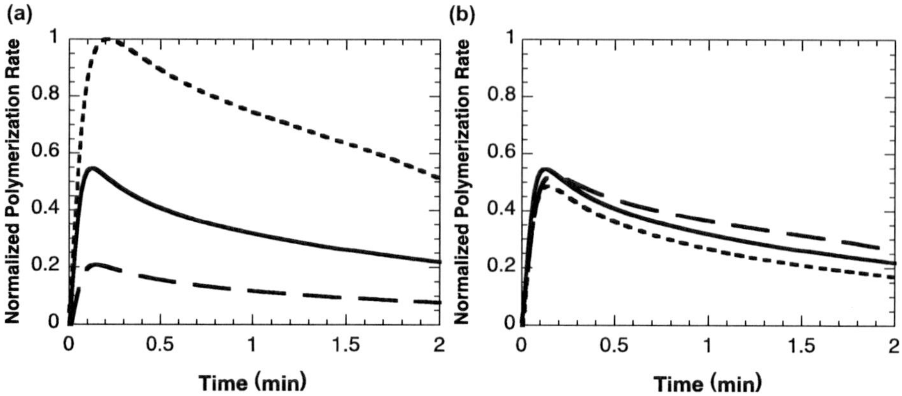


**Figure 1.9** Normalized polymerization rate *versus* irradiation time for hexanedithiol and trimethylolpropane diallyl ether in ethylene glycol diacetate. (a) 0.44 mol $L^{-1}$ allyl ether functional groups and 0.22 mol $L^{-1}$ (— —), 0.44 mol $L^{-1}$ (——), and 0.88 mol $L^{-1}$ (---) thiol functional groups. (b) 0.44 mol $L^{-1}$ thiol functional groups and 0.22 mol $L^{-1}$ (— —), 0.44 mol $L^{-1}$ (——), and 0.88 mol $L^{-1}$ (---) allyl ether functional groups. Samples contain 0.5 wt% benzophenone and are irradiated with UV light with principle output at 365 nm at 1.4 mW $cm^{-2}$ in a photo-DSC.
Figure adapted from ref. 62.

**Table 1.2** Kinetic parameters of thiol-ene systems determined from rotating sector experiments.[43,64]

| *Thiol-Ene* | $k_{psc}/L\,mol^{-1}\,s^{-1}$ | $k_{ct}/L\,mol^{-1}\,s^{-1}$ | $k_t/L\,mol^{-1}\,s^{-1}$ |
|---|---|---|---|
| Thiol-allyl ether | $2\times10^6$ | $2\times10^5$ | $2\times10^8$ |
| Thiol-norbornene | $3\times10^6$ | $3\times10^6$ | $4\times10^8$ |
| Thiol-vinyl ether | $2\times10^6$ | $3\times10^6$ | $4\times10^8$ |

parameter values are given in Table 1.2. Propagation kinetic parameters for thiyl radicals are on the order of $10^6\,L\,mol^{-1}\,s^{-1}$ and chain transfer kinetic parameters are on the order of $10^5$–$10^6\,L\,mol^{-1}\,s^{-1}$, where chain transfer kinetic parameters on the order of $10^6$ represent the more rapid thiol-norbornene and thiol-vinyl ether systems and chain transfer kinetic parameters of $10^5$ represent the less rapid thiol-allyl ether system. These kinetic parameters are in the same range as propagation kinetic parameters for chain growth systems,[28] with acrylate propagation kinetic parameters typically ranging from $10^4$–$10^6$. The termination kinetic parameters are much higher than typical termination kinetic parameters in chain growth systems. However, the termination kinetic parameters are similar to those in chain growth systems (on the order of $10^7$–$10^9\,L\,mol^{-1}\,s^{-1}$) during the very early stages of polymerization (<5% conversion) before the onset of diffusion limitations.[65] While termination kinetics rapidly decrease in chain growth polymerizations after the onset of diffusion limitations, the molecular weight evolution in thiol-ene systems is such that the propagation and chain transfer kinetic parameters remain relatively constant until much higher conversions are achieved.[59,62–64]

The mechanism of polymerization and knowledge of kinetic parameters enables the ability to accurately model and predict polymerization kinetics in thiol-ene systems over a range of polymerization conditions and functional group chemistries. This ability also extends to understanding and modeling of more complex ternary systems such as thiol-ene-ene and thiol-ene-acrylate.[64] In these systems simple changes in the relative concentrations and reactivity of the ene functional group have significant impact on the gel point, crosslink density, and network homogeneity.[59,61,64,66] In thiol-ene-ene systems where neither ene is homopolymerizable, the relative consumption of functional groups is proportional to the propagation kinetic parameters of the ene monomers and independent of the chain transfer kinetic parameter. In thiol-ene-acrylate systems where the ene monomer is not homopolymerizable, and in a mixed mode step-chain growth polymerization, the relative conversion of ene and acrylate functional groups depends on both the propagation and chain transfer kinetic parameters.

## 1.6 Volume Shrinkage and Shrinkage Stress

Unlike a typical chain growth radical mechanism, each ene double bond in a thiol-ene polymerization reacts with one thiol containing monomer instead of the two monomers that are generally coupled to each double bond. Thus as a thiol-ene polymer is formed, there is less contraction of the van der Waals distances separating the monomers as the polymer continues to be formed. The calculated shrinkage associated with a thiol-ene polymerization is between 12 and 15 $cm^3\ mol^{-1}$ per reacted double bond, in contrast with the shrinkage associated with a (meth)acrylate polymerization at 22–23 $cm^3\ mol^{-1}$. Thiol-ene photopolymerizations, by virtue of progressing *via* the step growth mechanism and thereby exhibiting delayed gelation, give rise to materials that have distinct advantages in terms of material properties. One of the significant advantages in terms of material properties is the low stress that is resultant from the delayed gelation during polymerization. Delayed gelation also results in a greater fraction of the volume shrinkage occurring prior to gelation, reducing the amount of post-gel shrinkage. Thiol-ene polymerizations also show reduced stress within the polymer network. Typically, the formation of a glassy polymer results in the development of stress within the network, as the post-gelation shrinkage that occurs within a highly crosslinked network results in permanent strain within the network as the network structure now resists deformation. However, in thiol-ene polymerizations, the reduced shrinkage and the delayed gel point of the step-growth mechanism result in reduced stresses within the network.

In a novel approach to further reducing the stress within thiol-ene networks, addition-fragmentation capable functional groups such as allyl sulfide moieties, which can react with the thiyl radicals formed at any point of the polymerization, were incorporated into the polymer backbone. By incorporating allyl sulfide moieties into multifunctional thiol and ene monomer systems a covalent adaptable network is created in which the allyl sulfide continuously participates in addition-fragmentation reactions such that the bond structure of the network is covalent, yet each individual crosslink can be broken and re-formed.[66–68] The ability to reversibly fragment and reform the backbone and crosslinked structure of the network serves to alleviate the stresses and strains within the network even after gelation. Shrinkage stress reductions of up to 75% have been demonstrated through incorporation of allyl sulfide moieties.[69] The use of allyl sulfides to reduce shrinkage stress has also been demonstrated in thiol-yne systems,[54] as well as high modulus thiol-ene-methacrylate systems.[70]

## 1.7 Material Properties

A few limited investigations have directly evaluated structure–property relationships in thiol-ene polymers and evaluated the effect of a range of monomer chemistries and functionalities on polymer properties such as modulus, $T_g$, hardness, *etc.*[71–74] Due to the chemical nature of thiol-ene systems, a broad range of different thiol and ene monomer structures, chemistries, and functionalities can be utilized. Judicious selection of these monomers can therefore be exploited to formulate systems with an incredibly broad range of polymerization kinetics, network structures, and material properties. In many cases the polymerization and material properties of thiol-ene systems are not achievable by any other classical photopolymerization processes.

There are numerous different classes of thiol-ene formulations, with the classical binary thiol-ene systems, utilizing non-homopolymerizable ene monomers, being the most common. The classical binary systems exhibit the well known advantageous properties of rapid polymerizations, no oxygen inhibition, and formation of homogeneous polymer networks. One of the drawbacks of these systems is that it is difficult to achieve high modulus and $T_g$. For many thiol-ene polymer networks, the available monomers and resultant flexible sulfide bonds results in low $T_g$, low modulus networks at ambient temperatures. For applications that desire high modulus and $T_g$ such as dental restoratives and automotive and aerospace resins, the low $T_g$ of thiol-ene networks is a detriment. Numerous approaches have been utilized to increase the mechanical properties of thiol-ene polymers. Multifunctional carbamates synthesized by sequentially reacting aliphatic and aromatic diisocyanates with a tetrafunctional thiol yield thiol monomers that result in systems with excellent hardness and impact properties along with $T_g$'s of up to 108 °C due to the extensive hydrogen bonding within the network.[75] In another approach, the use of norbornenes to form the thiol-ene network results in networks with $T_g > 80$ °C, largely due to the mobility restriction that norbornenes impart to the network.[74] The use of sequential/hybrid curing methodologies or hybrid systems have also been demonstrated to achieve thiol-ene materials with improved modulus and $T_g$.[76–79] Crosslinking linear polymers is a technique that can result in improved mechanical and physical properties of the crosslinked material. Several studies have shown that thiol-ene crosslinking reactions resulted in improved physical properties and were seen to be more efficient when compared with thermally catalyzed crosslinking reactions traditionally used to crosslink linear polymers.[80–83]

The use of thiol-acrylate and thiol-methacrylate systems is a simple way to expand the achievable material properties of thiol-ene systems to include high modulus and $T_g$.[10,11] These types of systems exhibit both step growth thiyl addition and chain transfer as well as (meth)acrylate chain growth homopolymerization. This step-chain growth mechanism is represented schematically in Figure 1.2, where the step growth thiyl radical addition and chain transfer mechanism is represented by the first cycle and the thiyl radical addition and chain growth mechanism is represented by the second cycle. These binary systems enable achievement of a broader range of material properties resultant from the combination of step and chain growth polymerization mechanisms, as well as the large number of commercially available (meth)acrylates and their increased crosslink density. The presence of the thiol in the (meth)acrylate system imparts a number of beneficial characteristics to the polymerization and polymer properties. Incorporation of the thiol monomer, and hence chain transfer step, into the formulation dramatically reduces the sensitivity of the film to oxygen inhibition. The use of low concentrations of thiols has been demonstrated to lead to systems readily curable with both UV and visible irradiation in the presence of oxygen.[31] Typically, 1–10 wt% thiol monomer is sufficient to reduce oxygen inhibition and form tack free polymers even in films that are only a few micrometres in thickness.[29] The reduced oxygen inhibition allows films to be cured with significantly less photoinitiator (and photoinitiator combinations) and reduced irradiation intensities while still maintaining rapid polymerization rates. In many cases, the polymerization rates are dramatically increased by up to 10 fold or even greater depending on the oxygen level. Additionally, the chain transfer step that is resultant from the presence of the thiol results in slower and more uniform molecular weight development that delays the gel point and reduces the amount of polymerization induced shrinkage stress. The delayed gel point also results in achievement of higher overall functional group conversions. For relatively small amounts of thiol, typically 5–20 wt% depending on the thiol and (meth)acrylate molecular structures, the modulus and glass transition temperature are not significantly impacted. Perhaps the one disadvantage of adding thiols to (meth)acrylates is that the amount of thiol that can be added is limited due to the nature of the polymerization.[35,84] The thiol monomer can only participate in the polymerization *via* chain transfer and subsequent thiyl radical addition to (meth)acrylate groups. If too much thiol is present, the (meth)acrylate groups will be completely consumed leaving unreacted thiol monomer in the final polymer. The specific ratios of thiol and (meth)acrylate that can be accommodated vary widely dependent upon the types of thiols and (meth)acrylates utilized and the desired polymer properties. Additionally, thiol-methacrylate polymerizations proceed relatively slowly often leading to low functional group conversions.

The limitations of binary thiol-ene and thiol-(meth)acrylate systems can be overcome to achieve an even broader range of properties by utilizing ternary (or greater) systems comprising thiols, enes, and (meth)acrylates. The advantages of thiol-ene systems are largely maintained when using these thiol-ene-(meth)acrylate ternary systems. With this approach, the mechanistic propagation pathways become much more complex providing the opportunity to manipulate the polymerization and polymer properties in ways not possible with binary systems.[64,85] Ternary thiol-ene systems enable greater control over material properties by incorporating multiple monomer types. Additionally, *via* the step growth mechanism the network architecture is also dictated by the relative reactivities of the different functional groups. One of the distinctive properties of thiol-ene systems is that a 1 : 1 ratio of thiol to ene functional groups is required for achievement of full conversion of both functional groups and maximum mechanical properties such as modulus and $T_g$. Even slight deviations from a 1 : 1 ratio of functional groups can result in a dramatic reduction these properties. The use of ternary systems, and especially thiol-ene-(meth)acrylate systems, provides a consistency of material properties over a range of functional group ratios, reducing the need for an exact 1 : 1 ratio of functional groups.

An example of ternary systems being used to achieve high modulus is seen in dental restorative materials. The rapid polymerizations to high conversion, lack of oxygen inhibition, and low shrinkage and stress is an ideal combination for application of thiol-enes as dental restorative materials.[86,87] In combination with methacrylate monomers in thiol-ene-meethacrylate systems, high modulus dental restorative materials are also achieved while maintaining the advantages of the thiol-ene polymerization.[53,87–89] Interestingly, reductions in shrinkage stress for thiol-ene-methacrylate systems are greater than for traditional thiol-ene systems. The increased reduction in shrinkage stress is resultant from a hybrid polymerization mechanism for thiol-enes and methacrylates. In these systems, the reaction often proceeds in two relatively distinct stages; a first stage that is dominated by methacrylate homopolymerization and a second stage that is dominated by thiol-ene polymerization.[90,91]

## 1.8 Applications

Resultant from the thiol-ene step growth mechanism, thiol-enes exhibit uniform network formation with low shrinkage and stress. Additionally, the gel point occurs at relatively high conversions relative to chain growth polymerizations followed by a rapid increase in crosslink density and modulus. Combined with the ability to overcome oxygen inhibition, these properties differentiate thiol-enes from other photopolymerization techniques and lead to significant polymerization and material property advantages for a wide range of applications.

The incorporation of sulfur into thiol-ene networks imparts a higher refractive index than comparable organic networks. The refractive index of thiol-ene polymers formed from typical thiol and ene monomers (such as pentaerythritol tetra(3-mercaptopropionate) and triallyl triazine trione) is commonly ~1.54 without the incorporation of any aromatic substituents. In fact, thiol-yne reactions have yielded polymers with refractive indexes up to 1.65 due to high concentrations of sulfur within the material.[92]

Much research has investigated the use of thiol-enes in the field of electro-optics. Thiol-ene systems have been used in applications that vary from flexible display components, photonic crystals, lens components, nanofiber high-oxygen-barrier networks, nanopillars and microbridges.[10] The uniform molecular weight development and delayed gel point of thiol-ene polymerizations result in the formation of PDLCs and HPDLCs with diffraction gratings that have excellent diffraction efficiency, high switching speeds and low switching voltages that make them desirable for use in photonic lasers, spectrometers for chemical and biological sensors and nanostructured devices. There are more than 50 research publications investigating the use of PDLCs and HPDLCs.[11] Thiol-ene polymerizations have been explored extensively as optical adhesives because of the outstanding optical clarity, high refractive index and ability to form nearly stress-free materials that do not cause optical distortion.[93]

Thiol-ene polymers have been demonstrated as ideal substrates for imprint lithographic applications.[33,94–98] Polymeric microdevices have been fabricated with improved feature qualities and high aspect ratios without warping or stress induced surface wrinkling.[99,100] Similarly, thiol-ene polymer derived ceramic devices have demonstrated improved curing and feature qualities.[101,102]

The uniform network formation of thiol-ene polymers results in networks that are very homogeneous and thereby exhibit glass transition regions that are much narrower than comparable chain growth systems. Uniquely, the narrow glass transition regions lead to extremely efficient energy absorption and dissipation for systems with glass transition temperatures near to ambient temperature.[103,104] The narrow glass transition region also leads to distinct and rapid shape memory polymer actuation and superior shape retention.[105] The uniform networks can also be tailored to serve as excellent barrier membranes for gas transport.[106]

## 1.9 Odor

Thiols are well known to exhibit a distinct odor that is often considered offensive. In some cases, the odor can be dealt with using traditional masking agents. Alternatively, eliminating thioglycolic acid or mercaptopropionic acid reduces the odor of thiols and a number of thiols are available with reported dramatic reductions in odor due to improved synthetic and purification procedures.[107,108] Thiols are also available from Yodo Chemicals with little to no odor. The odor issues associated with thiols are strongly correlated to the molecular weight of the monomers. Low molecular weight monomers such as ethane or butane dithiol exhibit extremely strong and offensive odors while higher molecular weight and commonly utilized trimethylolpropane tris(3-mercaptopropionate) (MW = 399) and pentaerythritol tetra(3-mercapto-propionate) (MW = 489) exhibit much milder odors, particularly when appropriately purified. Once polymerized, crosslinked thiol-ene polymers that achieve high conversion exhibit minimal if any residual odor.

## 1.10 Shelf Life

Thiol-ene monomer systems polymerize extremely rapidly and also exhibit minimal inhibition by oxygen and other traditional quinone based stabilizers. As such, thiol-ene systems can be rapidly polymerized with little to no inhibition period to form tack-free surfaces with low concentrations of photoinitiator and low irradiation intensities. While this feature is ideal in formulating ultra-fast and energy efficient systems, the drawback is that thiol-ene mixtures may exhibit poor shelf stability at room temperature. Unstabilized shelf lives of thiol-ene systems can range from hours to months with the most rapidly polymerizing thiol-ene systems (vinyl ethers and norbornenes) often exhibiting the poorest shelf stability. Depending on the particular thiol and ene monomers utilized, there are numerous mechanisms resulting in poor stability. The thiol-ene reactions can be base-catalyzed for enes substituted with electron withdrawing groups, with the base-catalyzed Michael addition reaction of thiols to acrylates being a very well known reaction. Decomposition of peroxide impurities results in generation of radicals that can initiate polymerization, while oxygen can initiate the polymerization through the ground-stage charge transfer complex.[37,38]

Numerous investigations have examined strategies for stabilizing thiol-ene systems including the use of vinyl radical scavengers,[109] phosphorous acid and hindered phenolic antioxidant stabilizers,[110] thermal stabilizers such as pyrogallol, hydroquinone, catechol, and 1,2,3-trihydroxybenzene,[111,112] sulfur and potassium iodine–iodine mixtures,[113] hindered phenolics,[114–118] phosphines and triarylphosphites,[107] and phosphoric acid.[119] Different inhibitors and combinations of inhibitors have varying effectiveness for different thiol-ene monomer systems and must be evaluated on a case-by-case basis. Perhaps the most effective and simplest stabilizer for a broad range of thiol-ene systems is the aluminium or ammonium salt of *N*-nitrosophenylhydroxylamine.[116,120]

Traditional stabilizers such as pyrogallol, hydroquinones, and catechol improve the stability of thiol-ene systems; however, rather than creating an initial polymerization inhibition period where the inhibitors are consumed before the reaction commences, they tend to act as polymerization retarders.[111] Additionally, stabilization often comes at the price of reduced polymerization kinetics and functional group conversion. Stabilizers are more effective in combination with acrylates, methacrylates, styrene derivatives, and conjugated 1,3-dienes.[10]

## 1.11 Oxidative and Photo Stability

Many polymer films degrade oxidatively when subjected to elevated temperature. Uniquely, thiol-ene polymer films have been found to exhibit excellent thermal stability.[121] Thiol-ene films exhibited no weight loss after being held in air at 200 °C for 40 min and the onset of thermal decomposition was greater than 250 °C with a temperature ramp of 10 °C $min^{-1}$. In thiol-ene networks, a thioether group is formed every time a thiol group adds across a double bond. Thioethers are known to act as antioxidant stabilizers for many commercial polymers when used in combination with radical scavengers.[122–124] In addition to oxidative stability thiol-enes have also demonstrated excellent optical stability with minimal discoloration after exposure to 50 °C for three years (Norland Optical Products Literature, www.norlandproducts.com), which correlates to greater than 20 years stability at ambient conditions. Thiol-ene films using monomers that do not contain aromatic groups have also been shown to exhibit low yellowing in simulated UV weathering experiments.[10]

## References


1. T. Posner, *Ber.*, 1905, **38**, 646.
2. J. v. Braun and R. Murjahn, *Ber.*, 1926, **59**, 1202.
3. M. S. Kharasch and F. R. Mayo, *Chem. Ind.*, 1938, **57**, 752.
4. D. D. Coffman, US Pat. 2 347 182, 1944.
5. H. C. Kolb, M. G. Finn and K. B. Sharpless, *Angew. Chem.*, 2001, **113**, 2056.
6. C. R. Morgan, F. Magnotta and A. D. Ketley, *J. Polym. Sci., Polym. Chem. Ed.*, 1977, **15**, 627.
7. C. R. Morgan and A. D. Ketley, *J. Polym. Sci., Polym. Lett. Ed.*, 1978, **16**, 75.
8. D. P. Gush and A. D. Ketley, *Mod. Paint Coat.*, 1978, November, 58.
9. A. F. Jacobine, in *Radiation Curing in Polymer Science and Technology III*, ed. J. D. Fouassier and J. F. Rabek, Elsevier, London, 1993, ch. 7, pp. 219–268.
10. C. E. Hoyle, T. Y. Lee and T. Roper, *J. Polym. Sci., Part A*, 2004, **42**, 5301.
11. C. E. Hoyle and C. N. Bowman, *Angew. Chem., Int. Ed.*, 2010, **49**, 1540.
12. E. Andrzejewska, *Photochemistry and UV Curing: New Trends*, 2006, 127.
13. A. B. Lowe, C. E. Hoyle and C. N. Bowman, *J. Mater. Chem.*, 2010, **20**, 4745.
14. M. J. Kade, D. J. Burke and C. J. Hawker, *J. Polym. Sci., Part A: Polym. Chem*, 2010, **48**, 743.
15. J. G. Kloosterboer, *Adv. Polym. Sci.*, 1988, **84**, 1.
16. T. Y. Lee, C. A. Guymon, E. S. Jonsson and C. E. Hoyle, *Polymer*, 2004, **45**, 6155.
17. C. Decker, *Surf. Coat. Int., Part B*, 2005, **88**, 9.
18. C. N. Bowman and C. J. Kloxin, *AICHE J.*, 2008, **54**, 2775.
19. T. M. Roper, T. Y. Lee, C. A. Guymon and C. E. Hoyle, *Macromolecules*, 2005, **38**, 10109.
20. L. J. Gou, B. Opheim, C. N. Coretsopoulos and A. B. Scranton, *Chem. Eng. Commun.*, 2006, **193**, 620.
21. C. N. Bowman and N. A. Peppas, *Macromolecules*, 1991, **24**, 1914.

22. H. Lu, J. W. Stansbury and C. N. Bowman, *Dent. Mater.*, 2004, **20**, 979.
23. G. P. Simon, P. E. M. Allen, D. J Bennett, D. R. G. Williams and E. H. Williams, *Macromolecules*, 1989, **22**, 3555.
24. A. R. Kannurpatti, J. W. Anseth and C. N. Bowman, *Polymer*, 1998, **39**, 2507.
25. A. R. Kannurpatti, K. J. Anderson, J. W. Anseth and C. N. Bowman, *J. Polym. Sci., Part B*, 1997, **35**, 2297.
26. H. H. Szmant, A. J. Mata, A. J. Namis and A. M. Panthananickal, *Tetrahedron*, 1976, **32**, 2665.
27. M. D. Goodner and C. N. Bowman, *Chem. Eng. Sci.*, 2002, **57**, 887.
28. G. Odian, *Principles of Polymerization*, John Wiley & Sons, New Jersey, 4th edn, 2004, ch. 3, pp. 256–263.
29. A. K. O'Brien, N. B. Cramer and C. N. Bowman, *J. Polym. Sci., Part A*, 2006, **44**, 2007.
30. N. B. Cramer, J. P. Scott and C. N. Bowman, *Macromolecules*, 2002, **35**, 5361.
31. M. R. Zonca, B. Falk and J. V. Crivello, *J. Macromol. Sci., Part A: Pure Appl. Chem.*, 2004, **41**, 741.
32. C. Nason, T. Roper, C. Hoyle and J. A. Pojman, *Macromolecules*, 2005, **38**, 5506.
33. L. M. Campos, K. L. Killops, R. Sakai, J. M. J. Paulusse, D. Damiron, E. Drockenmuller, B. W. Messmore and C. J. Hawker, *Macromolecules*, 2008, **41**, 7063.
34. C. Boyer, A. Granville, T. P. Davis and V. Bulmus, *J. Polym. Sci., Part A: Polym. Chem.*, 2009, **47**, 3773.
35. N. B. Cramer and C. N. Bowman, *J. Polym. Sci. Part A. Polym. Chem.*, 2001, **39**, 3311.
36. N. B. Cramer, S. K. Reddy, M. Cole, C. Hoyle and C. N. Bowman, *J. Polym. Sci., Part A*, 2004, **42**, 5817.
37. V. T. D'Souza, R. Nanjundiah, J. Baeza and H. H. Szmant, *J. Org. Chem.*, 1987, **52**, 1720.
38. V. T. D'Souza, V. K. Iyer and H. H. Szmant, *J. Org. Chem.*, 1987, **52**, 1725.
39. S. Sensfuss, M. Friedrich and E. Klemm, *Makromol Chem.*, 1991, **192**, 2895.
40. E. Klemm and S. T. Sensfuss, *J. Macromol. Sci. Chem.*, 1991, **A28**, 875.
41. D. D. Carlson and A. R. Knight, *Can. J. Chem.*, 1973, **51**, 1410.
42. N. P. Skerrett and N. W. Thompson, *Trans. Faraday Soc.*, 1938, **34**, 1238.
43. S. K. Reddy, C. N. Cramer and C. N. Bowman, *Macromolecules*, 2006, **39**, 3673.
44. T. F. Scott, C. J. Kloxin, R. B. Draughon and C. N. Bowman, *Macromolecules*, 2008, **41**, 2987.
45. T. M. Roper,C. E. Hoyle and D. H. Magers, in *Photochemistry and UV Curing: New Trends*, ed. J. P. Fouassier, Research Signpost, Kerala, 2006, pp. 253–264.

46. T. M. Roper, C. A. Guymon, E. S. Jonsson and C. E. Hoyle, *J. Polym. Sci., Part A: Polym. Chem.*, 2004, **42**, 6283.
47. H. Behringer, *Justus Liebigs Ann. Chem.*, 1949, **564**, 219.
48. K. Griesbaum, *Angew. Chem., Int. Ed. Engl.*, 1970, **9**, 273.
49. A. T. Blomquist and J. Wolinsky, *J. Org. Chem.*, 1958, **23**, 551.
50. J. A. Mampmeier and G. Chen, *J. Am. Chem. Soc.*, 1965, **87**, 2608.
51. A. A. Oswald, K. Griesbaum, B. E. Hudson and J. M. Bregman, *J. Am. Chem. Soc.*, 1964, **86**, 2877.
52. B. D. Fairbanks, T. F. Scott, C. J. Kloxin, K. S Anseth and C. N Bowman, *Macromolecules*, 2009, **42**, 211.
53. S. Ye, N. B. Cramer, I. R. Smith, K. R. Voigt and C. N. Bowman, *Macromolecules*, 2011, **44**, 9084.
54. H. Y. Park, C. J. Kloxin, T. F. Scott and C. N. Bowman, *Macromolecules*, 2010, **43**, 10188.
55. J. W. Chan, J. Shin, C. E. Hoyle, C. N. Bowman and A. B. Lowe, *Macromolecules*, 2010, **43**, 4937.
56. B. D. Fairbanks, E. A. Sims, K. S. Anseth and C. N. Bowman, *Macromolecules*, 2010, **43**, 4113.
57. D. Meziane, A. Barras, A. Kromka, J. Houdkova, R. Boukherroub and S. Szunerits, *Anal. Chem.*, 2012, **84**, 194.
58. M. Lomba, L. Oriol, R. Alcala, C. Sanchez, M. Moros, V. Grazu, J. L. Serrano and J. M. de la Fuente, *Macromol. Biosci.*, 2011, **11**, 1505.
59. S. K. Reddy, O. Okay and C. N. Bowman, *Macromolecules*, 2006, **39**, 8832.
60. O. Okay, S. K. Reddy and C. N. Bowman, *Macromolecules*, 2005, **38**, 4501.
61. O. Okay and C. N. Bowman, *Macromol. Theory Simul.*, 2005, **14**, 267.
62. N. B. Cramer, T. Davies, A. K. O'Brien and C. N. Bowman, *Macromolecules*, 2003, **36**, 4631.
63. N. B. Cramer, S. K. Reddy, A. K. O'Brien and C. N. Bowman, *Macromolecules*, 2003, **36**, 7964.
64. S. K. Reddy, C. N. Cramer and C. N. Bowman, *Macromolecules*, 2006, **39**, 3681.
65. S. Beuermann and M. Buback, *Prog. Polym. Sci.*, 2002, **27**, 191.
66. T. F. Scott, A. D. Schneider, W. D. Cook and C. N. Bowman, *Science*, 2005, **308**, 1615.
67. C. J. Kloxin, T. F. Scott, B. J. Adzima and C. N. Bowman, *Macromolecules*, 2010, **43**, 2643.
68. W. D. Cook, S. Chausson, F. Chen, L. Le Pluart, C. N. Bowman and T. F. Scott, *Polym. Int.*, 2008, **57**, 469.
69. C. J. Kloxin, T. J. Scott and C. N. Bowman, *Macromolecules*, 2009, **42**, 2551.
70. H. Y. Park, C. J. Kloxin, T. F. Scott and C. N. Bowman, *Dent. Mater.*, 2010, **26**, 1010.
71. H. Zhou, Q. Li, J. Shin and C. E. Hoyle, *Macromolecules*, 2009, **42**, 2994.
72. Q. Li, H. Zhou and C. E. Hoyle, *Polymer*, 2009, **50**, 2237.

73. K. Wutticharoenwong and M. D. Soucek, *J. Appl. Polym. Sci.*, 2009, **113**, 2173.
74. J. A. Carioscia, L. Schneidewind, C. O'Brien, R. Ely, C. Feeser, N. Cramer and C. N. Bowman, *J. Polym. Sci. Part A. Polym. Chem.*, 2007, **45**, 5686.
75. Q. Li, H. Zhou, D. A. Wicks and C. E. Hoyle, *J. Polym. Sci. Part A. Polym. Chem.*, 2007, **45**, 5103.
76. R. A. Ortiz, B. A. P. Urbina, L. V. C. Valdez, L. B. Duarte, R. G. Santos, A. E. G. Valdez and M. D. Soucek, J. Polym. Sci, *Part A*, 2007, **45**, 4829.
77. R. A. Ortiz, A. E. G. Valdez, L. B. Durante, R. G. Santos, L. R. O. Flores and M. D. Soucek, *Macromol. Chem. Phys.*, 2008, **209**, 2157.
78. R. A. Ortiz, B. A. P. Urbina, R. G. Santos, L. B. Duarte, A. E. G. Valdez and M. D. Soucek, *Macromol. Mater. Eng.*, 2008, **293**, 731.
79. O. Fichet, F. Vidal, J. Laskar and D. Teyssie, *Polymer.*, 2005, **46**, 37.
80. D. H. Lim, H. S. Do, H. J. Kim, J. S. Bang and G. H. Yoon, *J. Adhes. Sci. Technol.*, 2007, **21**, 589.
81. A. Taguet, B. Ameduri and B. Boutevin, in *Crosslinking in Materials Science*, vol. 184, Springer, Berlin, 2005, pp. 127–211.
82. A. C. Arsenault, D. P. Puzzo, I. Manners and G. A. Ozin, *Nat. Photonics*, 2007, **1**, 468.
83. D. B. Otts, E. Heidenreich and M. W. Urban, *Polymer*, 2005, **46**, 8162.
84. L. Lecamp, F. Houllier, B. Youssef and C. Bunel, *Polymer*, 2001, **42**, 2727.
85. S. Fan, F. Y. C. Boey and M. J. M. Abadie, *Eur. Polym. J.*, 2008, **44**, 2123.
86. H. Lu, J. A. Carioscia, J. W. Stansbury and C. N. Bowman, *Dent. Mater.*, 2005, **21**, 1129.
87. N. B. Cramer, C. L. Couch, K. M. Schreck, J. A. Carioscia, J. E. Boulden, J. W. Stansbury and C. N. Bowman, *Dent. Mater.*, 2010, **26**, 21.
88. N. B. Cramer, C. L. Couch, K. M. Schreck, J. E. Boulden, R. Wydra, J. W. Stansbury and C. N. Bowman, *Dent. Mater.*, 2010, **26**, 799.
89. J. E. Boulden, N. B. Cramer, K. M. Schreck, C. L. Couch, C. Bracho-Troconis, J. W. Stansbury and C. N. Bowman, *Dent. Mater.*, 2011, **27**, 267.
90. T. Y. Lee, J. Carioscia, Z. Smith and C. N. Bowman, *Macromolecules*, 2007, **40**, 1473.
91. T. Y. Lee, Z. Smith, S. K. Reddy, N. B. Cramer and C. N. Bowman, *Macromolecules*, 2007, **40**, 1466.
92. J. W. Chan, H. Zhou, C. E. Hoyle and A. B. Lowe, *Chem. Mater.*, 2009, **21**, 1579.
93. J. G. Woods and A. F. Jacobine, *RadTech '92 North Am. UV/EB Conf. Expo., Conf. Proc.*, 1992, vol. I, 173.
94. Y. W. Yi, V. Khire, C. N. Bowman, J. E. Maclennan and N. A. Clark, *J. Appl. Phys.*, 2008, **103**.
95. V. W. Khire, Y. Yi, N. A. Clark and C. N. Bowman, *Adv. Mater.*, 2008, **20**, 3308.
96. B. K. Long, B. K. Keitz and C. G. Willson, *J. Chem. Mater.*, 2007, **17**, 3575.

97. E. C. Hagberg, M. Malkoch, Y. B. Ling, C. J. Hawker and K. R. Carter, *Nano Lett.*, 2007, **7**, 2134.
98. I. W. Moran, A. L. Briseno, S. Loser and K. R. Carter, *Chem. Mater.*, 2008, **20**, 4595.
99. J. F. Ashley, N. B. Cramer, R. H. Davis and C. N. Bowman, *Lab Chip*, 2011, **11**, 2772.
100. M. D. Dickey, E. Collister, A. Raines, P. Tsiartas, T. Holcombe, S. V. Sreenivasan, R. T. Bonnecaze and C. G. Willson, *Chem. Mater.*, 2006, **18**, 2043.
101. S. K. Reddy, N. B. Cramer, T. Cross, R. Raj and C. N. Bowman, *Chem. Mater.*, 2003, **15**, 4257.
102. N. B. Cramer, S. K. Reddy, H. Lu, T. Cross, R. Raj and C. N. Bowman, *J. Polym. Sci., Part A: Polym. Chem.*, 2004, **42**, 1752.
103. A. F. Senyurt, H. Y. Wei, C. E. Hoyle, S. G. Piland and T. E. Gould, *Macromolecules*, 2007, **40**, 4901.
104. A. F. Senyurt, C. E. Hoyle, H. Y. Wei, S. G. Piland and T. E. Gould, *Macromolecules*, 2007, **40**, 3174.
105. D. P. Nair, N. B. Cramer, T. F. Scott, C. N. Bowman and R. Shandas, *Polymer*, 2010, **51**, 4383.
106. L. Kwisnek, S. Nazarenko and C. E. Hoyle, *Macromolecules*, 2009, **42**, 7031.
107. S. Davidson and C. Mead, *RadTech Eur. Tech. Conf. Proc.*, 2003, Vol. I, 579.
108. N. Rehnberg, U. Annby, C. A. Sjorgreen and R. S. Davidson, PCT Pat. Appl. WO 2001044377, 2001.
109. E. E. Stahly, U.S. Pat. 3 619 393, 1971.
110. J. L. Guthrie and F. J. Rendulic, U.S. Pat. 3 855 093, 1974.
111. E. Klemm, S. Sensfuss, U. Holfter and H. J. Flammersheim, *Angew. Makromol. Chem.*, 1993, **212**, 121.
112. G. Kuhne, J. S. Diesen and E. Klemm, *Angew. Makromol. Chem.*, 1996, **242**, 139.
113. J. G. Woods, *Radiation Curing: Science and Technology*, ed. S.P. Pappas, Plenum, New York, USA, 1992, 333.
114. D. M. Glaser, J. F. Jacobine and P. J. Grabek, U.S. Pat. 5 459 173, 1995.
115. J. P. Dowling, S. C. Richardson and K. Quigley, U.S. Pat. 5 358 976, 1994.
116. M. A. Rakas and A. F. Jacobine, *J. Adhes.*, 1992, **36**, 247.
117. C. L. Kehr and W. R. Wszolek, U.S. Pat. 3 661 744, 1992.
118. C. L. Kehr and W. R. Wszolek, *Proc. Am. Chem. Soc. Div. Org. Coat. Plast. Chem.*, 1973, **33**, 295.
119. C. E. Hoyle, R. D. Hensel and M. B. Grubb, *Polym. Photochem.*, 1984, **4**, 69.
120. B. K. Christmas, H. Ho, T. Ngo and D. Brenes, *Proc RadTech North Am. Tech. Conf. Proc.*, 2002, Vol. I, 712.
121. J. G. Woods, M. A. Rakas, L. M. Alberino, P.L. Kropp, D. M. Sutkaitis and S. T. Nakos, WO 95/00869, 1995.
122. G. Scott, *Dev. Polym. Stab.*, 1981, **4**, 1.
123. G. Scott, *Dev. Polym. Stab.*, 1983, **6**, 29.
124. J. R. Shelton, *Dev. Polym. Stab.*, 1981, **4**, 23.

CHAPTER 2

# End-group Functionalization of RAFT-prepared Polymers Using Thiol-X Chemistries

ANDREW B. LOWE

Centre for Advanced Macromolecular Design (CAMD), School of Chemical Engineering, University of New South Wales, Kensington, Sydney, New South Wales 2052, Australia
Email: a.lowe@unsw.edu.au

## 2.1 Introduction to RAFT Polymerization

Reversible addition-fragmentation chain transfer (RAFT) polymerization is an example of a reversible deactivation radical polymerization (RDRP) that operates on the principle of degenerative chain transfer.[1] The key to successful RAFT polymerization is the use of an appropriate chain transfer agent (CTA). Suitable RAFT CTAs are those based on thiocarbonylthio compounds such as dithioesters, trithiocarbonates, xanthates and dithiocarbamates, Figure 2.1.

The key structural features of these RAFT CTAs are the Z and R groups. These can, in principle, be of any chemical identity although not all structural variations are effective as RAFT agents or may, for example, be suitable for certain monomer classes and not others. Several excellent papers and reviews have discussed in detail the structural requirements of the Z and R groups for them to be effective and readers are directed to these manuscripts for a more detailed explanation.[2–6] However, we will briefly note that the Z group dictates the 'bulk' reactivity of the CTA (rates of addition to the C=S bond and

RSC Polymer Chemistry Series No. 6
Thiol-X Chemistries in Polymer and Materials Science
Edited by Andrew B. Lowe and Christopher N. Bowman

Published by the Royal Society of Chemistry, www.rsc.org

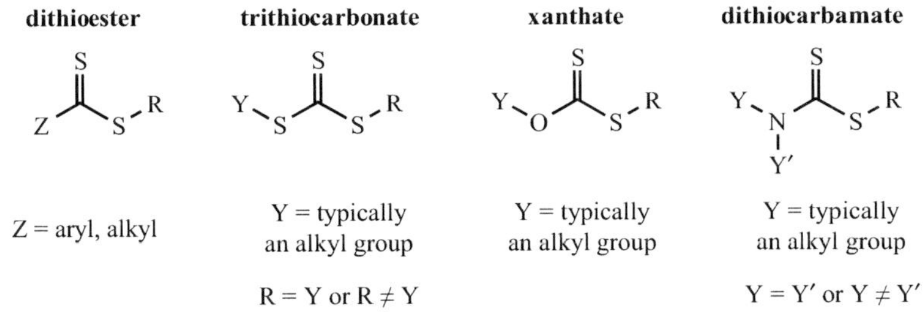


**Figure 2.1** General structures of common families of RAFT agents.

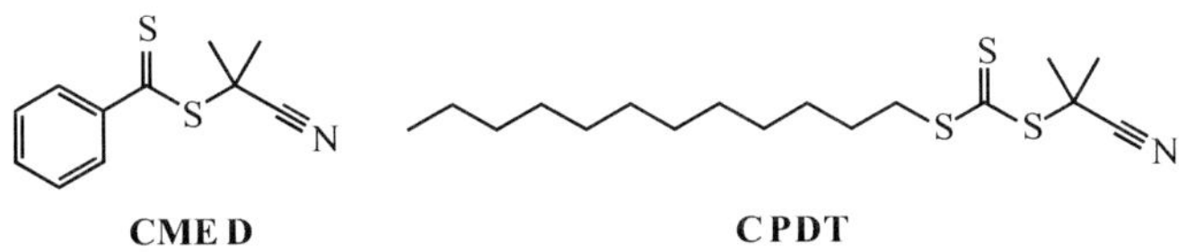


**Figure 2.2** Chemical structures of 1-cyano-1-methylethyl dithiobenzoate (**CMED**) and 2-cyano-2-propyldodecyl trithiocarbonate (**CPDT**) RAFT chain transfer agents.

stability/lifetime of the intermediate carbon centered radical, *vide infra*) and determines the general classification of the thiocarbonylthio compound. The R group, in contrast, allows for a fine-tuning of the CTA reactivity. The R group must be a good homolytic leaving group (and as such differences in its structure will also affect the lifetime of the intermediate radical to some extent) but must also be able to readily add to monomer, *i.e.* (re)initiate polymerization. For most practical purposes the use of certain dithiobenzoates, such as 1-cyano-1-methylethyl dithiobenzoate (**CMED**), or trithiocarbonates, such as 2-cyano-2-propyldodecyl trithiocarbonate (**CPDT**), derivatives, Figure 2.2, are sufficient for attaining a suitable level of control and facilitate the synthesis of target (co)polymers from an impressively broad range of monomers.

In the case of deactivated monomers such as vinyl esters and vinyl amides, the use of certain dithiocarbamates such as cyanomethyl methyl(phenyl)-carbamodithioate (**CMMPC**) is recommended. In addition to these substrate-specific RAFT agents, certain specific dithiocarbamate derivatives have recently been developed as so-called 'universal CTAs' being able to control the polymerization of *both* activated and deactivated monomer substrates. Examples of such CTAs include methyl 2-((methyl(pyridin-4-yl)-carbamothioyl)thio)propanoate (**MPCTP**) and 2-cyanopropan-2-yl methyl(pyridin-4-yl)carbamodithioate (**CPMPC**), Figure 2.3.

Finally, it should be pointed out that a series of the most commonly employed and effective RAFT CTAs are now commercially available including all the examples shown in Figures 2.2 and 2.3.

From a practical viewpoint it is noted that RAFT polymerizations are no more difficult to execute than a normal radical polymerization. Indeed, the basic components and range of experimental conditions under which RAFT

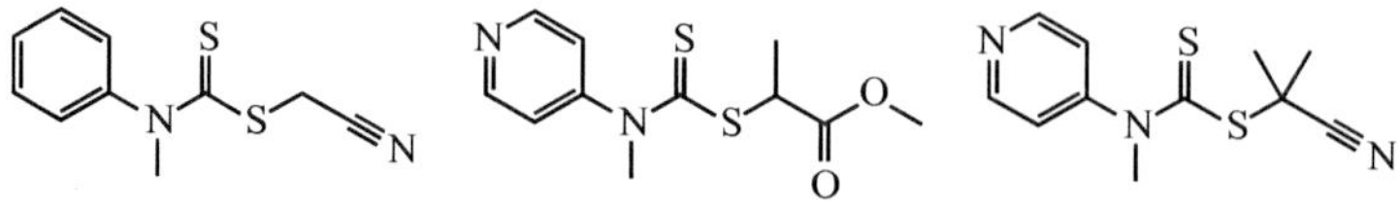


**Figure 2.3** Chemical structures of cyanomethyl methyl(phenyl)carbamodithioate (**CMMPC**), methyl 2-((methyl(pyridin-4-yl)carbamothioyl)thio)propanoate (**MPCTP**) and 2-cyanopropan-2-yl methyl(pyridin-4-yl)carbamodithioate (**CPMPC**).

can be successfully employed are as broad and versatile as normal radical polymerization, the only difference being the addition of a RAFT CTA. From a mechanistic viewpoint, therefore, it is not surprising that the basic polymerization process is also identical to a normal radical polymerization involving the basic steps of initiation, propagation and termination. However, superimposed on these elementary processes are those radical reactions involving the RAFT CTA, namely addition, fragmentation and reinitiation. The generally accepted RAFT mechanism is shown in Scheme 2.1.[4–7]

This warrants some discussion since a basic mechanistic picture of the RAFT process also explains why we are able to use RAFT-prepared (co)polymers in macromolecular thiol-based chemistries. As a radical polymerization, primary radicals (**1**, step I) must first be generated. This can be accomplished employing any of the standard methods for primary radical formation. The most common method, however, involves thermal generation of species **1** using common azo initiators such as 2,2′-azobis(2-methylpropionitrile) (AIBN). Neglecting cage reactions, **1** may react with monomer (M) to form the **1**-M adduct **3**, or add to RAFT agent, **2** (step II). In the case of the first option, adduct **3** can likewise add monomer, *i.e.* undergo propagation, or add to **2**. Given the extremely high chain transfer constants of species **2** it is very unlikely that more than a couple of monomer additions will occur before reaction with the RAFT agent. So, **3** is likely to add to **2** forming the intermediate carbon centered radical **5**. This reaction is, however, reversible and **5** may fragment reforming **2** and **3**. Alternatively, **5** can fragment to form a new RAFT agent **6** and the radical species **7**. This is the desired fragmentation pathway and is generally favoured with an appropriate choice of RAFT agent for a given monomer. Again, however, this process is reversible and **7** may add back to **6** *etc*. Provided **7** is a good reinitiating species then there is a high probability that **7** will add to monomer, forming a new propagating species **8**. The net result of this sequence of forward reactions is the generation of a new oligomer/polymer chain, **8**, initiated by the R group of the original RAFT agent, **2**, and the formation of a new thiocarbonylthio compound, **6**, that is also able to serve effectively as a RAFT agent. This entire sequence represents one addition-fragmentation cycle. The second option for the primary radical **1** is to add directly to the RAFT agent **2** to give the intermediate radical **4**. This reaction is reversible and, again, fragmentation in the forward direction is the preferred pathway. To favour forward fragmentation and formation of the new RAFT CTA **9** and the

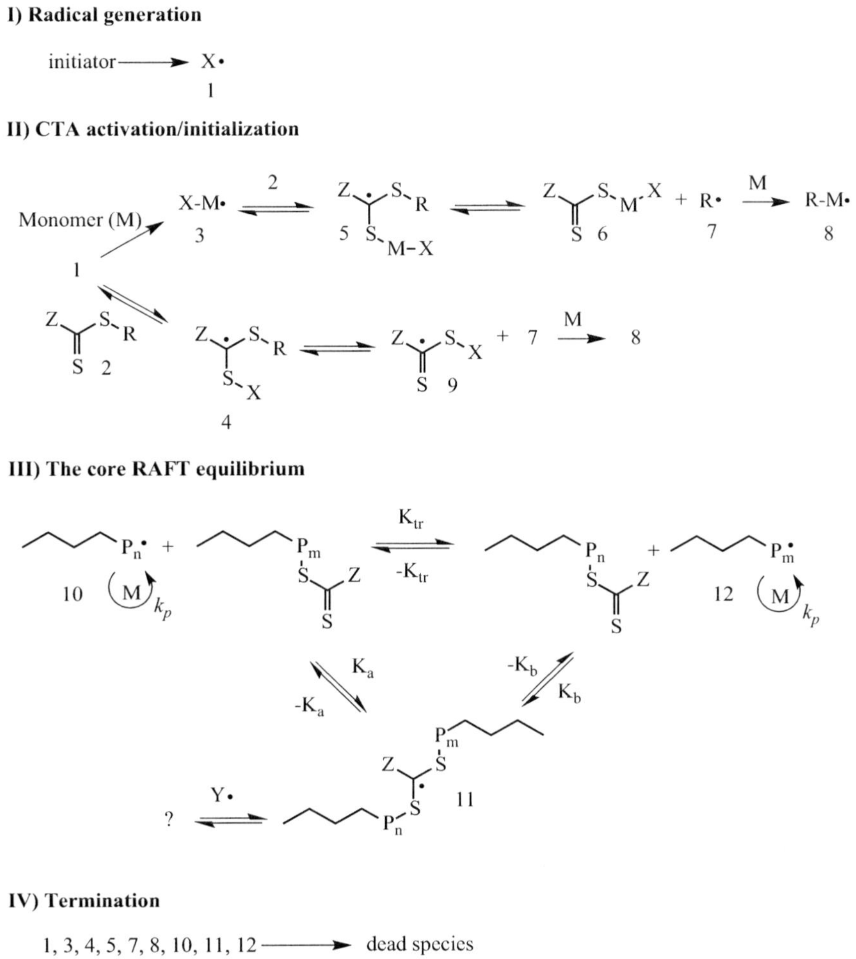


**Scheme 2.1** The RAFT mechanism.

radical **7**, the R group must be a better radical leaving group than **1**. The radical derived from the R group, **7**, will ideally add to a monomer initiating a new chain. Given the reversibility of many of these steps and the various reaction pathways available to the different generated radical species this part of the overall mechanism is fairly complex. However, the result of the series of reactions in step II, often termed the RAFT pre-equilibrium, is activation of the starting RAFT agent, **2**, to new oligomeric-type RAFT chain transfer agents. Once this pre-equilibrium is complete, the polymerization enters the main RAFT equilibrium (Step III Scheme 2.1). This is the stage that predominates in a RAFT polymerization and is the time period during which most of the monomer consumption, *i.e.* chain propagation, occurs. This stage involves the degenerative chain transfer of the thiocarbonylthio species between polymer chains **10** and **12** *via* the intermediate radical species **11**. Since RAFT is a radical

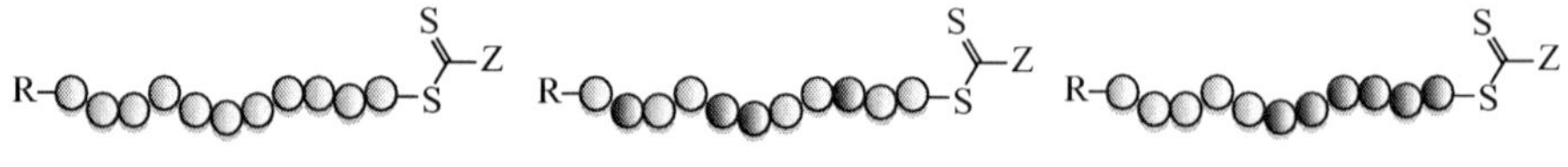


**Figure 2.4** Examples of accessible (co)polymer architectures with thiocarbonylthio end groups.

process and given that it is not possible to completely suppress all undesirable side reactions there will, inevitably be processes that occur that result in dead or inactive chains. However, assuming a RAFT polymerization proceeds 'smoothly', *i.e.* no thiocarbonylthio end groups are lost in such side reactions, then the vast majority (>95%) of (co)polymer chains will have thiocarbonylthio functional groups at the ω-termini, Figure 2.4, in the case where common RAFT CTAs are used.

In the early days of RAFT polymerization research, the presence of these thiocarbonylthio end-groups in the final polymeric materials was, in some instances, noted as a drawback. This was due, primarily, to the often intense colour associated with the RAFT CTAs (purples, reds, oranges and yellows are common for dithiobenzoates and trithiocarbonates), colour that is also observed in the final macromolecules since the polymers are simply polymeric versions of the thiocarbonylthio compounds. This led to researchers examining various methods for removing these groups from the polymer chain ends to obtain non-coloured products.

## Methods for Removing Thiocarbonylthio Groups from Polymer Chain Ends

As noted above, the presence of thiocarbonylthio end groups in (co)polymers, while crucial for successful RAFT (co)polymerizations, may be undesirable in the final product especially if coloured materials pose a potential problem. Fortunately, thiocarbonylthio compounds are inherently reactive and there are various chemistries that can be employed in their complete removal or transformation into a different chemical entity. Figure 2.5 highlights examples of these chemical and physical approaches to end-group removal and/or transformation. These reactions include radical based processes leading to the introduction of –H, –OH and –CN-based end groups, thermal elimination to give ene end-functional polymers (NB these four methods also result in the complete removal of S from the (co)polymer), heteroatom Diels–Alder functionalisation with a diene to give dithioacetal derivatives and nucleophilic cleavage, most commonly with a 1° or 2° amine or *via* hydrazinolysis, to yield the thiol, or thiolate, end-terminal macromolecule.

We will not discuss each of these modification routes in detail since there are a number of papers, highlight articles and mini-reviews that cover the fundamental characteristics of most of these reactions in detail including the use of newly introduced functionality in further chemical reactions.[8–12] Needless to say, however, not all of these approaches for chemical modification are as equally versatile or applicable. For example, the heteroatom Diels–Alder

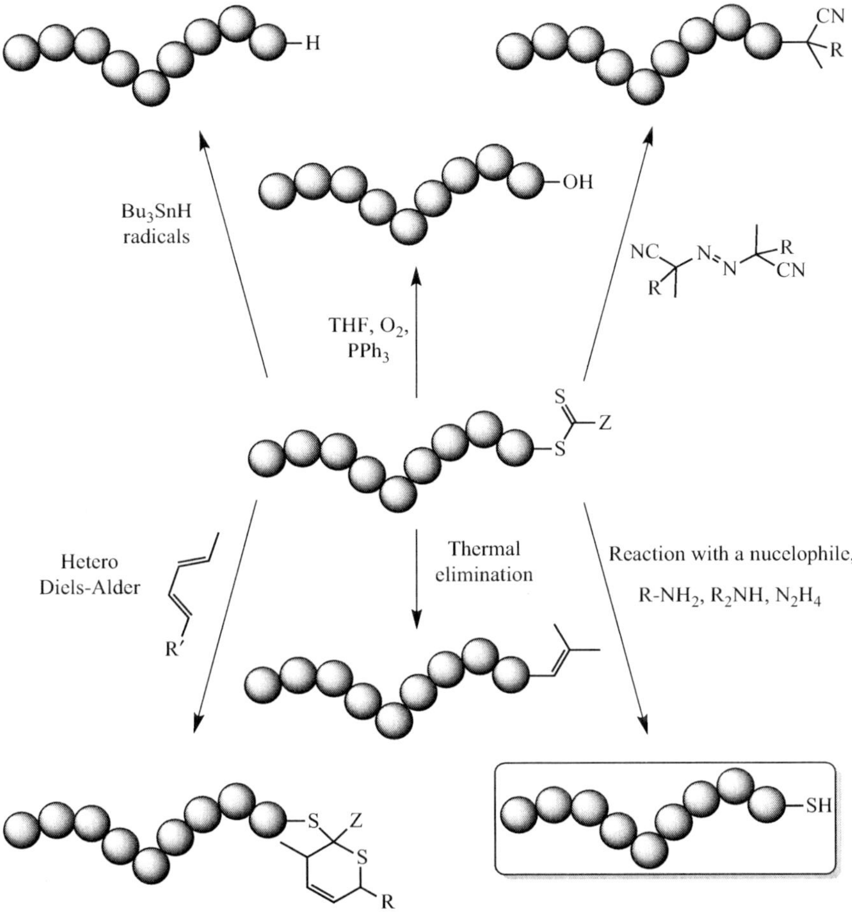


**Figure 2.5** Common literature reported methods for the removal or transformation of thiocarbonylthio end-groups in RAFT-prepared (co)polymers.

reaction with simple dienes is not universally applicable to all thiocarbonylthio species and thermal elimination can also, in principle, result in thiol-terminated species depending on the mechanism of elimination.[8] It is probably fair to state that the most commonly employed method for cleavage of thiocarbonylthio end-groups involves treatment with an appropriate primary or secondary amine (aminolysis) or by reaction with hydrazine (hydrazinolysis), both of which yield the corresponding macromolecular thiol.[8,12–14] This particular route for end-group modification is attractive for several reasons. Firstly, such reactions are well known and understood having been used routinely in biochemistry;[15,16] secondly, these processes are generally extremely rapid with reactions often being complete within seconds to a few minutes at room temperature; thirdly, cleavage results in the formation of a polymeric thiol and thus opens access to the extensive, and often facile, chemistry associated with the mercapto functional group.

In addition to aminolysis reactions as a means of cleaving RAFT end groups, the use of hydrazine has also attracted attention since the report from Shen *et al.*[13] In a comparative study on the rate and efficacy of the aminolysis of small molecule and polymer end-capped trithiocarbonates and dithioesters it was shown that the use of hydrazine, compared to *n*-butylamine, was beneficial on two counts. Firstly, the rates of thiocarbonylthio group cleavage, in both small molecules and polymers, were found to be significantly faster with hydrazine even when employed at significantly lower levels compared to *n*-butylamine. Secondly, and with respect to polymer end-group reactions, it was noted that the antioxidant properties of hydrazine allowed for the cleavage reactions to be performed in air without any undesirable oxidative coupling of the polymeric thiols. Such coupling reactions are known to occur readily and often precautions need to be taken (inert atmospheres and or the addition of reducing agents) to prevent such reactions occurring.

Finally, it is noted that other nucleophilic species can also be employed to cleave thiocarbonylthio end-groups to yield macromolecular thiols or thiolates. One of the earliest examples, reported by Lowe *et al.*,[17] involved the reaction of RAFT (co)polymers with $NaBH_4$ to give polymeric thiolate terminal (co)polymers that were used to stabilize colloidal gold. The same approach was used to prepare polymer modified planar gold surfaces[18] as well as gold nanorods.[19]

The chemistry of thiols is both rich and versatile. For basic information regarding physical and chemical properties of thiols readers are directed to two review articles by Koval[20,21] and the two-volume series edited by Patai.[22,23] Of particular relevance here, however, is the recent recognition that many 'common' thiol-based reactions may be accurately classified as click reactions or possess many of the desirable features associated with click and other highly efficient coupling/conjugation reactions.[24–28] The most common of these reactions are shown in Scheme 2.2 and include the thiol-ene,[24,27–30] thiol-yne,[25,31,32] thiol-isocyanate,[33] thiol-halogen,[34–37] thiol-epoxide[38–40] and thiol-thiol coupling reactions typically from reaction with a species containing an S–S bond.[12]

The fact that RAFT prepared (co)polymers, with their thiocarbonylthio end-groups, can be treated as protected macromolecular thiols that can be readily liberated by aminolysis reactions has resulted in many research groups examining this tool-box of thiol-based chemistries as a means of introducing new and specific chain end functionality as well as in the preparation of new and advanced structure (co)polymers. It is perhaps worth noting at this point that while many of these thiol chemistries are facile and high yielding there are some inherent problems with chemistry involving macromolecular thiols. As mentioned briefly above perhaps the best documented side-reaction involving polymeric thiols is oxidative coupling to give (co)polymers with twice the original molecular weight *via* disulfide bond formation. Such reactions can occur readily even when precautions, such as purging and use of inert atmospheres, are taken. As such, in order to obtain pristine macromolecular thiols it is advisable to perform the aminolysis step in the presence of a reducing agent. Commonly employed species include sodium dithionite ($Na_2S_2O_4$), or certain tertiary phosphines such as dimethylphenylphosphine (DMPP) or

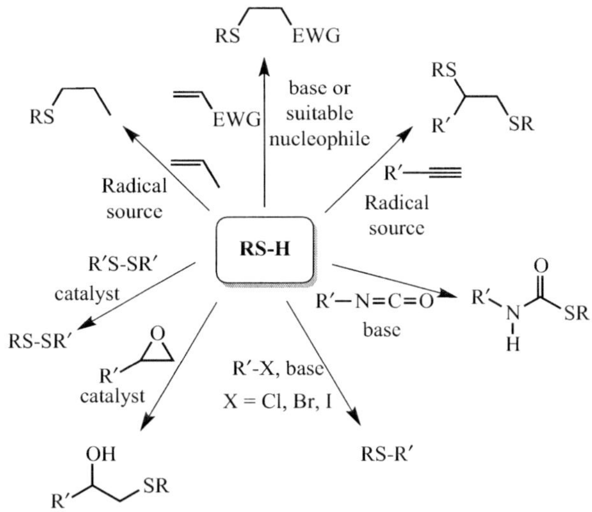


**Scheme 2.2** Common examples of highly efficient, thiol-based chemistries.

tris(carboxyethyl)phosphine (TCEP). Unfortunately, simple oxidative coupling is not the only reported side-reaction during aminolysis of thiocarbonylthio RAFT end groups and the nature of the repeat unit (styrenic *vs.* methacrylate for example) also have an influence on the outcome of such reactions. For example, Xu *et al.*[41] compared the products obtained from the cyclohexylamine-induced aminolysis of dithiobenzoate end-groups in low molecular weight polystyrene (PS) and poly(methyl methacrylate) (PMMA) and identified very different products. In the case of PS, aminolysis resulted in the formation of disulfides, the amount of which increased steadily over a period of 24 h, as judged by SEC. However, in the case of PMMA, while oxidative coupling to disulfide occurred the concentration of this species appeared to *decrease* with reaction time ultimately giving a material with the same approximate molecular weight as the starting material. While it is tempting to conclude that this species is the anticipated macromolecular thiol, it was shown using a combination of 2D NMR spectroscopy and MALDI-TOF MS that the polymer was composed, predominantly, of chains with thiolactone end groups formed *via* a back-biting reaction. Essentially identical observations were made for homopolymers of 2-(dimethylamino)ethyl methacrylate and lauryl methacrylate. While such cyclization reactions are clearly undesirable if the aim is to obtain free polymeric thiols it does offer a convenient route to thiolactone end-functional (co)polymers that may be subsequently used as macromonomers in ring opening polymerization or as reactive species for further functionalization in their own right (see Chapter 10).[42]

In a study on end group structure in poly(acrylic acid) (PAA) before and after base hydrolysis, a process in which thiol or thiolate end functional polymer chains would be expected, Llauro and co-workers[43] found that there was a significant amount of H-terminated chains formed by a transfer to water process as well as macromolecular disulfides and thiolactone end functional species and the authors cautioned the use of this route to thiol-terminated PAA (PAA-SH). These two examples alone highlight that care must be taken when reacting thiocarbonylthio end-groups with the intent of liberating the free macromolecular thiol. However, we also note that in these examples the authors intentionally attempted to isolate the cleavage products or allowed the cleavage reactions to proceed for an extended period of time. We will highlight below how many of these undesirable side reactions can, in fact, be avoided if the macromolecular thiols are not isolated but rather generated and either trapped or reacted immediately *in situ*.

In the remainder of this chapter we will highlight how researchers have employed these various thiol-X chemistries in (co)polymer modification with an emphasis on the chemistry of macromolecular thiols obtained from RAFT prepared precursor materials.

## 2.2 Thiol-ene Chemistry with Michael Acceptors

The thiol-ene reaction is an addition reaction that results in the hydrothiolation of a C=C bond.[24–26,29] As with many fundamental addition reactions, the thiol-ene reaction can proceed under a wide range of conditions and *via* different mechanistic pathways, including radical-mediated addition, traditional acid and base catalyzed addition and by nucleophilic initiation,[30] with the base and nucleophilic processes being, formally, examples of conjugate or heteroatom Michael addition reactions. In the context of the modification of end-groups in RAFT-prepared (co)polymers it is noted that the nucleophilic sulfa-Michael reaction is probably the most commonly employed due to the advantages associated with this chemistry as well as the apparent limitations of the radical thiol-ene reaction with polymeric thiols, at least in the case of polymer-polymer coupling reactions.[44]

One of the earliest reports describing the aminolysis of thiocarbonylthio end-groups in RAFT-prepared poly((meth)acrylate)s followed by Michael addition reactions was by Lima *et al.* in which the authors prepared a series of telechelic poly(meth)acrylates bearing either –OH or –$CO_2H$ end groups.[45] In the first instance, methyl methacrylate (MMA) was homopolymerized with the –OH functional RAFT agent 4-cyano-1-hydroxypent-4-yl dithiobenzoate in the presence of 4,4′-azobis(4-cyanopentanol) to give α-OH functional PMMA. Two different homopolymers with $M_n$'s of 2400 and 17 000 and polydispersity indices $\leq 1.21$ were prepared. Subsequent aminolysis of the homopolymers with hexylamine in THF at RT for 1 h followed by precipitation in heptane yielded the corresponding PMMA thiol. Analysis of this macromolecular thiol using a combination of size exclusion chromatography (SEC), critical liquid chromatography and matrix assisted laser-desorption ionization time of flight mass spectrometry (MALDI-TOF MS) revealed the presence of the target polymeric thiol but also H-terminated polymer chains and additional unidentified species. It is likely that at least one of these unidentified products was a thiolactone end-functional species, *vide supra*, the unambiguous formation of which was not verified until later studies. It is worth reiterating at this point that isolation of macromolecular methacrylic thiols is not recommended and will inevitably result in the presence of undesirable by-products. Nevertheless, the authors subsequently reacted the cleaved PMMA with 2-hydroxyethyl acrylate (HEA), under base catalyzed conditions, to give the Michael adduct and thus telechelic α,ω-OH functional PMMA. However, it was noted that due to the aforementioned side reactions and undesirable byproducts during aminolysis that the yield of the telechelic material was low.

In 2006, Qiu and Winnik,[46] reported the synthesis of telechelic poly(*N*-isopropylacrylamide) (PNIPAM) *via* a one-pot, two step aminolysis/Michael addition route. PNIPAM with isobutylsulfanylthiocarbonylsulfanyl groups at each chain end was dissolved in THF. *n*-Butylamine (5× molar excess based on thiocarbonylthio groups) and a small amount of TCEP were then added to the solution and the reaction stirred for 1 h at RT under a $N_2$ atmosphere. Aminolysis, followed by UV-Vis spectroscopy, was judged to be complete within this time frame after which HEA or butyl acrylate (BA) (10× molar excess) was added to the solution and reaction allowed to proceed for a further 10 h. Under these conditions, *n*-butylamine also serves as the nucleophilic catalyst for the conjugate addition step. Characterization of the modified polymer using a combination of SEC and $^1$H NMR and UV-vis spectroscopies confirmed ~99% end-group functionalization, and highlights the potential advantage(s) of the one-pot approach to thiol-Michael end group modification.

Since these earlier reports a number of research groups have examined the two step aminolysis/conjugate addition process with RAFT-prepared (co)polymers to produce new end-modified species. For example, Spruell *et al.*[47] described the chain end modification of dithiobenzoate-terminated PS with the series of acrylate Michael acceptors shown in Figure 2.6.

The authors evaluated a series of different one-pot, two step end-group cleavage/addition conditions with the intent of streamlining such end group modification reactions. Aminolysis, with *n*-propylamine, or end group cleavage with $NaBH_4$ in the presence of an acrylic species, yielded the Michael addition products in generally good to excellent yields. In both instances the formation of polymeric disulfide could be suppressed by the addition of $PBu_3$ – certain phosphines are well known to serve as both reducing agents and as initiators for

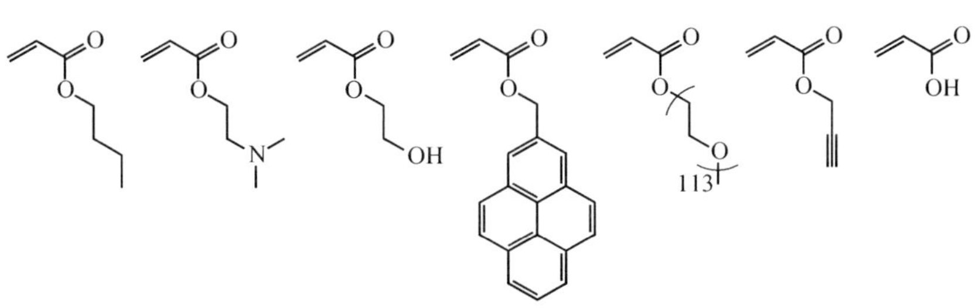

**Figure 2.6** Chemical structures of acrylic Michael acceptors employed in the end group modification of RAFT-prepared polystyrene.

thiol-Michael reactions. Interestingly, the use of $PBu_3$ could be avoided in the case of $NaBH_4$ reactions provided the reaction was thoroughly purged with argon. The approach was extended to AB diblock copolymers of PMMA-PS in which the nature of the ω-end group was shown to influence block copolymer assembly as demonstrated in the case of a cationically modified block copolymer thin film.

In a similar vein, Boyer *et al.*[48] described the one-pot thiol-Michael modification of a series of RAFT-prepared homopolymers with a somewhat broader range of ene Michael acceptors to prepare end-modified species, bioconjugates as well as macromonomers and advanced structure (co)polymers. These are summarized in Scheme 2.3.

Homopolymers of MMA, *N*-(2-hydroxypropyl) methacrylamide (HPMA) and NIPAM were prepared with dithioester (MMA and HPMA) or trithiocarbonate (NIPAM) RAFT CTAs to give materials with SEC measured molecular weights ≤ 25 000 and polydispersity indices ≤ 1.14. Subsequently, these polymers were subjected to a series of different one-pot aminolysis/thiol-Michael addition reactions to yield a range of new species. For example, PNIPAM and polyHPMA (PHPMA) were treated with a combination of hexylamine, triethylamine and biotin maleimide in a one-pot process to give the corresponding Michael adducts in very high yields as judged by a combination of $^1$H NMR and UV-vis spectroscopy (using Ellman's assay) and SEC. Consistent with previous reports, provided the aminolysis was performed in the presence of the maleimide derivative then oxidative coupling and disulfide formation as well as thiolactone formation was suppressed. Similarly,

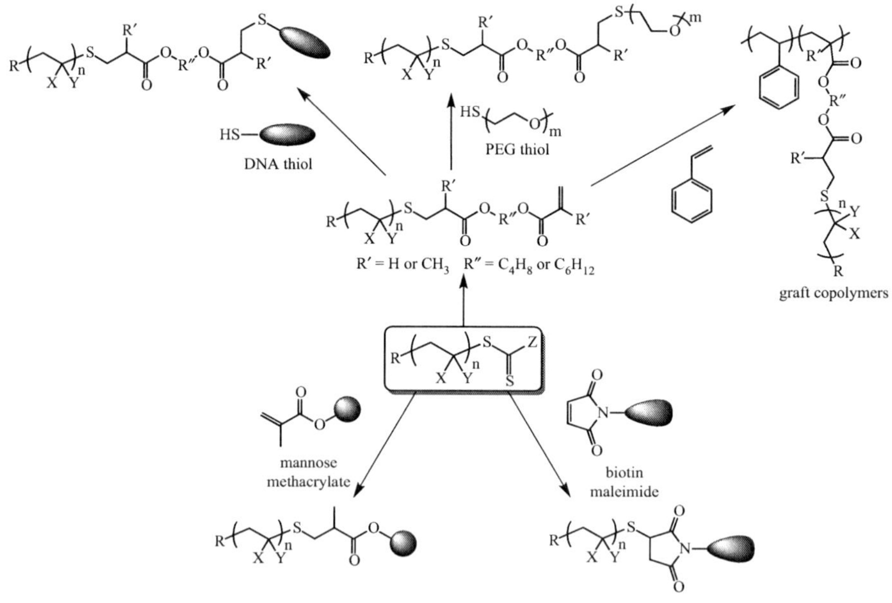


**Scheme 2.3** Thiol-Michael end-group modification of RAFT prepared (co)polymers.

PNIPAM was also modified with mannose methacrylate to give sugar end-capped PNIPAM with a conjugation efficiency of 90% as determined by $^1$H NMR spectroscopy. The use of difunctional Michael acceptors such as 1,6-hexanediol diacrylate or 1,4-butanediol dimethacrylate was also demonstrated and offers an attractive route to (meth)acrylic macromonomers. Such conjugations are performed with a significant excess of the di-(meth)acrylate in order to minimize or prevent the formation of double addition products. The products of such reactions are, clearly, (meth)acrylic end-functional homopolymers. These may be used as Michael acceptors or as polymerizable macromonomers. In the case of the former, conjugation with a thiol-terminated poly(ethylene glycol) yielded AB diblock copolymers in approximately 90% yield while reaction with a thiol-modified oligodeoxyribonucleotide gave PHPMA-DNA and PNIPAM-DNA bioconjugates in 60–70% yield. Copolymerization of PMMA or PNIPAM acrylate end functional homopolymers with styrene under conventional radical polymerization conditions gave graft copolymers with a high incorporation of the macromonomers. This report very nicely demonstrated the wide range of end-modified species that are readily accessible from RAFT-prepared (co)polymers and also highlighted how modifications are not necessarily limited to simple end-group transformation but also offers the ability to perform sequential reactions and prepare materials with more advanced architectures.

While specific modification of the RAFT end-group may be the primary motivation in a particular study it can also be utilized as a secondary consideration. For example, Yin *et al.*[49] recently described the preparation of glucose functionalized, polymeric micelles employing a combination of anionic and RAFT polymerizations. The RAFT component of the reported synthesis is shown in Scheme 2.4.

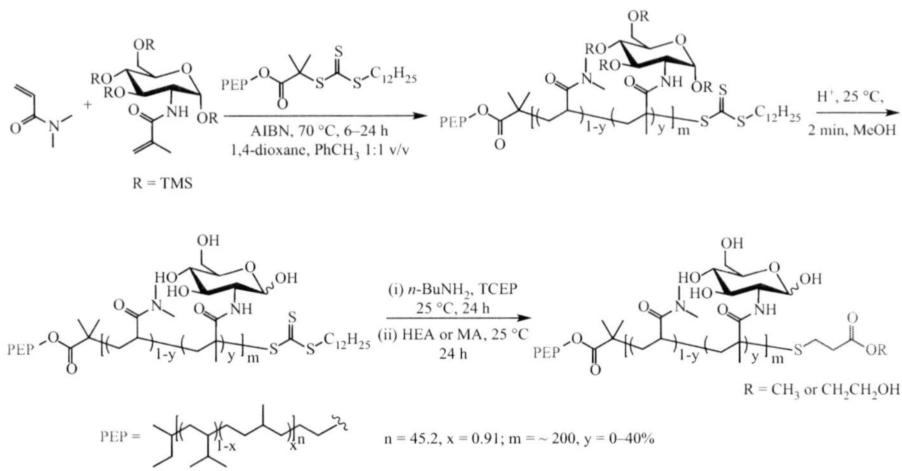


**Scheme 2.4** End group modification of a sugar-based amphiphilic block copolymer *via* sulfa-Michael addition.

A trithiocarbonate functional macro-RAFT CTA was first prepared by esterifying an appropriate trithiocarbonate with an ω-OH functional poly(ethylene-*alt*-propylene) (PEP) obtained *via* the hydrogenation of polyisoprene prepared *via* anionic polymerization. This CTA was used in the subsequent RAFT copolymerization of *N*,*N*-dimethylacrylamide (DMA) with α-2-deoxy-2-methacrylamido-1,3,4,6-tetra(*O*-trimethylsilyl)-D-glucopyranose (TMS-MAG). Subsequent removal of the protecting TMS groups yielded the target amphiphilic block copolymer. However, prior to an evaluation of the self-directed assembly of such copolymers in aqueous media the ω-thiocarbonylthio end groups were modified. This was done primarily to negate any possible undesirable effects on the self-assembly behaviour from the long hydrophobic C12 alkyl chain at the ω-terminus originating from the parent RAFT CTA. As such, prior to aqueous solution studies the authors cleaved the trithiocarbonate end group by treatment with *n*-butylamine followed by sulfa-Michael addition of the macromolecular thiol with methyl acrylate (MA) or 2-hydroxyethyl acrylate (HEA). Subsequent self-directed assembly in water indicated that these amphiphilic block copolymers undergo self-assembly forming spherical micelles as determined by a combination of dynamic light scattering and cryo-TEM. For example, Figure 2.7 shows a cryo-TEM image of a 1 wt% dispersion of a PEP-*block*-poly(DMA-*grad*-MAG) polymer with $M_n$PEP = 3K, $M_n$poly(DMA-*grad*-MAG) = 24K with 21 mol% MAG demonstrating the presence of aggregates with a spherical morphology and an average core radii of ~7 nm.

Roth and coworkers[50] recently described a study comparing the lower critical solution temperature (LCST) and upper critical solution temperature (UCST)

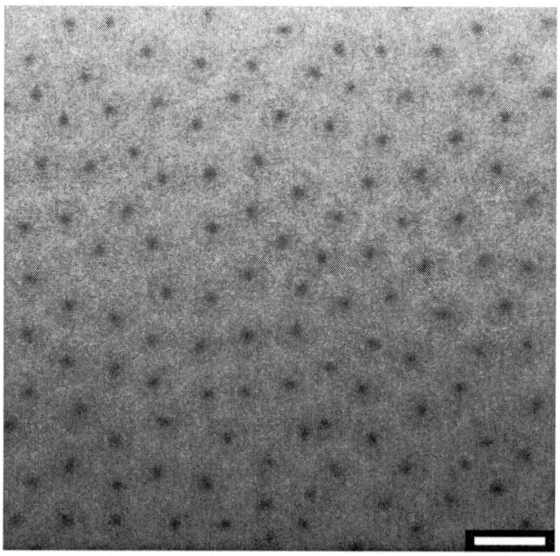

**Figure 2.7** Cryo-TEM image of the spherical self-assembled sugar-based amphiphilic structures formed from the block copolymer shown in Scheme 3.4.

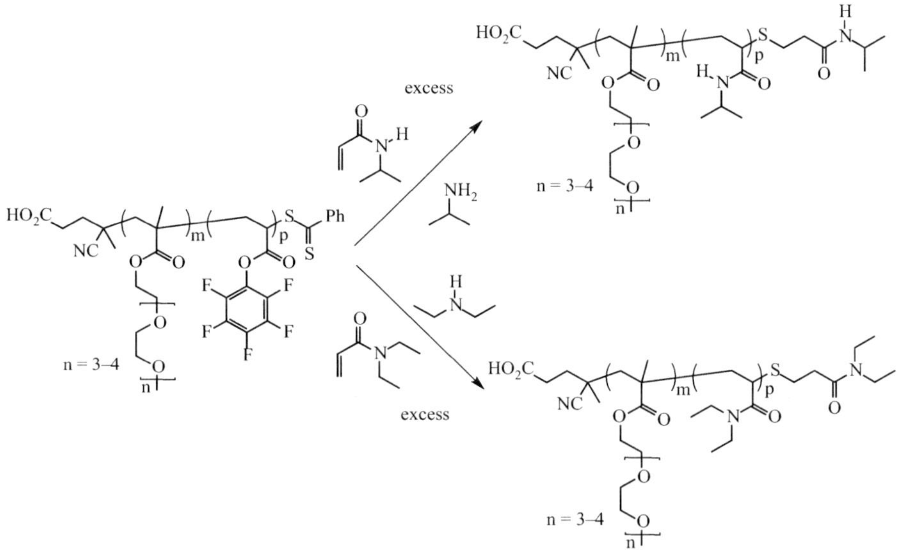


**Scheme 2.5** Simultaneous side-group and end-group modification of RAFT-prepared block copolymers of oligo(ethylene glycol) methyl ether methacrylate with pentafluorophenyl acrylate.

properties of doubly thermoresponsive AB diblock copolymers based on poly[oligo(ethylene glycol) methyl ether methacrylate] (POEGMA) with NIPAM or DEAM. Precursor AB diblock copolymers of POEGMA with pentafluorophenyl acrylate (PFPA) were initially prepared employing an acid-functional dithiobenzoate RAFT agent, Scheme 2.5. This yielded reactive precursor polymers that could be modified *via* acyl substitution reactions at the PFPA residues with primary or secondary amines yielding a library of functional copolymers with the same average degree of polymerization. In this instance the PFPA species were reacted with isopropylamine and diethylamine to introduce NIPAM and DEAM repeat units. However, such reactions will, of course, also result in the cleavage of the dithiobenzoate end groups and as such the acyl substitution reactions were conducted in the presence of NIPAM and DEAM monomers respectively.

This approach yielded the target POEGMA-NIPAM and POEGMA-DEAM AB diblock copolymers in which the ω-end groups were also converted to the Michael adducts and negated any potential disulfide formation and possible deleterious effects on the solution properties due to the presence of the dithiobenzoate end groups.

As noted above, the coupling of thiols to maleimides is generally very facile and, in fact, maleimides are generally known to be one of the more reactive family of Michael acceptors.[30] Sumerlin and co-workers[51,52] reported the synthesis of thermoresponsive, PNIPAM-based, end-functional polymers and polymer-bioconjugates *via* sequential thiol-maleimide/thiol-maleimide couplings, Scheme 2.6.

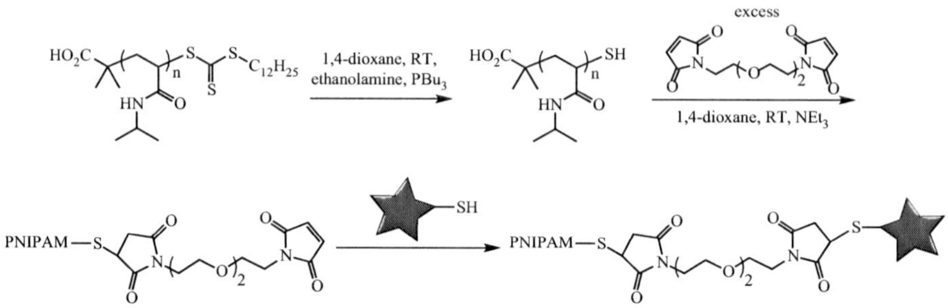


**Scheme 2.6** Sequential thiol-maleimide coupling reactions.

PNIPAM homopolymer with trithiocarbonate ω-end groups was first treated with ethanolamine in the presence of $PBu_3$ at RT in 1,4-dioxane to give the macromolecular thiol, PNIPAM-SH, which was actually isolated and analyzed in this particular instance. PNIPAM-SH was then reacted with an excess of bismaleimidodiethyleneglycol, with triethylamine as the base catalyst, at RT to give the maleimide ω-functional PNIPAM. It was subsequently demonstrated that this reactive species could be effectively coupled to small molecule thiols such as 4-methoxybenzyl mercaptan and 1-dodecanethiol under base mediated conditions as well as with a low molecular weight PS-SH thus demonstrating an effective modular route to AB diblock copolymers. In an essentially identical fashion, Li and co-workers also prepared polymer-bioconjugates of PNIPAM with bovine serum albumin (BSA) and ovalbumin (OVA) the only differences being the aminolysis reagent (hexylamine for protein conjugation *vs.* ethanolamine noted above) and the subsequent conditions employed for BSA and OVA conjugation. Specifically, phosphate buffer at pH 7.2 was employed for protein coupling. The successful formation of the PNIPAM-protein conjugates was confirmed using a combination of SDS-PAGE and aqueous SEC.

Kakwere and Perrier[53] described the synthesis, self-assembly and shell crosslinking of amphiphilic terpolymers of ethyl acrylate (EA) and 2-hydroxyethyl acrylate-*co*-*N*-acryloxysuccinimde (HEA-*co*-NAS). Self-assembly of the EA-*block*-(HEA-*co*-NAS) polymers in water yielded micelles with hydrodynamic diameters of 39–59 nm (depending on molar composition) as determined by dynamic light scattering. The NAS residues in the aggregate coronas were treated with hexamethylenediamine under dilute conditions to give the corresponding shell cross-linked particles with similar measured hydrodynamic diameters. The adopted synthetic approach ensured that the thiocarbonylthio end-groups were located on the periphery of the polymeric micelles and thus were available for subsequent modification. The end-groups were cleaved with $NaBH_4$ followed by treatment with TCEP to give polymeric micelles decorated with free thiol functional groups on the periphery. These were subsequently reacted with *N*-1-pyrenylmaleimide and biotin maleimide in thiol-Michael reactions to give fluorescent micelles as well as the biotin-functionalized particles that were shown to be able to react with avidin giving micelle-protein conjugates.

Others have also reported the coupling of thiol-terminated (co)polymers prepared by RAFT to maleimide functional species. Scales, Convertine and McCormick[54] described the synthesis of fluorescent-labeled PNIPAM *via* a three-step process. Trithiocarbonate end-capped PNIPAM was first converted to the corresponding macromolecular thiol by treatment with $NaBH_4$. Following purification by dialysis (which gave significant amounts of polymeric disulfide as an impurity) the product was treated with TCEP, in DMF, to reduce the disulfide to the free thiol that was subsequently reacted with pyrene maleimide in the presence of ethylenediamine to give the fluorescent tagged PNIPAM in approximately 83% yield. Delaittre *et al.*[55] reported the RAFT synthesis of acrylamide-based copolymers with novel photoreleasable functional groups. In a sequential thiol-maleimide/thiol-maleimide reaction sequence precursor copolymers were initially treated with ethanolamine to give the corresponding thiol-terminated species that were coupled with benzylmaleimide under basic conditions thus functionalizing the end-group near-quantitatively. Subsequently, the photoresponsive side groups were cleaved by irradiation at 366 nm giving pendant thiol-groups that were also coupled with benzylmaleimide, Scheme 2.7.

In another example of sequential thiol-based reactions, Yu and co-workers[56] reported a two-step sequence for preparing mono and bis end-functional PNIPAM, Scheme 2.8. Dithiobenzoate end-capped PNIPAM was first subjected to a one-pot aminolysis/thiol-Michael reaction with allyl methacrylate (ALMA) or propargyl acrylate (PRA) using *n*-octylamine and dimethylphenylphosphine (DMPP). Importantly, while ALMA and PRA contain two C–C multiple bonds it is only the activated (meth)acrylic C=C

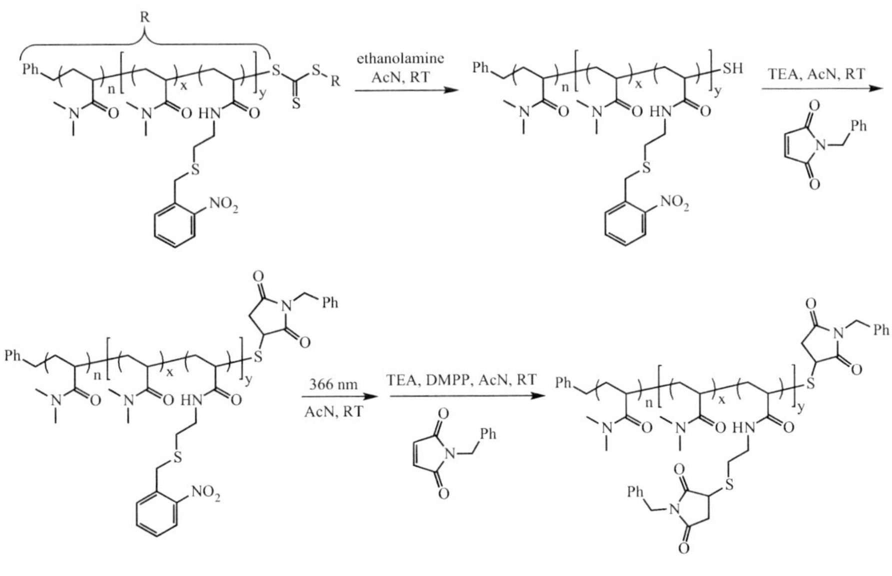


**Scheme 2.7** Sequential thiol-maleimide coupling reactions on end groups and side groups in a RAFT prepared copolymer.

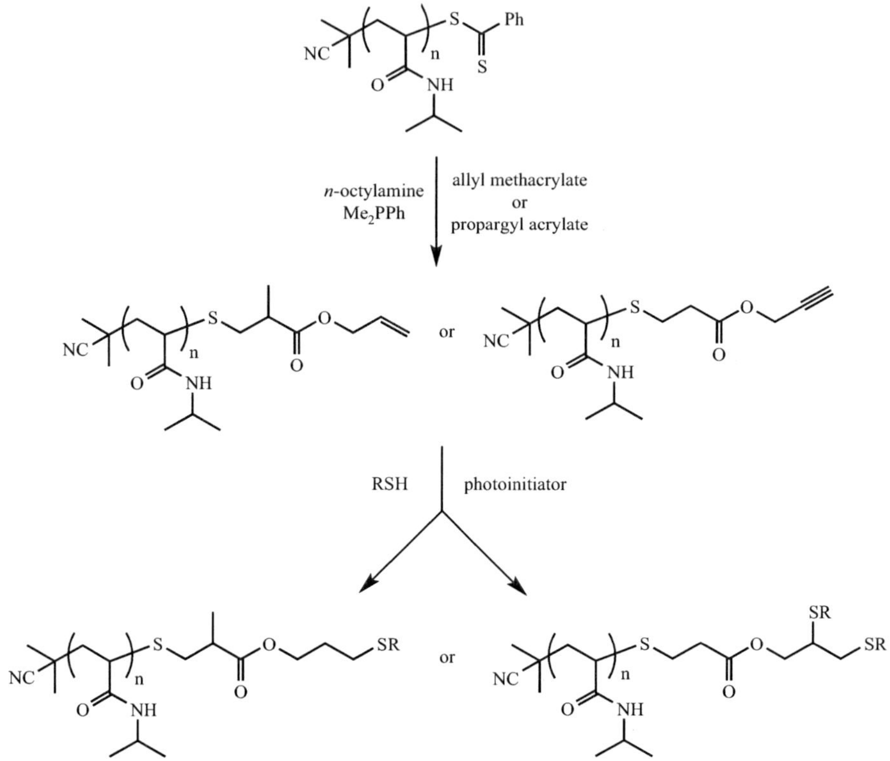


**Scheme 2.8** Sequential thiol-Michael/radical thiol-ene/yne reactions as a route to end functionalized poly(*N*-isopropylacrylamide).

bonds that are susceptible to Michael addition. This yielded the ene and yne end-functional PNIPAM, respectively, in quantitative yield as determined by $^{1}H$ NMR spectroscopy with little-to-no change in the molecular weight distribution as judged by SEC.

In a subsequent step the ene and yne end groups were treated with three different small molecule thiols, namely 6-mercaptohexan-1-ol, hexane-1-thiol and 3-mercaptopropylisobutyl-POSS, under UV irradiation in the presence of a photoinitiator. In the case of the ene end functional PNIPAM, this yields the mono addition thioether product while for the yne functional material double hydrothiolation is observed yielding the bis-1,2-dithioether. Quantitative formation of the addition products was confirmed using $^{1}H$ NMR spectroscopy.

In addition to simple chain end functionalization, the same types of chemistry can also be employed to prepare (co)polymers with more advanced architectures and or macromolecular species that can serve specific roles in other synthetic applications. For example, Chan and workers[57,58] reported the convergent synthesis of 3-arm star polymers based on poly(*N*,*N*-diethylacrylamide) (PDEA) and poly(*n*-butyl acrylate) (PnBA), *via* a sequential aminolysis/thiol-Michael addition with trimethylolpropane triacrylate, Scheme 2.9.

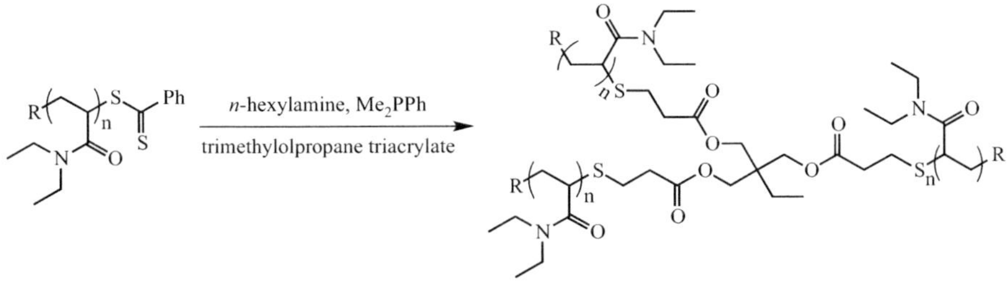


**Scheme 2.9** Convergent 3-arm star formation *via* sequential RAFT end-group aminolysis of poly(*N*,*N*-diethylacrylamide) homopolymers followed by thiol-Michael coupling with trimethylolpropane triacrylate.

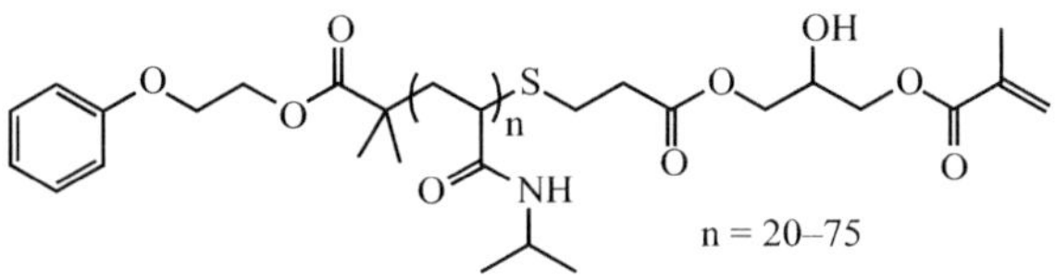


**Figure 2.8** Chemical structure of a methacrylate-end functionalized poly(*N*-isopropylacrylamide) macromonomer.

*N*,*N*-Diethylacrylamide (DEA) was first homopolymerized under typical RAFT conditions employing 1-cyano-1-methylethyl dithiobenzoate as the RAFT agent to give a PDEA homopolymer with an average $M_n$ of 4400 as determined by end-group analysis *via* $^1$H NMR spectroscopy and 4000 by MALDI-TOF MS. In a subsequent one-pot step, the PDEA homopolymer was reacted with hexylamine in the presence of DMPP and trimethylolpropane triacrylate. Under these conditions the dithiobenzoate end group is cleaved to give the PDEA macromolecular thiol (PDEA-SH), which then undergoes a thiol-Michael addition to the triacrylate to give the target 3-arm star polymers. Successful formation of the star polymers was confirmed using MALDI-TOF MS. Generally the reaction proceeded smoothly although the presence of a small amount of 2-arm star as well as unreacted PDEA homopolymer was detected by MALDI-TOF MS.

In another recent example, McKee *et al.*[59] described the synthesis of sterically stabilized polystyrene latexes employing well-defined PNIPAM macromonomers that had been prepared *via* RAFT and subsequently modified. The trithiocarbonate end groups of the PNIPAM homopolymers were reacted with hexylamine (cleavage agent) in the presence of triethylamine and 3-(acryloyloxy)-2-hydroxypropyl methacrylate (AHPMA). AHPMA contains both acrylic and methacrylic functionality but the former is far more reactive towards thiol-Michael addition and as such the approach yields the PNIPAM-based methacrylic macromonomers (PNIPAM-MA) shown in Figure 2.8.

The PNIPAM-MA macromonomers were subsequently used as reactive polymeric steric stabilizers in methanolic dispersion polymerizations of styrene at 70 °C. The authors very nicely demonstrated the need for methacrylate

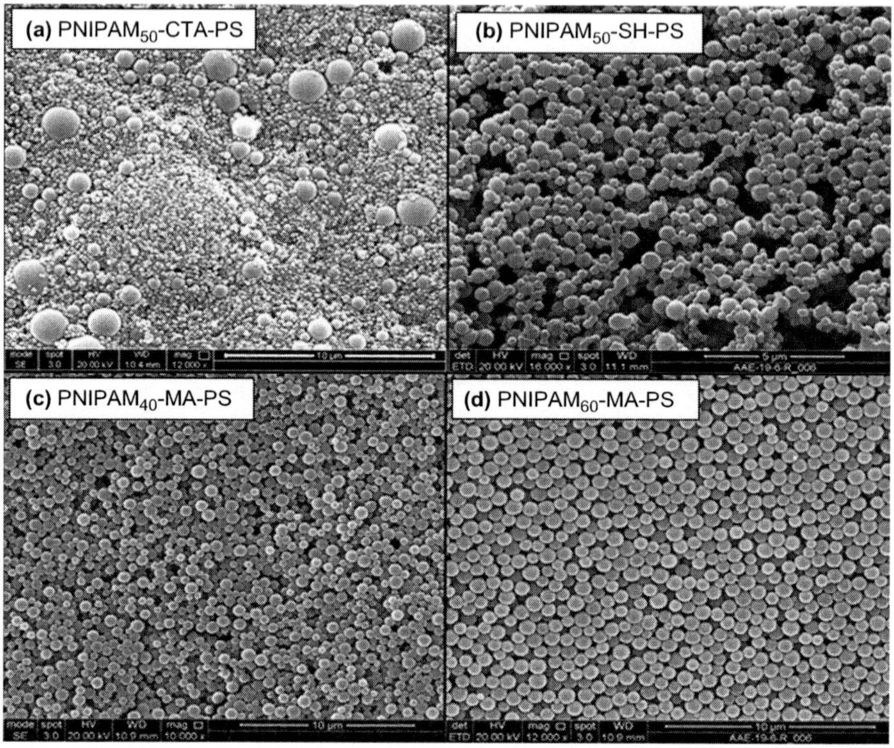


**Figure 2.9** Scanning electron micrographs of sterically stabilized polystyrene latexes prepared by alcoholic dispersion polymerization in methanol at 70 °C with the parent $PNIPAM_{50}$ macroCTA (a), the $PNIPAM_{50}$-SH macromolecular thiol (b), a $PNIPAM_{50}$-MA macromonomer (c), and a $PNIPAM_{60}$-MA macromonomer (d).
Reprinted with permission from J. R. McKee, V. Ladmiral, J. Niskanen, H. Tenhu and S. P. Armes, *Macromolecules*, 2011, **44**, 7692–7703. 

end-groups to obtain narrow size distribution latexes. For example, the use of a parent trithiocarbonate PNIPAM homopolymer ($n = 50$) or the cleaved polymeric thiol (PNIPAM-SH) resulted in latexes with either a broad size distribution (Figure 2.9 (a)) and or interparticle fusion (Figure 2.9 (b)) whereas those prepared with PNIPAM-MA yielded well-defined narrow size distribution particles as judged by scanning electron microscopy (SEM) (Figure 2.9 (c) and (d)).

While these colloidal dispersions are perfectly stable in MeOH, the well-documented lower critical solution temperature (LCST) of PNIPAM may be exploited to give thermoresponsive latexes when dispersed in water. For example, Figure 2.10 (a) and (b) shows pictures of a 1 wt% solution of the $PNIPAM_{50}$-MA macromonomer at 20 and 50 °C. As expected the PNIPAM-MA homopolymer is soluble at the lower temperature but insoluble at 50 °C – a temperature significantly higher than the reported LCST (32 °C) for PNIPAM.

**Figure 2.10** 1 wt% aqueous dispersion of a poly(*N*-isopropylacrylamide) methacrylate macromonomer below (a) and above (b) the lower critical solution temperature of poly(*N*-isopropylacrylamide) and a polystyrene latex stabilized with poly(*N*-isopropylacrylamide) below (c) and above (d) the lower critical solution temperature.
Reprinted with permission from J. R. McKee, V. Ladmiral, J. Niskanen, H. Tenhu and S. P. Armes, *Macromolecules*, 2011, **44**, 7692–7703. Copyright 2011 American Chemical Society.

Similarly, Figure 2.10 (c) and (d) show pictures of a 1 wt% aqueous dispersion of a $PNIPAM_{50}$-MA stabilized polystyrene latex at the same two temperatures. At 20 °C the latex is colloidally stable whereas at 50 °C the dispersion flocculates due to the coil-to-globule transition of the stabilizing PNIPAM chains at this higher temperature. This behaviour is completely reversible.

While these examples above highlight the utility of (meth)acrylic and maleimide activated substrates as useful Michael acceptors in modifying RAFT polymer end-groups, other activated substrates can also be employed. For example, Grover and co-workers[60] described the dithiobenzoate-mediated polymerization of poly(ethylene glycol acrylate) (PEGA) yielding homopolymers with SEC molecular weights ranging from 6700 to 26 500 and polydispersity indices $\leq$ 1.24. Treatment of these polyPEGA homopolymers with *n*-butylamine in MeOH in an oxygen free environment for 10 min followed by reaction with divinylsulfone in the presence of TCEP and phosphate buffer yielded the semitelechelic vinylsulfone end-capped polymers in essentially quantitative yield. The presence of the activated vinylsulfone group at the ω-chain terminus clearly allows for the subsequent addition of another nucleophilic species. In this particular instance, the authors conjugated the end functional homopolymers to bovine serum albumin (BSA) *via* free protein-bound thiol groups. Successful conjugation was confirmed using sodium dodecyl sulfate polyacrylamide gel electrophoresis (SDS-PAGE). An important consideration in the preparation of polymer-protein conjugates relates to the subsequent activity of the protein species. In the case of BSA it is known that changes to the secondary or tertiary structure of the native protein results in a loss of esterase activity. As such, it is important to evaluate whether polymer conjugation has an impact on fundamental activity. In the above example the authors evaluated the esterase activity of modified and unmodified BSA with 4-nitrophenyl acetate by monitoring the appearance of 4-nitrophenylate at 405 nm – BSA is readily acetylated by the former. Results indicated that the polymer-protein conjugate retained approximately 92% esterase activity compared to unmodified BSA, demonstrating that the formation of the conjugate has a very minimal effect on activity.

Ho and co-workers[61] reported a three-step sequential process for the modification of the ω-chain ends in PNIPAM. Cleavage of the trithiocarbonate end groups was first achieved with *n*-hexylamine in the presence of DMPP, Scheme 2.10. In the second step the PNIPAM-SH was reacted with vinyl azlactone (VAZ) employing DMPP as the nucleophilic catalyst in this instance. The successful modification of the end group was confirmed using standard techniques such as $^{1}$H NMR spectroscopy and MALDI-TOF MS. The VAZ group is a useful reactive group that can be readily ring-opened with primary amines for example.[62,63] In the above example, the VAZ end modified PNIPAM was reacted with *p*-fluorobenzylamine to demonstrate the ability to further react the VAZ functionality. Successful reaction was confirmed using $^{1}$H NMR spectroscopy and UV detection in SEC. In the case of the former, new aromatic peaks were clearly visible at 7.0 and 7.4 ppm while a new peak at 263 nm was observed in the SEC-UV detector.

We have seen from several of the examples above that the ability to perform sequential conjugation reactions has offered a route to series of novel functional materials, whether it be by thiol-maleimide/thiol-maleimide, thiol-Michael/thiol-ene, thiol-Michael/thiol-yne, thiol-divinylsulfone/thiol-vinylsulfone or thiol-Michael with VAZ followed by nucleophilic ring-opening. The examples presented are not an exhaustive representation of the literature but hopefully serve to demonstrate that the combination of RAFT (co)polymer synthesis with mono or sequential end group modification reactions give access to a near-infinite number of new and interesting materials with potentially novel properties.

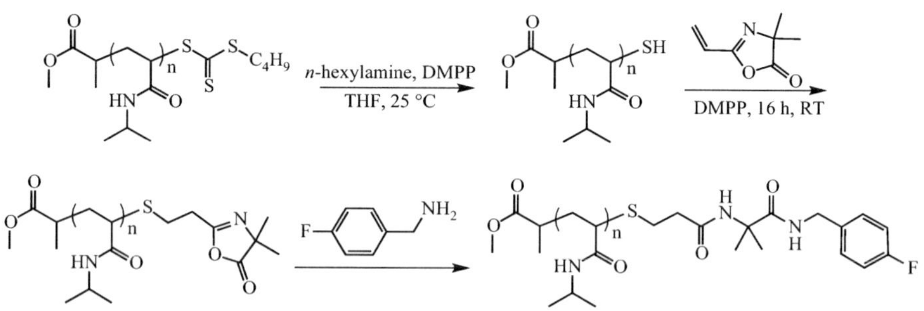


**Scheme 2.10** Aminolysis of poly(*N*-isopropylacrylamide) end groups followed by thiol-Michael reaction with vinyl azlactone with subsequent nucleophilic ring-opening as a two step route to end functional materials.

## 2.3 Thiol-yne Chemistry

The radical-mediated thiol-yne reaction can be thought of as a sister reaction to the more common and widely employed radical thiol-ene reaction, although we note that its use is becoming more generally widespread in the polymer science community.[31,32,64] The addition of two equivalents of thiol, under radical mediated conditions, results in the double hydrothiolation of a C≡C bond with 1,2-regioselectivity. Other modes of addition are, of course, possible and these are discussed elsewhere.[64]

In terms of the use of thiol-yne chemistry specifically in the modification of RAFT-polymer end-groups there are currently very few examples in the literature. However, it is noted that there are a significant number of yne-functional RAFT CTAs that have been described in the literature to prepare α (R-group)-functional copolymers that have subsequently been combined with small molecule and polymeric azides in Cu-catalyzed alkyne-azide cycloaddition reactions.[65] Such materials, of course, require no post-polymerization modification chemistry to introduce the desired yne functional groups. In addition, there are numerous examples where the thiol-yne reaction has been employed to modify side chains as well as end groups in non-RAFT prepared materials.[64]

Yu *et al.*,[56] as described above (Scheme 2.8), prepared an yne-end modified PNIPAM *via* a one-pot aminolysis and thiol-Michael addition with propargyl acrylate. The yne end functionalized PNIPAM was then treated with 6-mercapto-1-hexanol, 1-hexanethiol and 3-mercaptopropyl polyhedral oligomeric silsesquioxane (100% excess based on yne functional groups) in the presence of Irgacure 651 photoinitiator. This resulted in the quantitative formation, as judged by $^{1}$H NMR spectroscopy, of the corresponding bis-ω-thioether adducts.

Kumar, Bousquet and Stenzel[66] described the RAFT homopolymerization of *tert*-butyl acrylate (TBA) in the presence of benzyl pyridine-2-carbodithioate – a RAFT agent known to undergo heteroatom Diels–Alder reactions with dienes. Reaction of the resulting polyTBA (PTBA) homopolymer with hexa-2,4-dien-1-yl prop-2-yn-1-yl succinate, in a [4+2] cycloaddition process, yielded the yne end functional PTBA homopolymer, Scheme 2.11.

This yne functional species was then treated with thioglycerol under radical conditions in the presence of a photoinitiator yielding the thiol-yne double addition product – a four functional alcohol. Esterification of the OH groups with 4-oxo-4-(prop-2-ynyloxy)butanoic anhydride then yielded the corresponding four functional yne species, which was treated with β-thioglucose under radical conditions to give the 8-functional end modified dendron-like polymer.

## 2.4 Thiol-isocyanate Chemistry

The base-mediated reaction of a thiol with an isocyanate gives thiocarbamates. The reaction is generally very fast and efficient and is not plagued by the same side reactions often observed in alcohol-isocyanate coupling. As with the radical thiol-yne reaction, there are very few examples of thiol-isocyanate

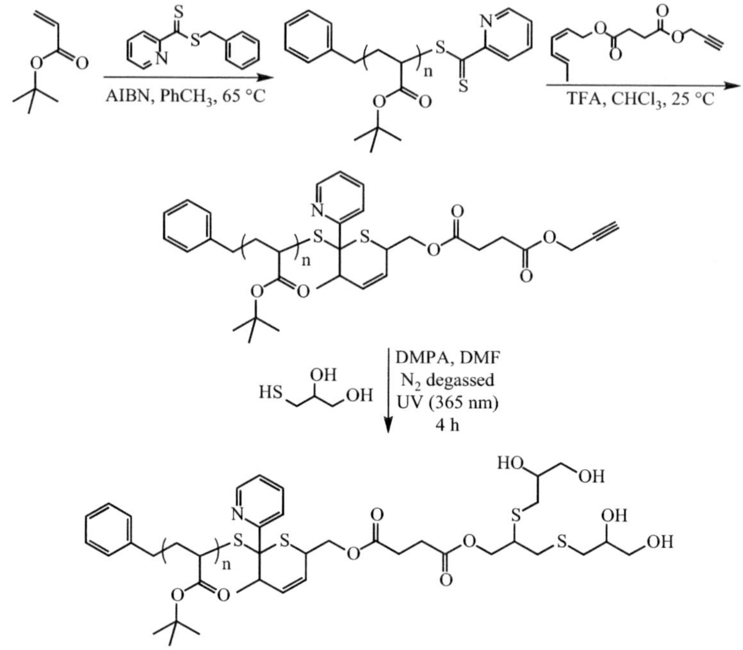


**Scheme 2.11** Sequential heteroatom Diels-Alder/radical thiol-yne reactions to give a 4-functional, OH end-modified PTBA homopolymer.

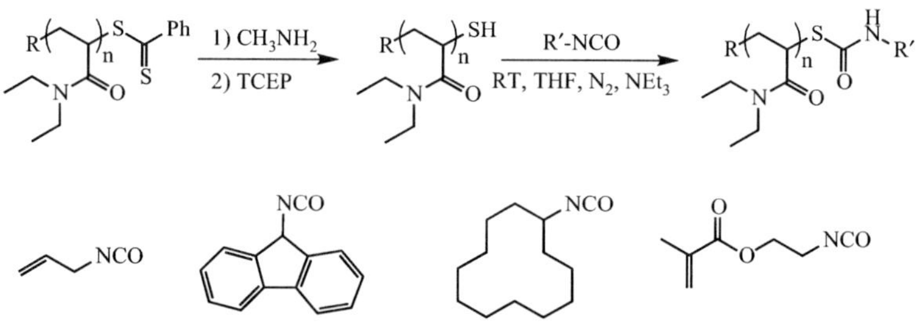


**Scheme 2.12** End group modification of a RAFT-prepared poly(*N*,*N*-diethylacrylamide) homopolymer *via* reaction of the macromolecular thiol with commercially available small molecule isocyanates.

coupling specifically employed with the aim of modifying end groups in RAFT-prepared polymers, although there are examples in the literature where it has been employed in side chain modification of free isocyanate groups,[67] or surface-bound polymers bearing isocyanate groups.[68]

Li and coworkers[33] described the modification of dithiobenzoate end-groups in a low molecular weight PDEAM homopolymer employing a combination of end-group cleavage and thiol-isocyanate coupling with a wide range of commercially available small molecule isocyanates, Scheme 2.12.

PDEAM was first treated with methylamine and then TCEP to give the thiol-terminated PDEAM (PDEAM-SH). This macromolecular thiol was isolated

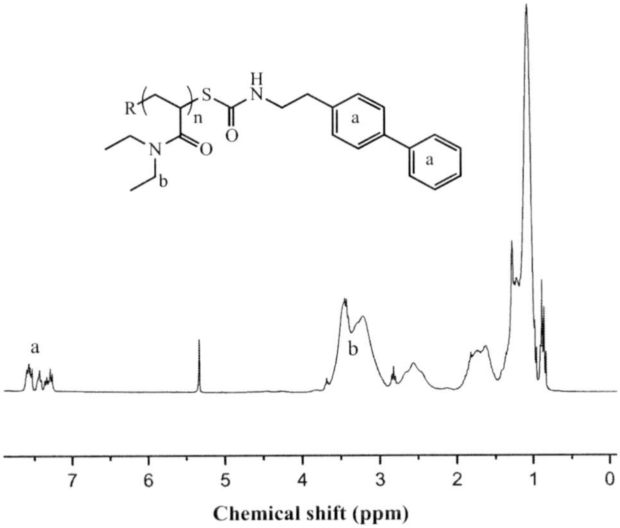


**Figure 2.11** $^1$H NMR spectrum of the product obtained from the reaction of a poly(*N*,*N*-diethylacrylamide) macromolecular thiol with 4-(2-isocyanatoethyl)-1,1′-biphenyl.

and characterized, and its formation verified using a combination of $^1$H NMR spectroscopy and UV-vis spectrophotometry. Reaction of PDEAM-SH with 14 different isocyanates, examples of which are shown in Scheme 2.12, under basic conditions yielded the corresponding thiocarbamate end-functional PDEAM species in generally excellent-to-quantitative yields. As a representative example, Figure 2.11 shows the $^1$H NMR spectrum obtained from the reaction of PDEAM-SH with 4-(2-isocyanatoethyl)-1,1′-biphenyl.

Successful coupling is confirmed by the presence of the aromatic phenyl end-groups at $\delta \sim 7.5$ ppm (labeled **a**). A ratio of the integrals associated with these aromatic species with the species labeled **b** (the methylene hydrogens bound directly to the amide N in the polymer main chain) indicate a near-quantitative formation of the thiocarbamate derivative. PDEAM is documented to possess an LCST of ~33 °C,[69] essentially identical to that of the more commonly employed PNIPAM. Given the low molecular weight of the PDEAM homopolymer employed in this study (2540 as determined by size exclusion chromatography with narrow molecular weight PMMA standards) it was anticipated that the introduction of hydrophobic end groups *via* the thiol-isocyanate coupling reaction would have a measureable effect on the LCST temperature. Indeed, the experimentally determined LCSTs of the modified PDEAM homopolymer were up to 10 °C lower than PDEAM-SH (LCST = 34 °C) further reinforcing the well-documented effect that polymer end groups can have on solution properties in low molecular weight species.

In the above study, one of the commercially available isocyanates that was employed to modify the PDEAM end-group was 2-(isocyanato)ethyl

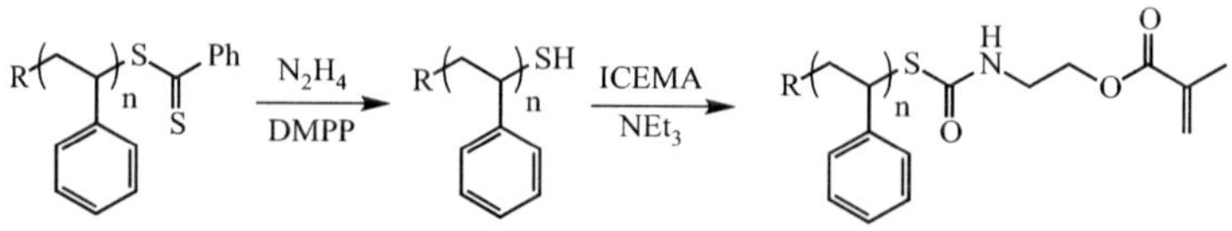


**Scheme 2.13** Synthesis of polystyrene-based methacrylic macromonomers *via* sequential hydrazinolysis and thiol-isocyanate coupling with 2-(isocyanato)ethyl methacrylate.

methacrylate (ICEMA). As demonstrated extensively earlier in this chapter, polymeric thiols can undergo Michael addition to methacrylic bonds and, as noted above, react readily with isocyanate groups. However, there is a significant difference in the reactivity of these two functional groups towards thiols, with, for example, methacrylic species actually being one of the least reactive substrates towards thiol-Michael reactions. In contrast, reactions of thiols with isocyanates are extremely rapid. As such, is it possible to react thiols with isocyanates in the presence of methacrylic species and get exclusive reaction with the isocyanate functional group. The reaction of PDEAM-SH with ICEMA, therefore, proceeds exclusively at the –NCO functional group in ICEMA and was shown to be an efficient and facile route to PDEAM-based methacrylic macromonomers. Indeed, this represents an alternative, but complementary, route to such methacrylic macromonomers, other examples of which were highlighted above and described by Boyer *et al.*[48] and Armes and coworkers.[59]

Recently, Harvison and Lowe[70] described the synthesis and copolymerization of polystyrene-based methacrylic macromonomers taking advantage of this orthogonal reactivity of thiols with ICEMA, Scheme 2.13.

Low-to-intermediate molecular weight precursor homopolymers of styrene were first prepared *via* thermal initiation in the presence of a dithiobenzoate RAFT agent. Hydrazinolysis of the end groups in the presence of DMPP yielded the corresponding polystyrene-thiols, which when treated with ICEMA under basic conditions gave the target methacrylic macromonomers in high yield. These were subsequently copolymerized with methacrylic comonomers such as MMA and lauryl methacrylate (LMA).

## 2.5 Thiol-epoxide Chemistry

The ring-opening reaction of epoxides with thiols under basic or Lewis acid catalyzed conditions can be a facile means of generating alcohols. In the original 'click' article by Kolb, Finn and Sharpless[71] the ring-opening of spring loaded substrates such as epoxides and aziridines by nucleophiles was listed as a reaction that could meet the criteria to be accurately described as a click reaction. Indeed, there is extensive literature dealing with small molecule reactions that justifies that classification. Epoxides have, of course, also formed the basis of an extensive body of work in the polymer literature as highly reactive functional species.

As with the thiol-yne and thiol-isocyanate chemistries, there is currently very little in the open literature detailing the combination of macromolecular thiols obtained from RAFT prepared copolymers with ring opening reactions on epoxides. In fact, the only example, to the best of our knowledge, was described by Harvison *et al.*[38] Homopolymers of polystyrene and PDEAM were initially prepared *via* dithiobenzoate-mediated RAFT. Subsequently, two different routes were examined for affecting macromolecular thiolysis of a series of small molecule commercially available epoxides. In the first route, the thiocarbonylthio end-groups in polystyrene were first cleaved *via* hydrazinolysis and the intermediate macromolecular thiol isolated. Reaction with 1,2-epoxy-5-hexene under either base or Lewis acid catalysis gave the ring-opened products with conversions in the range of 74–89%. In the second route, a one-pot procedure was employed in which end-group cleavage was accomplished with $NaBH_4$ in the presence of an epoxide. In the case of PDEAM, end group conversions >99% were obtained.

## 2.6 Thiol-halogen Chemistry

The reaction between a thiol and an alkyl halide is a nucleophilic substitution reaction yielding thioethers. Such chemistry is straightforward and has been employed to prepare a range of functional materials. For example, Nakayama and Okano[72] prepared amphiphilic AB diblock copolymers comprised of a hydrophobic poly(benzyl methacrylate) (PBzMA) block with a thermoresponsive NIPAM-*co*-DMA block *via* dithiobenzoate-mediated RAFT. In one instance, the end-groups were cleaved *via* aminolysis followed by reaction with 2-iodoethanol to give the corresponding OH terminal copolymer. In a similar manner, Segui, Qiu and Winnik[73] reported the synthesis of α,ω-pyrene and cholesterol modified PNIPAM prepared from HS-PNIPAM-SH (obtained *via* aminolysis of a parent PNIPAM prepared with a difunctional RAFT CTA) and 4-(1-pyrenyl)butyl iodide or cholest-5-en-3β-yl 6-iodohexyl ether. In both instances, near-quantitative incorporation of the desired functional groups was obtained. The introduction of these bulky hydrophobic end-groups resulted in the self-directed assembly of the now amphiphilic species, as determined employing a combination of $^1H$ NMR and fluorescence spectroscopies.

The reaction of ω-thiols with α-bromoesters at the α end-group of the same polymers was reported by Xu *et al.*[74] Methyl acrylate (MA) was polymerized with RAFT CTAs carrying two 2-bromopropionate or two 2-bromoisobutyrate groups in the R fragment. Aminolysis of the dithioester ω end-groups liberated thiols that reacted *in situ* with the α bromo groups. In model reactions, the 2-bromopropionate groups were found to react faster with thiols than the 2-bromoisobutyrate groups. The substitution of one of the two 2-bromopropionates at the α end group of the telechelic polymers thus influences the reactivity of the second one more dramatically than in the case of slower reacting 2-bromoisobutyrates. In the case of the former polymer, a thiol preferentially reacts with an α end that has not yet been substituted, producing multiblock polymers. In the case of the 2-bromoisobutyrates, a terminal thiol reacts with α end-groups on which the first bromide has already been substituted producing hyperbranched polymers, Scheme 2.14.

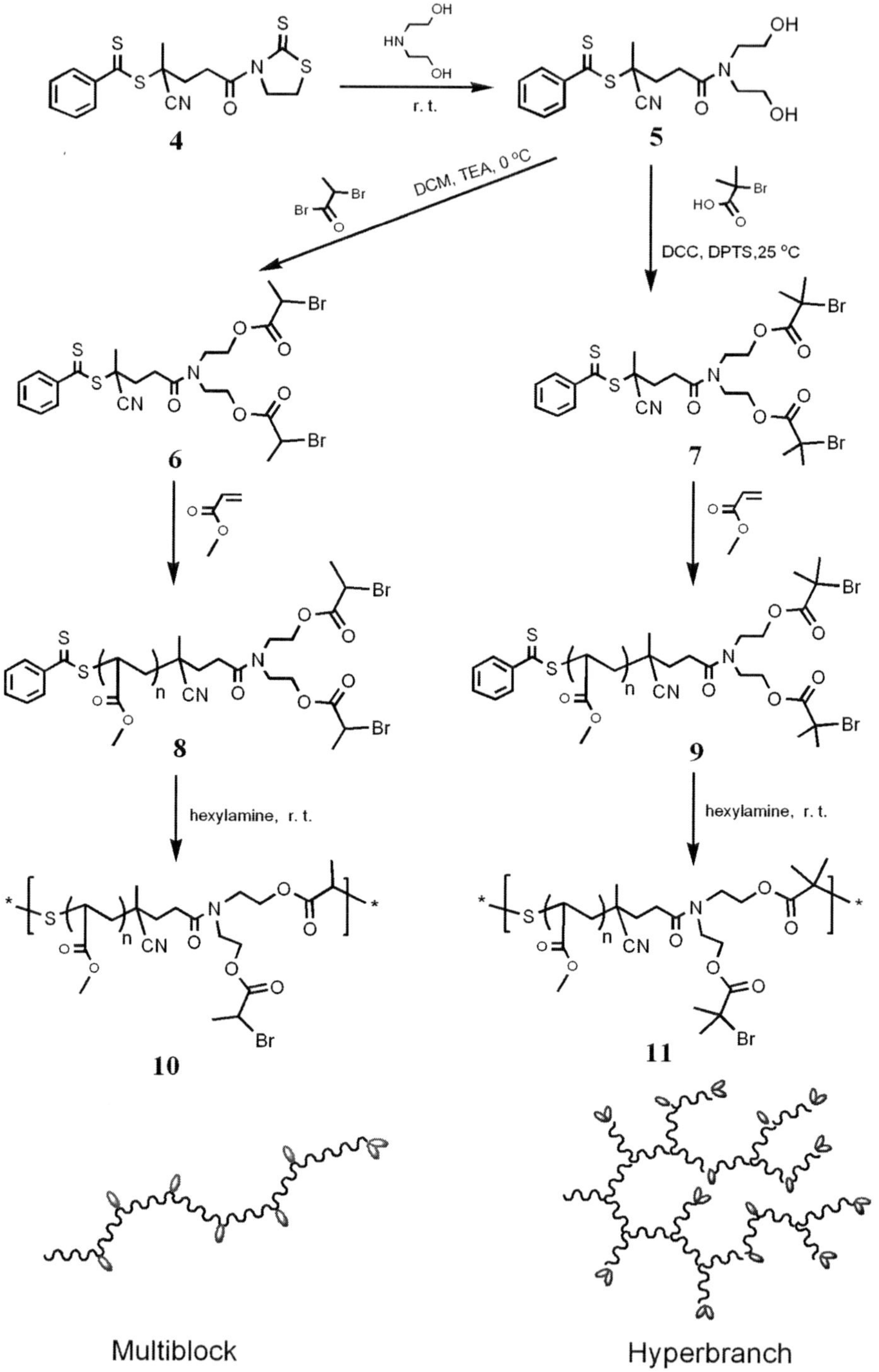


**Scheme 2.14** Formation of multiblock and hyperbranched polymers *via* inter- and intramolecular thiol-bromo substitution reactions.

## 2.7 Thiol-thiol Coupling Chemistry

In addition to the thiol-based chemistries highlighted above, there are a series of thiol-thiol coupling processes that have also emerged and proven to be synthetically useful reactions in trapping intermediate polymeric thiols or in the formation of novel functional species. These include thiol coupling/exchange reactions with pyridyl disulfide (PDS) and methyl methanethiosulfonate (MMTS) and its derivatives. The application of these reactions will not be discussed here since Roth and Theato highlight such chemistries in Chapter 5 and a recent review discusses the application of pyridyl disulfide chemistry in combination with RAFT-prepared (co)polymers.[12]

## 2.8 Summary and Outlook

In this chapter a series of reactions involving macromolecular thiols obtained from RAFT-prepared parent (co)polymers with a variety of substrates has been highlighted. Examples based on thiol-ene, thiol-yne, thiol-isocyanate, thiol-epoxide and thiol-halogen chemistries have been presented and demonstrate the rich, versatile and often selective chemistry of thiol functional groups even when instilled at a (co)polymer chain end(s). The resurgence in interest in thiol-based chemistries, some of which may be accurately described as click chemistries under appropriate conditions, has enabled synthetic polymer chemists to prepare a wide range of new and interesting materials with novel and, in some cases, unique properties. The continued and growing interest in the application of simple chemistries for the facile modification of (co)polymer end groups will no doubt lead to many more exciting reports and applications of such materials prepared from the thiol-X toolbox.

## References

1. J. Chiefari, Y. K. Chong, F. Ercole, J. Krstina, J. Jeffrey, T. P. T. Le, R. T. A. Mayadunne, G. F. Meijs, C. L. Moad, G. Moad, E. Rizzardo and S. H. Thang, *Macromolecules*, 1998, **31**, 5559.
2. R. T. A. Mayadunne, E. Rizzardo, J. Chiefari, J. Krstina, G. Moad, A. Postma and S. H. Thang, *Macromolecules*, 2000, **33**, 243–245.
3. M. Benaglia, J. Chiefari, Y. K. Chong, G. Moad, E. Rizzardo and S. H. Thang, *J. Am. Chem. Soc.*, 2009, **131**, 6914–6915.
4. G. Moad, E. Rizzardo and S. H. Thang, *Aust. J. Chem.*, 2005, **58**, 379–410.
5. G. Moad, E. Rizzardo and S. H. Thang, *Aust. J. Chem.*, 2006, **59**, 669–692.
6. G. Moad, E. Rizzardo and S. H. Thang, *Aust. J. Chem.*, 2009, **62**, 1402–1472.
7. C. Barner-Kowollik, M. Buback, B. Charleux, M. L. Coote, M. Drache, T. Fukuda, A. Goto, B. Klumperman, A. B. Lowe, J. B. McLeary, G. Moad, M. J. Monteiro, R. D. Sanderson, M. P. Tonge and P. Vana, *J. Polym. Sci., Part A: Polym. Chem.*, 2006, **44**, 5809–5831.
8. G. Moad, E. Rizzardo and S. H. Thang, *Polym. Int.*, 2011, **60**, 9–25.

9. M. Chen, G. Moad and E. Rizzardo, *J. Polym. Sci., Part A: Polym. Chem.*, 2009, **47**, 6704–6714.
10. Y. K. Chong, G. Moad, E. Rizzardo and S. H. Thang, *Macromolecules*, 2007, **40**, 4446–4455.
11. M. A. Harvison, P. J. Roth, T. P. Davis and A. B. Lowe, *Aust. J. Chem.*, 2011, **64**, 992–1006.
12. P. J. Roth, C. Boyer, A. B. Low and T. P. Davis, *Macromol. Rapid Commun.*, 2011, **32**, 1123–1143.
13. W. Shen, Q. Qiu, Y. Wang, M. Miao, B. Li, T. Zhang, A. Cao and Z. An, *Macromol. Rapid Commun.*, 2010, **31**, 1444–1448.
14. M. Deletre and G. Levesque, *Macromolecules*, 1990, **23**, 4733–4741.
15. G. Levesque, P. Arsene, V. Fanneau-Bellenger and T.-N. Phan, *Biomacromolecules*, 2000, **1**, 387–399.
16. G. Levesque, P. Arsene, V. Fanneau-Bellenger and T.-N. Phan, *Biomacromolecules*, 2000, **1**, 400–406.
17. A. B. Lowe, B. S. Sumerlin, M. S. Donovan and C. L. McCormick, *J. Am. Chem. Soc.*, 2002, **124**, 11562–11563.
18. B. S. Sumerlin, A. B. Lowe, P. A. Stroud, P. Zhang, M. W. Urban and C. L. McCormick, *Langmuir*, 2003, **19**, 5559–5562.
19. J. W. Hotchkiss, A. B. Lowe and S. G. Boyes, *Chem. Mater.*, 2007, **19**, 6–13.
20. I. V. Koval, *Russ. J. Org. Chem.*, 2005, **41**, 631–648.
21. I. V. Koval, *Russ. J. Org. Chem.*, 2007, **43**, 319–346.
22. *The Thiol Group: Volume 1*, ed. S. Patai, John Wiley & Sons Ltd., 1974.
23. *The Thiol Group: Volume 2*, ed. S. Patai, John Wiley & Sons Ltd., 1974.
24. A. B. Lowe, *Polym. Chem.*, 2010, **1**, 17–36.
25. C. E. Hoyle, A. B. Lowe and C. N. Bowman, *Chem. Soc. Rev.*, 2010, **39**, 1355–1387.
26. C. E. Hoyle and C. N. Bowman, *Angew. Chem., Int. Ed.*, 2010, **49**, 1540–1573.
27. M. J. Kade, D. J. Burke and C. J. Hawker, *J. Polym. Sci., Part A: Polym. Chem.*, 2010, **48**, 743–750.
28. A. Dondoni, *Angew. Chem., Int. Ed.*, 2008, **47**, 8995–8997.
29. C. E. Hoyle, T. Y. Lee and T. Roper, *J. Polym. Sci., Part A: Polym. Chem.*, 2004, **42**, 5301–5338.
30. J. W. Chan, C. E. Hoyle, A. B. Lowe and M. Bowman, *Macromolecules*, 2010, **43**, 6381–6388.
31. A. B. Lowe, C. E. Hoyle and C. N. Bowman, *J. Mater. Chem.*, 2010, **20**, 4745–4750.
32. R. Hoogenboom, *Angew. Chem., Int. Ed.*, 2010, **49**, 3415–3417.
33. H. Li, B. Yu, H. Matsushima, C. E. Hoyle and A. B. Lowe, *Macromolecules*, 2009, **42**, 6537–6542.
34. B. M. Rosen, G. Lligadas, C. Hahn and V. Percec, *J. Polym. Sci., Part A: Polym. Chem.*, 2009, **47**, 3931–3939.
35. B. M. Rosen, G. Lligadas, C. Hahn and V. Percec, *J. Polym. Sci., Part A: Polym. Chem.*, 2009, **47**, 3940–3948.

36. C. R. Becer, K. Babiuch, D. Pilz, S. Hornig, T. Heinze, M. Gottschaldt and U. S. Schubert, *Macromolecules*, 2009, **42**, 2387–2394.
37. J. Xu, L. Tao, C. Boyer, A. B. Lowe and T. P. Davis, *Macromolecules*, 2010, **43**, 20–24.
38. M. A. Harvison, T. P. Davis and A. B. Lowe, *Polym. Chem.*, 2011, **2**, 1347–1354.
39. A. Brandle and A. Khan, *Polym. Chem.*, 2012, **3**, 3224–3227.
40. I. Gadwal and A. Khan, *Polym. Chem.*, 2013, **4**, 2440–2444.
41. J. Xu, J. He, D. Fan, X. Wang and Y. Yang, *Macromolecules*, 2006, **39**, 8616–8624.
42. E. Pieter, F. Goethals and F. E. Du Prez, *J. Am. Chem. Soc.*, 2011, **133**, 1678–1681.
43. M.-F. Llauro, J. Loiseau, F. Boisson, F. Delolme, C. Ladaviere and J. Claverie, *J. Polym. Sci., Part A: Polym. Chem.*, 2004, **42**, 5439.
44. S. P. S. Koo, M. M. Stamenović, R. A. Prasath, A. J. Inglis, F. E. Du Prez, C. Barner-Kowollik, W. Van Camp and T. Junkers, *J. Polym. Sci., Part A: Polym. Chem.*, 2010, **48**, 1699–1713.
45. V. Lima, X. Jiang, J. Brokken-Zijp, P. J. Schoenmakers, B. Klumperman and R. Van Der Linde, *J. Polym. Sci., Part A: Polym. Chem.*, 2005, **43**, 959–973.
46. X.-P. Qiu and F. M. Winnik, *Macromol. Rapid Commun.*, 2006, **27**, 1648–1653.
47. J. M. Spruell, B. A. Levy, A. Sutherland, W. R. Dichtel, J. Y. Cheng, J. F. Stoddart and A. Nelson, *J. Polym. Sci., Part A: Polym. Chem.*, 2009, **47**, 346–356.
48. C. Boyer, A. Granville, T. P. Davis and V. Bulmus, *J. Polym. Sci., Part A: Polym. Chem.*, 2009, **47**, 3773–3794.
49. L. Yin, M. C. Dalsin, A. Sizvovs, T. M. Reineke and M. A. Hillmyer, *Macromolecules*, 2012, **45**, 4322–4332.
50. P. J. Roth, T. P. Davis and A. B. Lowe, *Polym. Chem.*, 2012, **3**, 2228–2235.
51. M. Li, P. De, S. R. Gondi and B. S. Sumerlin, *J. Polym. Sci., Part A: Polym. Chem.*, 2008, **46**, 5093–5100.
52. M. Li, P. De, H. Li and B. S. Sumerlin, *Polym. Chem.*, 2010, **1**, 854–859.
53. H. Kakwere and S. Perrier, *J. Am. Chem. Soc.*, 2009, **131**, 1889–1895.
54. C. W. Scales, A. J. Convertine and C. L. McCormick, *Biomacromolecules*, 2006, **7**, 1389–1392.
55. G. Delaittre, T. Pauloehrl, M. Bastmeyer and C. Barner-Kowollik, *Macromolecules*, 2012, **45**, 1792–1802.
56. B. Yu, J. W. Chan, C. E. Hoyle and A. B. Lowe, *J. Polym. Sci., Part A: Polym. Chem.*, 2009, **47**, 3544–3557.
57. J. W. Chan, B. Yu, C. E. Hoyle and A. B. Lowe, *Chem. Commun.*, 2008, 4959–4961.
58. J. W. Chan, B. Yu, C. E. Hoyle and A. B. Lowe, *Polymer*, 2009, **50**, 3158–3168.
59. J. R. McKee, V. Ladmiral, J. Niskanen, H. Tenhu and S. P. Armes, *Macromolecules*, 2011, **44**, 7692–7703.

60. G. N. Grover, S. N. S. Alconcel, N. M. Matsumoto and H. D. Maynard, *Macromolecules*, 2009, **42**, 7657–7663.
61. T. H. Ho, M. Levere, J.-C. Soutif, V. Montembault, S. Pascual and L. Fontaine, *Polym. Chem.*, 2011, **2**, 1258–1260.
62. M. E. Buck and D. M. Lynn, *Polym. Chem.*, 2012, **3**, 66–80.
63. S. M. Heilmann, J. K. Rasmussen and L. R. Krepski, *J. Polym. Sci., Part A: Polym. Chem.*, 2001, **39**, 3655–3677.
64. A. B. Lowe and J. W. Chan, in *Functional Polymers by Post-Polymerization Modification: Concepts, Guidelines, and Applications*, ed. P. Theato and H.-A. Klok, Wiley-VCH Verlag GmbH & Co. KGaA, 2013, pp. 87–118.
65. M. A. Harvison and A. B. Lowe, *Macromol. Rapid Commun.*, 2011, **32**, 779–800.
66. J. Kumar, A. Bousquet and M. A. Stenzel, *Macromol. Rapid Commun.*, 2011, **32**, 1620–1626.
67. J. D. Flores, J. Shin, C. E. Hoyle and C. L. McCormick, *Polym. Chem.*, 2010, **1**, 213–220.
68. R. M. Hensarling, S. B. Rahane, A. P. LeBlanc, B. J. Sparks, E. M. White, J. Locklin and D. L. Patton, *Polym. Chem.*, 2011, **2**, 88–90.
69. I. Idziak, D. Avoce, D. Lessard, D. Gravel and X. X. Zhu, *Macromolecules*, 1999, **32**, 1260–1263.
70. M. A. Harvison, *PhD Thesis*, University of New South Wales, 2013, in production.
71. H. C. Kolb, M. G. Finn and K. B. Sharpless, *Angew. Chem., Int. Ed.*, 2001, **40**, 2004–2021.
72. M. Nakayama and T. Okano, *Biomacromolecules*, 2005, **6**, 2320–2327.
73. F. Segui, X. P. Qiu and F. M. Winnik, *J. Polym. Sci., Part A: Polym. Chem.*, 2008, **46**, 314–326.
74. J. Xu, L. Tao, C. Boyer, A. B. Lowe and T. P. Davis, *Macromolecules*, 2010, **43**, 20–24.

CHAPTER 3

# Thiol-X Chemistries for the Production of Degradable Polymers

KATIE L. POETZ, OLIVIA Z. DURHAM AND DEVON A. SHIPP*

Department of Chemistry & Biomolecular Science, Clarkson University, Potsdam New York, 13699-5810, U.S.A.
*Email: dshipp@clarkson.edu

## 3.1 Introduction

Thiol-X chemistries incorporate a variety of reactions that may be utilized in polymerization, functionalization, and crosslinking.[1–3] These reactions may involve the combination of thiol species with alkenes, alkynes, epoxies, isocyanates and other functional molecules. Consequently, they utilize an assortment of reaction mechanisms, and thus provide a range of products. These chemistries are expanded on in other parts of this volume. However, three of the most well known chemistries associated with thiol-X chemistry are: thiol-ene, thiol-yne and thiol-Michael additions. Each of these chemistries can be employed for polymerizations (in propagation and crosslinking) and functional modifications.[1–3]

The widespread use of thiol-X chemistries, and thiol-ene in particular, has in part been because of the interest in the concept of "click chemistry", into which many thiol-X chemistries fall.[4–6] To be labeled a "click" chemistry, a given reaction must embody most if not all of the following features: high yield

RSC Polymer Chemistry Series No. 6
Thiol-X Chemistries in Polymer and Materials Science
Edited by Andrew B. Lowe and Christopher N. Bowman

Published by the Royal Society of Chemistry, www.rsc.org

of a single product, region- and stereo-selectivity, simple reaction conditions involving benign or no solvent, and little or no purification. Thiol-X reactions typically exemplify many of these features, and thus have been widely used in functionalization and polymerization processes. Such is the utility of thiol-X and click chemistries that many chemical-based problems in bio-related applications can now be tackled with greater certainty than in the past. The subject matter of this chapter is degradable polymers, particularly those for use in biomedical applications, and is one area that benefits from these developments in synthetic chemistry.

Mechanistically, the three common thiol-X chemistries (thiol-ene, thiol-yne and thiol-Michael additions) can be grouped into two categories. The first grouping consists of thiol-ene and thiol-yne, which are radical-mediated addition processes. Thiol-Michael additions, on the other hand, involve base-catalyzed nucleophilic additions of thiolate anions to electrophilic $\beta$-carbon atoms in carbon–carbon double bonds. The mechanisms of these reactions are discussed in more detail below.

Thiol-ene chemistry is a prominent polymerization method that utilizes the reaction between a bifunctional or multifunctional thiol and a bifunctional or multifunctional alkene.[7,8] The generally accepted mechanism for thiol-ene polymerization is shown in Scheme 3.1. Thiol-ene polymerization is commonly initiated photochemically, although other initiating processes can be used, such as redox and thermal initiating. A thiyl radical generated by H-abstraction by an initiator-derived radical adds to an alkene. This radical adduct undergoes another H-abstraction from another thiol, regenerating a thiyl radical species. This two-step process makes up the propagation stage and results in polymer chain growth until a termination reaction occurs. Although the reaction is a

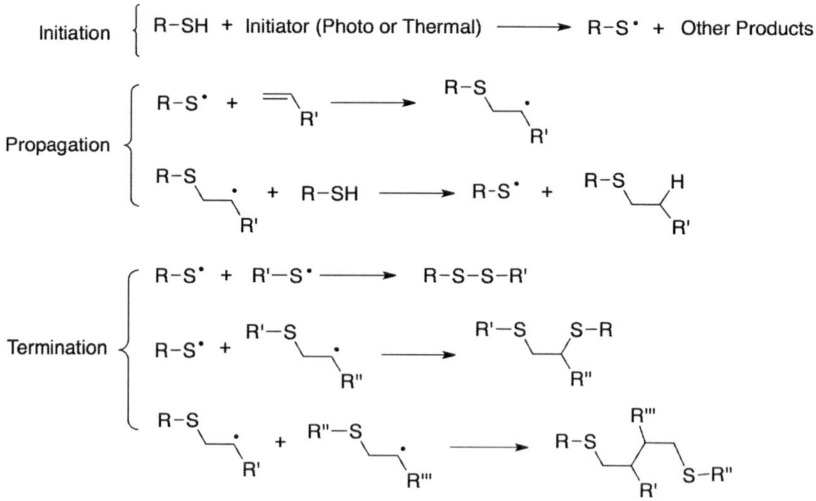


**Scheme 3.1** Thiol-ene polymerization mechanism.

radical-based chain reaction, molecular weight growth in thiol-ene polymerizations follows a step-growth rather than a chain growth pattern.[7]

The mechanism of thiol-yne polymerizations[9] is similar to that of thiol-ene, with the major exception being that once a thiyl radical has added across a C≡C triple bond, the resulting thioether still contains a C=C double bond that is susceptible to attack from a second thiyl radical. Hence, each -yne has a functionality of two, whereas -enes have a functionality of one in these polymerizations.

The other common mechanism of thiol-X reactions is thiol-Michael addition (Scheme 3.2).[10–12] The efficiency of these reactions is practically 100%. The high efficiency can be attributed to a number of reasons, including the high reactivity of the thiolate anion toward conjugate addition and the low $pK_a$ value of thiols, both factors resulting in these reactions typically being completed in minutes. In these reactions, a base is used to create a thiolate anion which subsequently attacks the β-carbon of a C=C bond. The carbanion formed then deprotonates the conjugate acid of the base catalyst, yielding the final product and regenerating the catalyst.

The utilization of thiol-X chemistries in biomedical applications, such as drug delivery and cell targeting, has become increasingly evident.[13] This is, in part, because thiol-X chemistries provide the required synthetic control that allows specific molecular architecture and functionality to be introduced into materials of interest. In order to apply polymer materials as tools for drug design or as drug delivery devices *in vivo*, they must be able to fulfill several biological demands of fundamental importance. Polymeric materials used in the application of drug delivery systems should be non-toxic, non-immunogenic, able to cross bio-barriers (like intestines, blood brain barrier, and cellular membranes), able to target specific structures, and must be able to stay in circulation in the body for the time needed to yield a clinical effect. These properties relate back to molecular architecture and functionality.

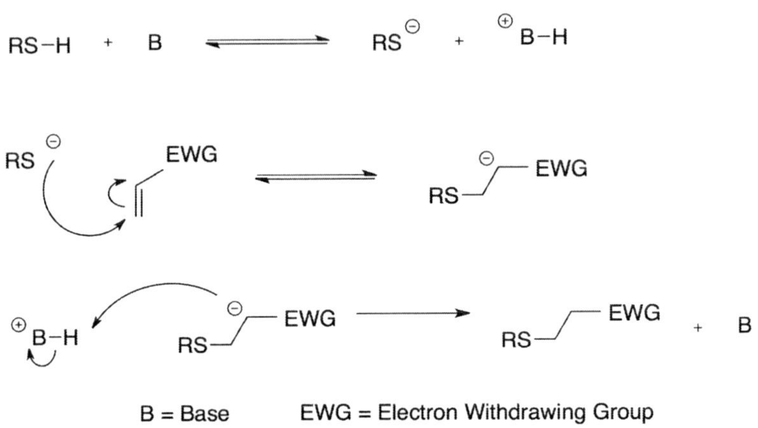


**Scheme 3.2** Thiol-Michael addition mechanism.

Drug delivery is a prime application for degradable polymers. There are a significant number of factors and requirements that must be met for a degradable polymer to be used as a drug delivery device. In concurrence with material properties, control over drug release kinetics is essential for obtaining the most efficient and safe administration and circulation of drug materials. Drug delivery systems must demonstrate easy fabrication, practical application, safety, optimal performance, and improved efficacy and targeted delivery in order to be considered for the treatment of different illnesses or diseases. As such, significant efforts have been made to synthesize well-tailored macromolecular assemblies using different polymer materials.

This chapter discusses the use of thiol-X chemistries in the synthesis of a range of degradable polymers, many of which have potential for drug delivery applications. The focus is on thiol-ene radical and thiol-Michael chemistries in particular as these have been the most widely used in this area.

## 3.2 Thiol-X Chemistries Applied to Degradable Polymer Synthesis

### 3.2.1 Generalities

Synthetic polymeric hydrogels synthesized using thiol-X chemistries offer promising advancements in a wide range of applications that require a well-defined molecular architecture that is similar to the natural structure of the extracellular matrix (ECM) as well as possessing easily tunable physical and chemical properties.[13] The two common covalent gelation mechanisms used for hydrogel fabrication are (1) radical step-growth photopolymerization of thiols and -enes, and (2) the conjugate Michael addition of multifunctional thiol and activated -ene precursors. The following discussion is based on these two methods and their application to the production of degradable gel materials with specific bio-applications.

### 3.2.2 Thiol-ene Radical Polymerizations

Thiol-ene photopolymerizations have become an imperative fabrication mechanism for the preparation of biopolymers that exhibit orthogonality and high efficiency. Thiol-ene chemistry and biomolecule synthesis can be utilized in various biomedical applications including the development of protein biochips, drug delivery, cellular microarrays, fabrication of hydrogels, and the creation of dental restorative materials. Radical thiol-ene polymerizations have several advantages over their thiol-Michael cousins, including the ability to be photoinitiated, thereby providing spatial and temporal control, and possessing an impressively wide range of enes and thiols that can be used. These unique capabilities allow for the production of degradable polymers with various functionalities, such as polyanhydrides and polyesters, the former of which would be incompatible with the thiol-Michael addition because of the susceptibility of the anhydride to nucleophilic attack by the thiolate anion.

Rydholm *et al.* have studied the degradation behavior of polyethers made using an ester-containing dithiol and PEG-based diallyl ethers.[14] The ester functionality contained in the thiol monomers provides the degradability. In the preparation of different polymeric materials, studies have examined the effects of solvent concentration on degradation rates. In the synthesis of thiol-allyl ether polymers, it was shown that as the amount of solvent increased, the rate of degradation increased. As degradation continued, the modulus of the materials decreased nearly exponentially. In addition, the equilibrium swelling ratio increased as the solvent concentration was increased, which is expected due to the decrease in crosslink density of the polymer material. Additional studies with the same material examined the effect of monomer molecular weight. It was determined that as the monomer molecular weight increases, the degradation rate increases, which in turn causes the crosslink density to decrease. The thiol concentration and its affect on degradation, swelling, and modulus were also measured. From this study, it was demonstrated that as the thiol content increased, the degradation rate, along with the equilibrium swelling ratio, increased while the modulus decreased. As the amount of non-degradable thiol content increased, the degradation rate decreased, and in most cases, the moduli tended to increase as the amount of non-degradable thiol content decreases. These changes in non-degradable thiol content lead to small changes in swelling ratios, and due to the decreased amount of ester hydrolysis, this causes increased swelling. When there are fewer degradable linkages, the amount of swelling over time will vary minimally due to fewer changes in the polymer network structure.[14] This demonstrates how changes in the process of polymer synthesis can have a significant impact on different important chemical and mechanical properties of the material.

Shipp *et al.*[15,16] have used thiol-ene polymerizations where the diene monomer contains anhydride functionality in order to make crosslinked and linear polyanhydrides (Scheme 3.3). Examination of the degradation rate of the crosslinked polyanhydrides with various crosslink densities in the network structure has demonstrated that the degradation follows a surface erosion profile. Thermal and mechanical properties of these polymers were monitored as a function of crosslink density; the monomers used were 4-pentenoic anhydride, pentaerythritol tetrakis(3-mercaptopropionate) and 3,6-dioxa-1,8-dithiooctane.[16] The glass transition temperature ($T_g$) increased as a function of crosslink density from −55 to −15 °C. The Young's modulus was also dependent on crosslink density, and ranged from ~ 10 to 4 MPa at 37 °C as the crosslink density decreased from 0.91 to 0.55 M.

Thiol-ene chemistry has also become a useful reaction mechanism for the development of PEG-based hydrogels for biomedical applications.[13] PEG-based hydrogels can be formed through various mechanisms. However, there are significant limitations to the suitability of different mechanisms due to applications that require *in situ* fabrication of hydrogels. PEG hydrogels formed by thiol-ene photopolymerizations offer many advantages that

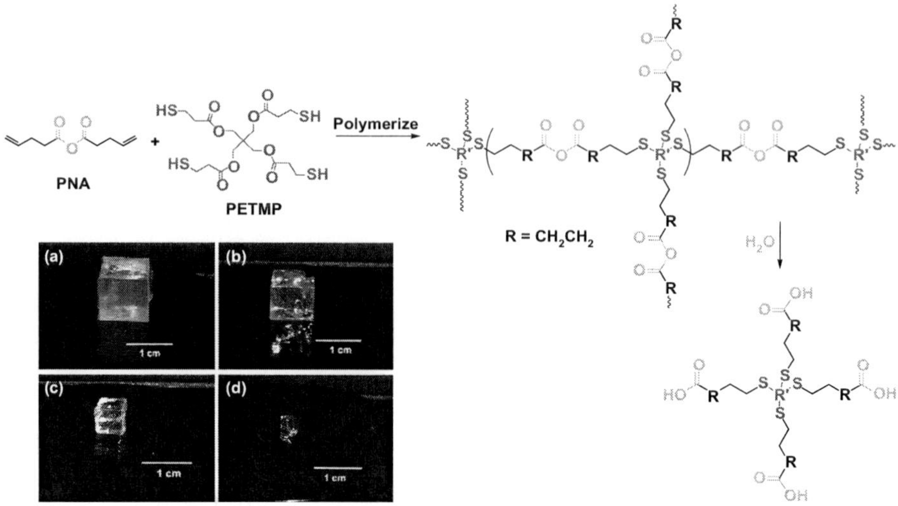


**Scheme 3.3** Outline of the synthesis of polyanhydrides from thiol-ene polymerization. The photos show a 1 $cm^3$ sample during the erosion process in PBS solution at 25 °C (after a: 0 h, b: 18 h, c: 48 h, d: 72 h).
Adapted and reprinted with permission from ref. 15. Copyright 2009 The Royal Society of Chemistry.

overcome difficulties associated with *in situ* development of the hydrogel scaffold. The step-growth mechanism of thiol-ene chemistry contributes to the cytocompatability of the reaction as well as resulting in relatively homogeneous thiol-ene network structures and enabling enhanced control over physical properties of the hydrogel material and its constituents. Coupling thiol-ene chemistry and photopolymerization as a gelation mechanism for hydrogel development allows PEG-based hydrogels to be synthesized with spatiotemporally defined properties that influence common cellular activities including cell spreading, migration, and differentiation.

The use of peptide crosslinkers in PEG-based hydrogels has been shown to allow hydrogels to undergo enzymatic degradation, which can be used in conjunction with the release of drug molecules, proteins, or even the promotion of cell generation and differentiation. One example from Lin *et al.*[17] used a PEG hydrogel made from a multi-arm PEG-norbornene macromer (PEG4NB) that was crosslinked by matrix metalloproteinase (MMP)-cleavable peptides flanked with bis-cysteines (Scheme 3.4). PEG-norbornene hydrogels have been shown to successfully encapsulate different biomolecules including fibroblasts, valvular interstitial cells, mesenchymal stem cells, and fibrosarcoma cells. In addition, similar methodologies have allowed for the development of enzyme-sensitive, surface eroding thiol-ene hydrogels.[18] In particular, enzyme responsive PEG hydrogels have been used for the controlled release of proteins. In studies involving release of proteins, it is important to understand physiological conditions and chemicals that influence controlled release of the protein from the hydrogel material. Human neutrophil elastase (HNE) is an

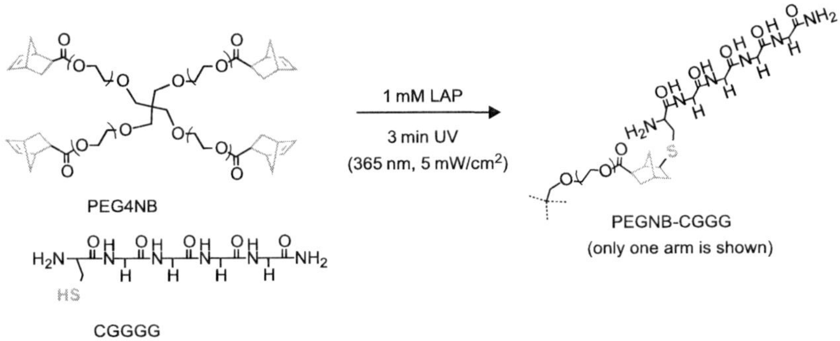


**Scheme 3.4** Schematics of step-growth thiol-ene click reaction using PEG4NB and mono-cysteine peptide (CGGGG) to form non-crosslinked PEG-peptide conjugates. Photoinitiator: 1 mM LAP (lithium arylphosphanate). Reprinted with permission from ref. 17. Copyright 2011 Elsevier Ltd.

enzyme that is released as inflammation occurs, and Aimetti *et al.*[18] have shown that a network of PEG hydrogels, which contain a peptide chain, has a controlled release of a protein under inflammatory conditions in the presence of HNE. The thiol-ene hydrogel investigated was formed by 4-arm PEG-norbornene ($PEG_{5k}$-norbornene) and bis-cysteine in the presence of a photoinitiator and HNE responsive peptide.

Benton *et al.* used a four-arm PEG that was functionalized with norbornene that was polymerized with peptides that contained cysteine to form a 3D hydrogel that would culture valvular interstitial cells (VICs).[19] A crosslinking unit composed of MMP was used because VICs secrete an enzyme that allows for cell-dictated gel remodeling. This degradable system had adhesion proteins incorporated into this linkage to study the morphology, differentiation proliferation, and migration. The VICs showed the ability to undergo cell division by the use of live cell imaging. The spread morphology could be predicted by dictating the crosslink density of the hydrogel. A lower crosslink density showed the spread morphology could be achieved earlier than in higher crosslink density gels. The peptide RGD showed an increased bioactivity by allowing the cell to become more elongated. This finding shows that these cells have the ability to respond to their environment by sensing other peptides.[19] Thiol-ene hydrogels have also been used to encapsulate cells for growth in cell proliferating medium. The amount of MMP was varied and the behavior of human mesenchymal stem cells (hMSCs) was monitored to investigate how these environments affected the hMSCs. It was found that the cells survive equally well in degradable environments as well as non-degradable environments. In addition, it has been shown that hMSCs are capable of remodeling their surroundings, and have improved differentiating capability in degradable environments.[20]

Ortiz *et al.* used sucrose as a building block for thiol-ene polymerization that resulted in a sucrose-based polymer that was subsequently degradable in acidic media.[21,22] Sucrose was modified to introduce two allyl groups along with small percentages of mono and triallyl derivatives. These modified sucrose materials were photopolymerized with dithiothreitol, a difunctional thiol.

Thiol-ene chemistry can also be used as a method to link two polymers each containing functionalized end groups that have the ability to undergo polymerization. Preformed hydrogels containing synthetic peptide sequences have been incorporated into a crosslinked hydrogel. DeForest *et al.*[23] have demonstrated the ability to control the amount of thiol-ene functionalization by varying the amount of initiator and the amount of time spent under the UV lamp. These variations have allowed for the development of patterned hydrogel materials. After the hydrogel is removed from the light source, the unreacted thiol monomers are able to diffuse through the gel network.

Thiol-ene chemistry has also been applied to the development of synthetic polymeric materials that incorporate natural molecules or polymers. A recent area of interest includes hydrogels that contain networks of hyaluronic acid. Hyaluronic acid (HA) is a glycosaminoglycan material that naturally occurs in all animals and some bacteria species. HA is commonly found in extracellular and pericellular matrices. This particular material is of significant interest in the application of hydrogel formation due to its biocompatibility and easy degradation by mammalian enzymes. Three different families of HA-based photopolymerized PEG diacrylate (PEGDA) hydrogels were compared in the analysis of network architecture on the mechanical properties and degradability of the hydrogels.[24] Through additional analysis of these hydrogels, the materials that utilized thiolated HA as a chain transfer agent during PEGDA polymerization have been found as the best candidate for applications involving tissue engineering and cell spreading. These hydrogels can be made and tailored to possess a constant density of adhesion peptides that promote natural cell activity.

### 3.2.3 Thiol-Michael Additions

Thiol-Michael additions have been used extensively in recent years to create several types of degradable polymers with a wide range of biomedical uses.[11,25–58] Because of the bio-application emphasis, these polymerizations usually involve acrylated polymers, often PEGs, which are crosslinked with thiolated compounds that are often bio-sourced or biodegradable. This approach is normally used to make a crosslinked network structure for tissue engineering, drug delivery, adhesives, cell growth matrices, and other related biomedical applications.

The significant activity began with a seminal paper by Hubbell *et al.*[11] in 2001 in which they developed the thiol-Michael reaction for the crosslinking of diacrylated PEG with α- and ω-cysteine terminated oligopeptides. The same authors followed up with a second paper that used vinyl sulfone-terminated PEGs instead of acrylate-based PEGs (Scheme 3.5) where it was found that the use of vinyl sulfone also yielded crosslinked hydrogels.[25] There are several advantages to the use of thiol-Michael addition. Perhaps the two primary attributes are that the reaction is fast and goes to completion (thus falls under

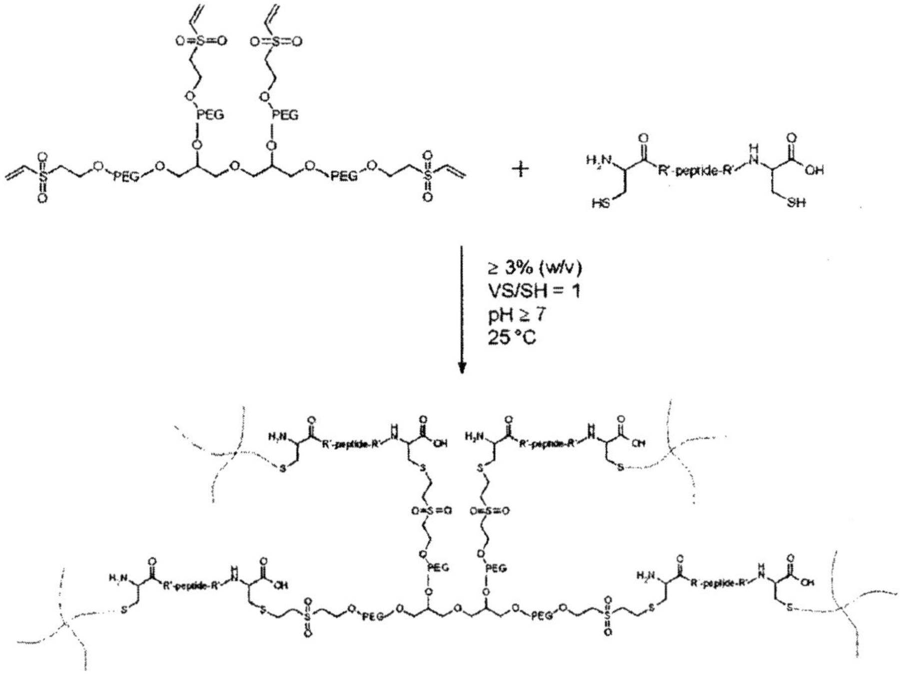


**Scheme 3.5** Synthesis of PEG and peptide-based hydrogels using thiol-Michael addition as developed by Lutolf and Hubbell.
Reprinted with permission from ref. 25. Copyright 2003 American Chemical Society.

the purview of click chemistry from that perspective) without the need for catalysts, and can be carried out under physiological conditions. In addition, it was shown that the thiol-Michael reaction is sensitive to various parameters, such as pH, stereoelectronics of the thiol, and the p$K_a$ of the thiol (which can be influenced by nearby amino acids).[12] Many subsequent studies have utilized the availability of acrylated and other vinyl-functionalized PEGs in making polymers (both crosslinked and non-crosslinked). The availability of cysteine-containing peptides allows the easy incorporation of complex biological materials into a polymer matrix. Furthermore, other biopolymers that are readily thiolated can be used; a common example is that of thiolated HA.

Degradation of gels made by thiol-Michael addition may occur *via* several mechanisms, depending of the functionality of the materials used in their synthesis. Those containing esters, such as those based on PEG-acrylates, may undergo ester hydrolysis. Early work with such materials showed that the gels degrade over several days or even weeks, depending on the degree of cross-linking and degradation conditions.[11] The inclusion of peptide sequences that may undergo enzymatic degradation has also been examined. While this approach allows for specificity as a result of enzymatic degradation, it is also more expensive to fabricate and may not be stored for long periods of time.[25,59]

As stated above, the thiol-Michael addition reaction as applied to the synthesis of degradable gels has received significant attention in recent years. Below are several examples of the use of thiol-Michael addition in the synthesis of such materials to illustrate the utility of this reaction in the production of biomedical devices and materials.

Cellesi *et al.*[60] focused on the crosslinking of telechelic linear poly(propylene glycol)-*block*-poly(ethylene glycol)-*block*-poly(propylene glycol) (Pluronic®) with two sets of terminal groups including thiol groups and acrylate groups. The gelation process is based on the occurrence of both physical and chemical mechanisms (Scheme 3.6). This work optimized the molecular architecture and functionality of precursor species for the use of encapsulated materials designed for calcium alginate. Degradation was monitored by measuring the swelling of the hydrogel beads and is based on the hydrolysis of ester bonds that are incorporated into the crosslinked network with Michael-type addition reactions.

Synthetic biomimetic materials, primarily degradable hydrogels, have been used in order to establish proliferating environments for cell cultures. PEG hydrogels that have been functionalized with vinyl sulfones will actively undergo Michael addition reactions with cysteine containing species. Commonly, the cysteine group will be incorporated into a peptide sequence, which allows for further functionalization to occur in order to promote cell adhesion. Zisch *et al.*[61] used a 4-arm PEG hydrogel functionalized with vinyl sulfone end groups and crosslinked it with a peptide that contains a cysteine unit. Vascular endothelial growth factor (VEGF) was covalently incorporated through a cysteine amino acid from pendant peptide units, which allowed for controlled release of VEGF. Local and controlled release of VEGF was provided through the presence of MMP peptides. In a similar study by Pratt *et al.*,[62] a di-functionalized PEG unit was functionalized with vinyl sulfone and crosslinked through a tricysteine peptide.[62] The longevity of the hydrogels was investigated and it was found that the gels were insoluble after four hours and maintained properties for at least four months. Degradation studies showed significant increase in the presence of plasmin. Furthermore, cellular activity

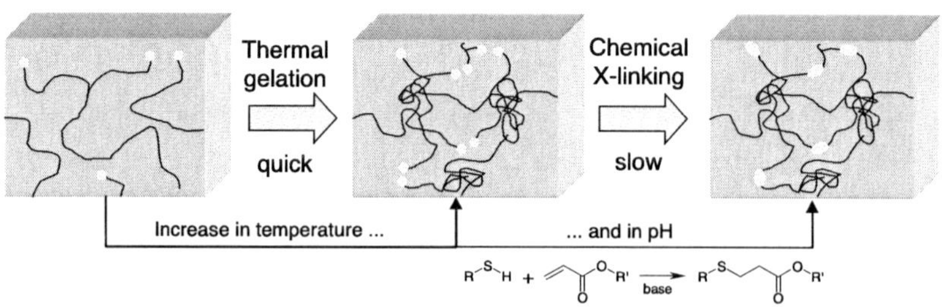


**Scheme 3.6** Outline of the thermal gelation and thiol-Michael crosslinking of Pluronic® materials.
Reprinted with permission from ref. 60. Copyright 2003 Elsevier Ltd.

increased significantly in the presence of RGD as opposed to studies without RGD, and also increased in the presence of L-cysteine over D-cysteine.

There has been significant interest in utilizing thiol-Michael addition reactions for the production of extracellular matrix (ECM) mimetics and hydrogel scaffolds for tissue engineering in general. As such, several workers have examined the combination of acrylate- and vinyl sulfone-based PEGs with thiolated HA. An early example was that of Park *et al.*,[63] who reacted thiolated HA into the PEG-based hydrogels in order to create a hydrolytically sensitive ester bond. Censi *et al.*[64] utilized methylacylate bearing ABA-triblock copolymers of a PEG-based structure containing thermosensitive blocks that were cured by Michael type addition with thiolated HA to yield injectable materials that would undergo gelation *in situ* under physiological conditions. Lin *et al.*[65] used thiolated dextran and thermosensitive acrylate/vinyl sulfone modified Pluronic 127 copolymers to form hydrogel materials *in situ via* a Michael addition reaction.

The materials discussed immediately above for tissue engineering are passive degradable materials (*i.e.* they degrade at a rate dictated by the hydrolysis of the ester bonds in the acrylate-PEG). However, for hydrogel materials to be used for tissue engineering or cell seeding it is necessary to promote cell–cell interactions at a certain point in time or at a specific location. In order to achieve this feat, it is necessary to utilize chemistries that allow for the controlled degradation of hydrogel materials in order to release cells for appropriate biological interactions. Such hydrogel materials can be made to incorporate enzymatically degradable moieties. More specifically, Kloxin *et al.*[66] showed that the incorporation of photodegradable crosslinkers with an enzymatically cleavable thiol-containing peptide could be constructed using Michael addition. The composition of such photodegradable hydrogels involves the use of PEG-based materials. To demonstrate the impact of hydrogel materials for cell migration, the erosion of these photodegradable hydrogels was examined through studies that involved degradation using photolithography. The use of photolithography allowed for the formation of microscopic or macroscopic patterns into the hydrogel structure. In the presence of encapsulated cells, these photodegradable hydrogels create a platform for cell–cell interactions and connectivity. It was found that cell migration occurred exclusively in areas of the hydrogel structure that exhibited complete or near-complete gel erosion. In addition, gel crosslinking density was examined in order to observe the effects on cell spreading. It was also found that when the crosslinking density of hydrogel materials was reduced through photodegradation, cell spreading was promoted.

Patterson and Hubbell[67] developed PEG-based hydrogels made from thiol-Michael additions that are enzymatically or proteolytically sensitive through the incorporation of MMP substrate sequences in order to promote cellular responses and cellular migration into the polymeric hydrogel scaffold. Their work developed cell-responsive hydrogels from multi-arm PEG-based materials terminated with vinyl-sulfone groups. These materials are then subsequently functionalized with important ligands and growth factors with cysteine-containing groups. After gelation, contact between the hydrogel material and cells, *in vitro* or *in vivo*, resulted in local degradation of the hydrogel as the cells respond to biological cues promoted by the hydrogel material. It was observed that the degree of degradation of the hydrogel material influenced the rate of cell spreading. More degradation allowed for decreased crosslinking density, which in turn promoted cell spreading and cell invasion *in vitro*. Similarly, Miller *et al.*[68] used PEG diacrylate with bis-cysteine MMP-sensitive peptides, which yield a product that is biodegradable and a photoactive macromer in the form of acrylate-PEG-(peptide-PEG)-acrylate. PEG-based hydrogel materials have also been utilized for the development of materials with well-defined molecular architectures that allow for the tailoring of desired mechanical and chemical properties. In the development of these materials, especially in applications such as cellular scaffolds, the mechanical properties must be tailored to provide sufficient strength for cell growth, cell migration, and the development of new blood vessels during angiogenesis of regions undergoing cell restoration.

Buwalda *et al.*[69] prepared chemically crosslinked hydrogels using a Michael addition reaction between an acrylated poly(ethylene glycol)-poly(lactide) (PEG-PLA) star block copolymer and a thiol-terminated multi-arm PEG molecule. The resulting PEG-PLA hydrogel exhibited degradation through the hydrolysis of the PLA chains in the block copolymers. Through rheological studies, the PEG-PLA hydrogels were found to possess robust mechanical properties when formed *in situ*. Degradation studies conducted on these materials indicated that PEG-PLA hydrogels exhibit adequate *in vitro* stability with 30% mass loss after 30 days.

Injectable hydrogels have become increasingly more prominent due to improved control over different physical and chemical properties of the material. In addition, injectable biomaterials are able to conform to complex shapes at the site of injury and can adhere to tissue to promote cell growth and regeneration. More importantly, *in situ* gelation of biomaterials allows for the minimization of surgical invasiveness and allows for possible release of cells that can secrete ECM that can adhere to the hydrogels to stimulate cell activity and repair. Pritchard *et al.*[70] have developed an injectable PEG-based hydrogel based on poly(ethylene glycol) diacrylate (PEG-400-DA) and ethoxylated trimethylolpropane tri-3-mercaptopropionate (ETTMP) (Scheme 3.7). The degradation of the hydrogels was examined in PBS (pH 7.4) at 37 °C and the

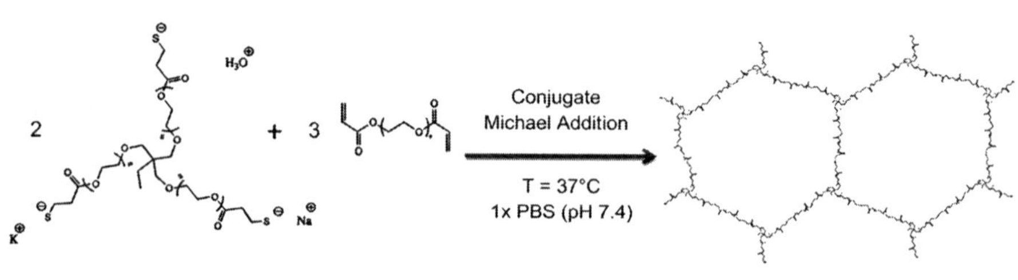


**Scheme 3.7** Outline of the formation of an injectable PEG-based hydrogel. Reprinted with permission from ref. 70. Copyright 2010 Elsevier Ltd.

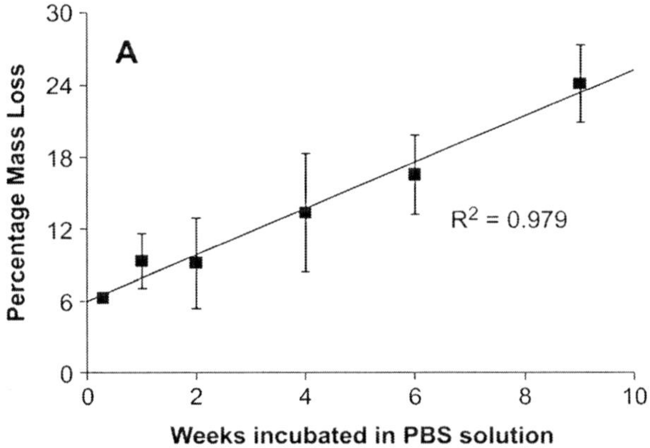


**Figure 3.1** Degradation of injectable PEG-based hydrogel based on the chemistry shown in Scheme 4.7.
Reprinted with permission from ref. 70. Copyright 2010 Elsevier Ltd.

gels were found to reach equilibrium within 48 hours of incubation with no appreciable mass loss, with degradation occurring linearly over nine weeks *in vitro* (Figure 3.1). For the examination of cell adhesion, the hydrogels were functionalized through the incorporation of an oligolysine peptide (Cys-($Lys_{14}$)-Cys).[70]

Another approach to the fabrication of injectable *in situ* cross-linked hydrogels involves the modification of chitosan by the use of Michael addition.[71] Porous structures in hydrogel fabrication offer a potential scaffold for cell infiltration and growth. The particular hydrogel examined was synthesized through the Michael addition of thiol-modified chitosan (CS-NAC) and PEG diacrylate (PEGDA). It was found that important mechanical and swelling properties of the injectable biomaterial were influenced by the content of free thiols in CS-NAC, temperature, and concentration of each of the precursors. Consequently, these hydrogels offer great potential as a biocompatible material for biomedical applications such as drug delivery and tissue engineering.

## 3.3 Conclusions

As presented in this chapter, thiol-X chemistries have been developed and widely applied in the synthesis of a range of degradable polymers over the past decade or so. In particular, thiol-ene and thiol-Michael chemistries have shown significant promise because of their ability to be high yielding, orthogonal and functional group tolerant. The reactions also exhibit the capacity to be conducted in physiological conditions, thus allowing them to be utilized *in vivo*. Furthermore, a wide range of thiol-containing systems, such as cysteine residues in peptides, can be incorporated into the polymer material, thereby providing functionality that may be specifically tailored, *e.g.* for targeted delivery, degradation, adhesion or cell growth. Degradation of such polymers may be achieved through a number of bond scission processes, often hydrolysis (*e.g.* of ester linkages) but may also be catalyzed by enzymes. Many examples in the literature are hydrogels, particularly based on PEG, which can be rendered functional post-polymerization to provide specific responses or induce reactions that lead to desired results. Continued interest in the development of new materials for tissue engineering and the delivery of therapeutics is bound to further the involvement of thiol-X chemistries, and should see them become mainstays in the production of biomedical devices in the near future.

## References

1. A. B. Lowe and M. A. Harvison, *Aust. J. Chem.*, 2010, **63**, 1251–1266.
2. C. E. Hoyle, A. B. Lowe and C. N. Bowman, *Chem. Soc. Rev.*, 2010, **39**, 1355–1387.
3. B. S. Sumerlin and A. P. Vogt, *Macromolecules*, 2010, **43**, 1–13.
4. H. C. Kolb, M. G. Finn and K. B. Sharpless, *Angew. Chem., Int. Ed.*, 2001, **40**, 2004–2021.
5. C. J. Hawker and K. L. Wooley, *Science*, 2005, **309**, 1200–1205.
6. R. K. Iha, K. L. Wooley, A. M. Nystrom, D. J. Burke, M. J. Kade and C. J. Hawker, *Chem. Rev.*, 2009, **109**, 5620–5686.
7. C. E. Hoyle, T. Y. Lee and T. Roper, *J. Polym. Sci. Part A: Polym. Chem.*, 2004, **42**, 5301–5338.
8. C. E. Hoyle and C. N. Bowman, *Angew. Chem., Int. Ed.*, 2010, **49**, 1540–1573.
9. A. B. Lowe, C. E. Hoyle and C. N. Bowman, *J. Mater. Chem.*, 2010, **20**, 4745–4750.
10. M. Friedman, J. F. Cavins and J. S. Wall, *J. Am. Chem. Soc.*, 1965, **87**, 3672–3682.
11. D. L. Elbert, A. B. Pratt, M. P. Lutolf, S. Halstenberg and J. A. Hubbell, *J. Controlled Release*, 2001, **76**, 11–25.
12. A. Metters and J. Hubbell, *Biomacromolecules*, 2005, **6**, 290–301.
13. A. M. Kloxin, C. J. Kloxin, C. N. Bowman and K. S. Anseth, *Adv. Mater.*, 2010, **22**, 3484–3494.
14. A. E. Rydholm, S. K. Reddy, K. S. Anseth and C. N. Bowman, *Polymer*, 2007, **48**, 4589–4600.
15. D. A. Shipp, C. W. McQuinn, B. G. Rutherglen and R. A. McBath, *Chem. Commun.*, 2009, 6415–6417.
16. B. G. Rutherglen, R. A. McBath, Y. L. Huang and D. A. Shipp, *Macromolecules*, 2010, **43**, 10297–10303.
17. C. C. Lin, A. Raza and H. Shih, *Biomaterials*, 2011, **32**, 9685–9695.
18. A. A. Aimetti, A. J. Machen and K. S. Anseth, *Biomaterials*, 2009, **30**, 6048–6054.
19. J. A. Benton, B. D. Fairbanks and K. S. Anseth, *Biomaterials*, 2009, **30**, 6593–6603.

20. S. B. Anderson, C. C. Lin, D. V. Kuntzler and K. S. Anseth, *Biomaterials*, 2011, **32**, 3564–3574.
21. R. A. Ortiz, A. Y. R. Martinez, A. E. G. Valdez and M. L. B. Duarte, *Carbohydr. Polym.*, 2010, **82**, 822–828.
22. R. A. Ortiz, A. E. G. Valdez, M. G. M. Aguilar and M. L. B. Duarte, *Carbohydr. Polym.*, 2009, **78**, 282–286.
23. C. A. DeForest, E. A. Sims and K. S. Anseth, *Chem. Mater.*, 2010, **22**, 4783–4790.
24. S. Ouasti, R. Donno, F. Cellesi, M. J. Sherratt, G. Terenghi and N. Tirelli, *Biomaterials*, 2011, **32**, 6456–6470.
25. M. P. Lutolf and J. A. Hubbell, *Biomacromolecules*, 2003, **4**, 713–722.
26. J. W. Bae, E. Lee, K. M. Park and K. D. Park, *Macromolecules*, 2009, **42**, 3437–3442.
27. H. H. Bearat, B. H. Lee, J. Valdez and B. L. Vernon, *J. Biomater. Sci., Polym. Ed.*, 2011, **22**, 1299–1318.
28. B. Blakely, B. H. Lee, C. Riley, R. McLemore, C. P. Pathak and B. L. Vernon, *J. Biomed. Mater. Res., Part B*, 2010, **93B**, 9–17.
29. F. Brandl, M. Henke, S. Rothschenk, R. Gschwind, M. Breunig, T. Blunk, J. Tessmar and A. Gopferich, *Adv. Eng. Mater.*, 2007, **9**, 1141–1149.
30. F. Brandl, F. Kastner, R. M. Gschwind, T. Blunk, J. Tessmar and A. Gopferich, *J. Controlled Release*, 2010, **142**, 221–228.
31. K. Chawla, T. B. Yu, S. W. Liao and Z. B. Guan, *Biomacromolecules*, 2011, **12**, 560–567.
32. M. Davidovich-Pinhas and H. Bianco-Peled, *J. Mater. Sci.: Mater. Med.*, 2010, **21**, 2027–2034.
33. M. Davidovich-Pinhas and H. Bianco-Peled, *Acta Biomater.*, 2011, **7**, 625–633.
34. T. Dispinar, W. Van Camp, L. J. De Cock, B. G. De Geest and F. E. Du Prez, *Macromol. Biosci.*, 2012, **12**, 383–394.
35. J. W. DuBose, C. Cutshall and A. T. Metters, *J. Biomed. Mater. Res. A*, 2005, **74A**, 104–116.
36. C. Hiemstra, L. J. van der Aa, Z. Y. Zhong, P. J. Dijkstra and J. Feijen, *Macromolecules*, 2007, **40**, 1165–1173.
37. C. Hiemstra, L. J. van der Aa, Z. Y. Zhong, P. J. Dijkstra and J. Feijen, *Biomacromolecules*, 2007, **8**, 1548–1556.
38. G. A. Hudalla, T. S. Eng and W. L. Murphy, *Biomacromolecules*, 2008, **9**, 842–849.
39. R. Jin, L. S. M. Teixeira, P. J. Dijkstra, C. A. van Blitterswijk, M. Karperien and J. Feijen, *Biomaterials*, 2010, **31**, 3103–3113.
40. Y. S. Jo, J. Gantz, J. A. Hubbell and M. P. Lutolf, *Soft Matter*, 2009, **5**, 440–446.
41. J. M. Jukes, L. J. van der Aa, C. Hiemstra, T. van Veen, P. J. Dijkstra, Z. Y. Zhong, J. Feijen, C. A. van Blitterswijk and J. de Boer, *Tissue Eng., Part A*, 2010, **16**, 565–573.

42. Y. G. Lei, Q. K. T. Ng and T. Segura, *Microsc. Res. Tech.*, 2010, **73**, 910–917.
43. Y. G. Lei and T. Segura, *Biomaterials*, 2009, **30**, 254–265.
44. Y. Y. Liang, T. W. Jensen, E. J. Roy, C. Y. Cha, R. J. DeVolder, R. E. Kohman, B. Z. Zhang, K. B. Textor, L. A. Rund, L. B. Schook, Y. W. Tong and H. Kong, *Biomaterials*, 2011, **32**, 2004–2012.
45. S. Q. Liu, Q. A. Tian, J. L. Hedrick, J. H. P. Hui, P. L. R. Ee and Y. Y. Yang, *Biomaterials*, 2010, **31**, 7298–7307.
46. S. Q. Liu, Q. A. Tian, L. Wang, J. L. Hedrick, J. H. P. Hui, Y. Y. Yang and P. L. R. Ee, *Macromol. Rapid Comm.*, 2010, **31**, 1148–1154.
47. K. N. Minh, T. H. Cong, G. H. Gao, J. H. Kim, P. H. Dai, S. Y. Chae, K. C. Lee and D. S. Lee, *Soft Matter*, 2011, **7**, 2994–3001.
48. W. L. Murphy, W. S. Dillmore, J. Modica and M. Mrksich, *Angew. Chem., Int. Ed.*, 2007, **46**, 3066–3069.
49. J. Patterson and J. A. Hubbell, *Biomaterials*, 2011, **32**, 1301–1310.
50. E. A. Phelps, N. O. Enemchukwu, V. F. Fiore, J. C. Sy, N. Murthy, T. A. Sulchek, T. H. Barker and A. J. Garcia, *Adv. Mater.*, 2012, **24**, 64–70.
51. S. C. Rizzi and J. A. Hubbell, *Biomacromolecules*, 2005, **6**, 1226–1238.
52. S. A. Robb, B. H. Lee, R. McLemore and B. L. Vernon, *Biomacromolecules*, 2007, **8**, 2294–2300.
53. A. Shikanov, R. M. Smith, M. Xu, T. K. Woodruff and L. D. Shea, *Biomaterials*, 2011, **32**, 2524–2531.
54. W. Steinhauer, H. Keul and M. Moller, *Polym. Chem.*, 2011, **2**, 1803–1814.
55. Z. C. Wang, X. D. Xu, C. S. Chen, L. Yun, J. C. Song, X. Z. Zhang and R. X. Zhuo, *ACS Appl. Mater. Inter.*, 2010, **2**, 1009–1018.
56. Y. Yu and Y. Chau, *Biomacromolecules*, 2012, **13**, 937–942.
57. Y. X. Yu, C. Deng, F. H. Meng, Q. Shi, J. Feijen and Z. Y. Zhong, *J. Biomed. Mater. Res. A*, 2011, **99A**, 316–326.
58. S. P. Zustiak and J. B. Leach, *Biomacromolecules*, 2010, **11**, 1348–1357.
59. S. P. Zustiak and J. B. Leach, *Biotechnol Bioeng*, 2011, **108**, 197–206.
60. F. Cellesi, N. Tirelli and J. A. Hubbell, *Biomaterials*, 2004, **25**, 5115–5124.
61. A. H. Zisch, M. P. Lutolf, M. Ehrbar, G. P. Raeber, S. C. Rizzi, N. Davies, H. Schmokel, D. Bezuidenhout, D. V, P. Zilla and J. A. Hubbell, *FASEB J.*, 2003, **17**, 2260–2262.
62. A. B. Pratt, F. E. Weber, H. G. Schmoekel, R. Muller and J. A. Hubbell, *Biotechnol. Bioeng.*, 2004, **86**, 27–36.
63. Y. Park, N. Tirelli and J. Hubbell, *Biomaterials*, 2003, **24**, 893–900.
64. R. Censi, P. J. Fieten, P. di Martino, W. E. Hennink and T. Vermonden, *Macromolecules*, 2010, **43**, 5771–5778.
65. C. Lin, P. Zhao, F. Li, F. Guo, Z. Li and X. Wen, *Mater. Sci. Eng. C*, 2010, **30**, 1236–1244.
66. A. M. Kloxin, A. M. Kasko, C. N. Salinas and K. S. Anseth, *Science*, 2009, **324**, 59–63.

67. J. Patterson and J. A. Hubbell, *Biomaterials*, 2010, **31**, 7836–7845.
68. J. S. Miller, C. J. Shen, W. R. Legant, J. D. Baranski, B. L. Blakely and C. S. Chen, *Biomaterials*, 2010, **31**, 3736–3743.
69. S. Buwalda, P. Dijkstra and J. Feijen, *J. Controlled Release*, 2011, **152**, 329–428.
70. C. D. Pritchard, T. M. O'Shea, D. J. Siegwart, E. Calo, D. G. Anderson, F. M. Reynolds, J. A. Thomas, J. R. Slotkin, E. J. Woodard and R. Langer, *Biomaterials*, 2011, **32**, 587–597.
71. D. Y. Teng, Z. M. Wu, X. G. Zhang, Y. X. Wang, C. Zheng, Z. Wang and C. X. Li, *Polymer*, 2010, **51**, 639–646.

CHAPTER 4

# *Thiol–Thiosulfonate Chemistry in Polymer Science: Simple Functionalization of Polymers via Disulfide Linkages*

PETER J. ROTH*[a] AND PATRICK THEATO*[b]

[a] Centre for Advanced Macromolecular Design (CAMD), School of Chemical Engineering, University of New South Wales, UNSW Sydney, NSW 2052, Australia; [b] Institute for Technical and Macromolecular Chemistry, University of Hamburg, Bundesstrasse 45, D-20146 Hamburg, Germany

*Email: p.roth@unsw.edu.au; theato@chemie.uni-hamburg.de

## 4.1 Introduction

In recent years, post-polymerization modification of polymers has been revisited and has found broad application in combination with controlled polymerization methods. The possibility to prepare architecturally well-defined polymers and the precise introduction of functional groups in the prepared polymer represents an ideal synthetic combination. As such, post-polymerization modification techniques are versatile and complementary. Among the different chemistries employed in post-polymerization modifications, functionalization utilizing thiols have found broad application. Due to the ubiquitous presence of thiol groups in proteins, *i.e.* cysteine moieties, selective functionalization chemistries have been developed in biochemistry and

RSC Polymer Chemistry Series No. 6
Thiol-X Chemistries in Polymer and Materials Science
Edited by Andrew B. Lowe and Christopher N. Bowman

Published by the Royal Society of Chemistry, www.rsc.org

are used intensively in the conjugation of polymers with proteins/polypeptides.[1–5] These chemical concepts have also found application in polymer chemistry[6–23] and as such, the book at hand on "Thiol-X Chemistries in Polymer and Materials Science" represents a comprehensive compilation of the various aspects of thiol chemistries employed in the preparation and functionalization of polymers. The present chapter covers the particularly useful reaction of thiols with thiosulfonates, yielding asymmetric disulfides. Because of the ease of cleaving disulfides on demand by mild reduction, these reactions offer the possibility of reversible functionalisation.[21,24,25]

Methanethiosulfonates (MTS) have found broad utilization in biochemistry. Due to their highly selective reactivity toward thiols in water and fast reaction rates, functional MTS reagents have particularly been exploited for the site-specific modification of cysteine residues of proteins,[26–28] including fluorescent labeling,[29] glycosylation,[30–32] and biotinylation.[33] In particular, ion channel proteins have been targeted with charged MTS reagents, giving insights into their structure and function by way of the substituted cysteine accessibility method (SCAM).[29,34–42] Site directed spin labeling with a radical modified MTS reagent and measurement of the electron paramagnetic resonance spectrum has been used to map the topography of membrane proteins and to determine secondary structures.[43–45]

The versatility of MTS reactions has recently been adapted to polymer chemistry and represents a sub-class of the manifold thiol-X reactions employed in the functionalization of polymers. When functional thiosulfonates are used, the respective functional groups can be installed in polymers. In the following sections, first, the reactivity and synthesis of functional thiosulfonates and their utilization will be described, followed by a section on polymeric thiosulfonates. Next, special emphasis will be laid on the modification of polymers that were prepared by the RAFT technique. Subsequently, developments on the use of MTS chemistries in self-assembled monolayers will be covered. Finally, this chapter will end with an experimental section providing the reader with exemplary recipes.

## 4.2 Reactivity and Synthesis of Functional Thiosulfonates

### 4.2.1 Reactivity

*S*-Alkyl alkanethiosulfonates consist of a functional group (denoted $R^1$ in Scheme 4.1) attached to an electrophilic sulfur atom carrying a sulfinyl leaving group with residue $R^2$. Commonly, $R^2$ is a methyl or phenyl group making up the classes of *S*-substituted methanethiosulfonates (MTS) and benzenethiosulfonates (BnTS). The electrophilic sulfur atom presents high selectivity toward soft nucleophiles, especially thiols, R–SH. Reaction rates of MTS reagents with mercaptoethanol in aqueous buffer (pH 7) have been shown to be in the order of $10^3$ to $10^4\,mol^{-1}\,s^{-1}$, depending on the chemical nature of the functional group.[46] Even at low concentrations below 1 mM in water, the

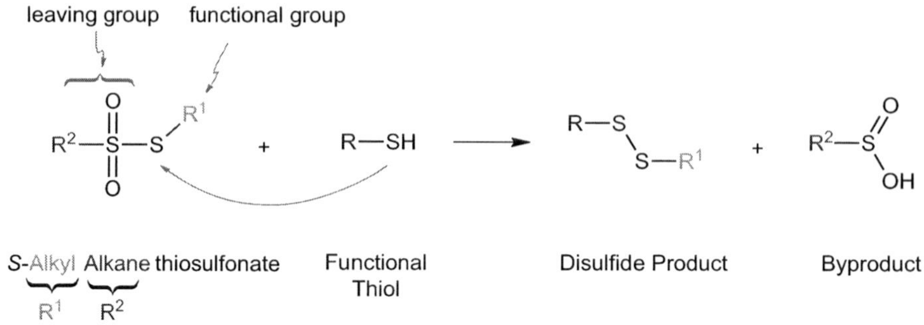


**Scheme 4.1** General mechanism of the reaction of an alkyl alkanethiosulfonate with a thiol and definitions of groups. The most common reagents are methanethiosulfonates (MTS, $R^2$ = methyl) and benzenethiosulfonates (BnTS, $R^2$ = phenyl).

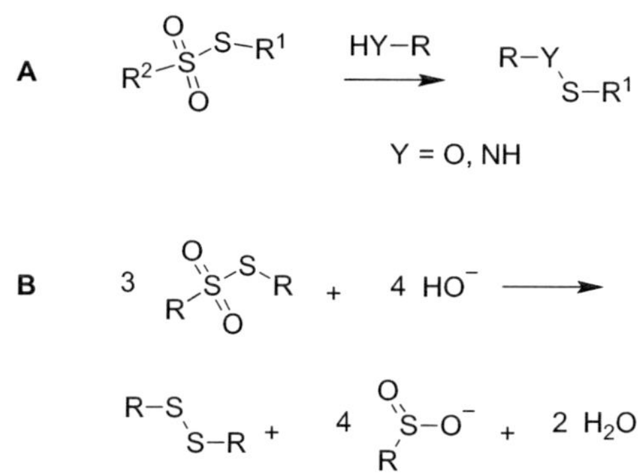


**Scheme 4.2** Reaction of thiosulfonates with alcohols, amines (A) and aqueous base (B).

reaction of methanethiosulfonates with cysteine residues of proteins have been shown to proceed to completion within seconds.[39,40] Thiolates have been shown to react up to $10^7$ times faster than thiols.[47] These substitutions yield the asymmetric disulfides RS-SR$^1$.

The reaction with hard nucleophiles such as alcohols or amines (R-OH, R-$NH_2$) typically also takes place on the –S(R$^1$) atom, as with thiols, giving the corresponding sulfenylated products (Scheme 4.2A), similar to a reaction of alcohols or amines with a sulfenyl chloride (R$^1$S-Cl).[48–50] Sulfenates (R$^1$SO$^-$) produced by reacting a thiosulfonate with aqueous base are generally not stable. Disproportionation can give thiols which react with the thiosulfonate starting materials yielding disulfides as final products (Scheme 4.2B).[48] Hydrolysis occurs at neutral pH with half times ranging from 10 min to 2 hours[51,52] and decreasing with increasing pH. To avoid hydrolysis, reactions in organic media are preferred which typically proceed smoothly. Under Lewis acid catalysis, electron rich aromatics can be used as nucleophiles which similarly substitute the sulfinate leaving group on the electrophilic sulfur yielding functional arylthioethers, $R^1S$-Ar.[53,54]

### 4.2.2 Synthesis of Functional Thiosulfonates

Many functional MTS (or BnTS) reagents are commercially available. A selection of common synthetic pathways for their preparation is depicted in Scheme 4.3. Typically, sodium methanethiosulfonate (NaMTS) (left reactant in reactions A and B, $R^2 = CH_3$) is used as a nucleophilic starting material to replace leaving groups such as bromide[27,55–57] or mesylate[58] on the functional substrate $R^1$. NaMTS is available either through reaction of sodium sulfinate with sulfur,[55,56] which gives sufficiently pure product in high yield after a short reaction time, or through reaction of sodium sulfide (hydrate or nonahydrate) with methanesulfonyl (mesyl) chloride,[56,59,60] which involves cheaper reactants. Other techniques involve the reaction of two different sulfur containing compounds. Polymeric thiosulfonates have been prepared by reacting dithiols with alkanedi(sulfonyl chloride) in the presence of a base (Scheme 4.3C) to give *S*-substituted alkanethiosulfonate linkages.[61] Furthermore, sulfinates have been reacted with *S*-nitrosothiols formed *in situ* to give substituted alkanethiosulfonates (Scheme 4.3D).[62]

Symmetrical thiosulfonates, among them most prominently *S*-methyl methanethiosulfonate, MMTS ($R^1 = R^2 = CH_3$), can be obtained by selective oxidation of a symmetrical organic disulfide, *e.g.* dimethyldisulfide (Scheme 4.4). Recent oxidant/catalyst systems suitable for this reaction include dimethylsulfoxide/rhenium complexes[63] and iodine/ceric ammonium nitrate.[64] Further syntheses of symmetric and asymmetric organic thiosulfonates are

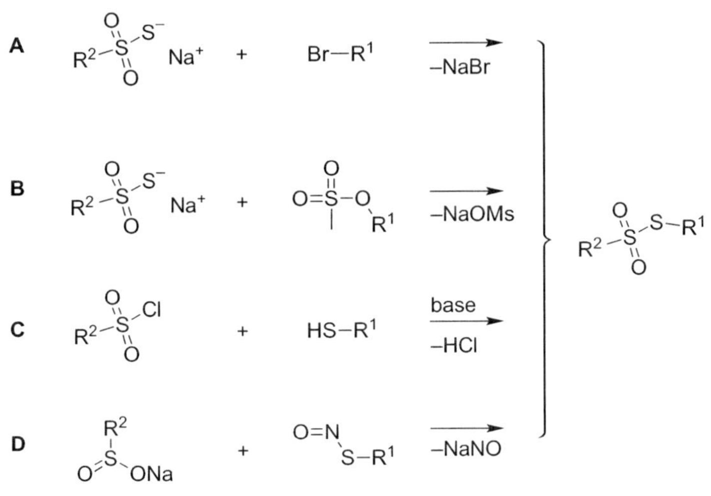


**Scheme 4.3** Selection of common synthetic pathways to *S*-functional alkanethiosulfonates. Ms = mesyl, $CH_3SO_2$–.

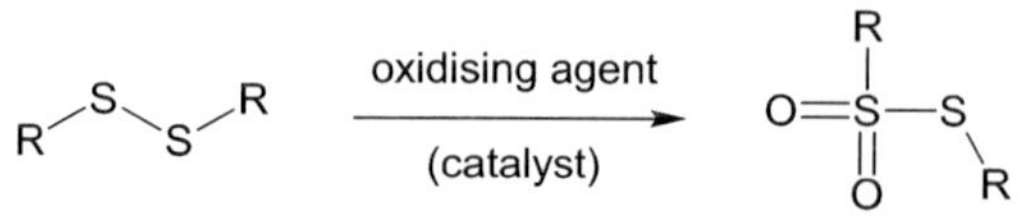


**Scheme 4.4** Selective oxidation of symmetrical disulfides to yield S-substituted thiosulfonates.

presented in a comprehensive but slightly dated review by Zefirov *et al.* from 1993.[48]

# 4.3 Polymeric Thiosulfonates

In this section, polymeric thiosulfonates are reviewed and grouped according to which of the residues of functional group ($R^1$) or leaving group ($R^2$), (as defined in Scheme 4.1) is polymeric.

## 4.3.1 Polymeric Leaving Group ($R^2$ = Polymer)

Raija *et al.*[62] proposed functional polystyrene (PS) immobilized thiosulfonate reagents to produce functional low molecular weight disulfides with thiols. In this case, the sulfinate leaving group was attached to the polymeric support. The removal of the sulfinate side product could thus be accomplished by simple filtration of the matrix. Chlorosulfonated PS was reacted with *S*-substituted *S*-nitrosothiols affording a polystyrene support carrying up to 1.86 mmol $g^{-1}$ PS of immobilized thiols. These reagents were used in 2-fold excess toward thiols (cysteine, acetyl cysteine, reduced glutathione) to produce the functional disulfide derivatives in yields above 92%.

## 4.3.2 Polymeric Functional Group ($R^1$ = Polymer)

*S*-Polymeric methanethiosulfonates yield functional polymeric disulfides upon exposure to thiols. This architecture is advantageous as the resulting disulfide linkages can be easily cleaved under physiological conditions with intracellular glutathione,[24,25] providing the opportunity for delivery applications. Thiosulfonate functionality can be introduced into polymers by substitution of halides or mesylates, respectively.

Zhao *et al.*[65] prepared the thiol-reactive methoxy poly(ethylene glycol) (mPEG) derivatives mPEG-MTS and mPEG-BnTS. Hydroxy terminated mPEG was converted into a bromo species using $PPh_3/CBr_4$, followed by treatment with NaMTS. The thiol containing proteins papain and recombinant human granulocyte colony-stimulating factor were site specifically PEGylated using mPEG-MTS and mPEG-BnTS.

Fang and Bentley[58] employed multifunctional PEG stars with MTS or BnTS terminated arms to produce crosslinked hydrogels. Thiosulfonate groups were introduced using either NaMTS/NaBnTS to replace mesylate groups or by way of amino functionalized MTS species that were coupled to the arms using activated esters. Addition of base to MTS/BnTS terminated stars caused hydrolysis and disulfide crosslinking of the PEG stars (see Scheme 4.2B).

A recent study by Boyer *et al.*[66] demonstrated that MTS chemistry can give access to versatile thiol–disulfide and thiol–ene reactions on polymers prepared by atom transfer radical polymerization (ATRP). Using $CuBr_2$ termination, poly(*tert*-butyl acrylate) carrying bromide end groups was obtained. Substitution of terminal bromides using NaMTS produced MTS functional polymer with end group functionality above 90%. The resulting polymer was coupled quantitatively with functional thiols *via* disulfide bonds (Scheme 4.5A). Heating of the MTS terminated polymer in DMF containing triethylamine and traces of water was reported to produce thiol end groups. These were successfully reacted with fluorescein acrylate in a thiol-ene reaction.

The possibility of introducing MTS functionality along a polymer main chain is to (co)polymerise *p*-chloromethyl styrene (Cl-St) (Scheme 4.5B). This approach was employed by Duong *et al.*[67] in the production of drug delivery vehicles. Poly[oligo(ethylene glycol) acrylate] was chain extended by activated ester monomers and Cl-St. MTS functionality, introduced by substitution of the chlorides by NaMTS, was used to couple model drugs to the copolymers *via*

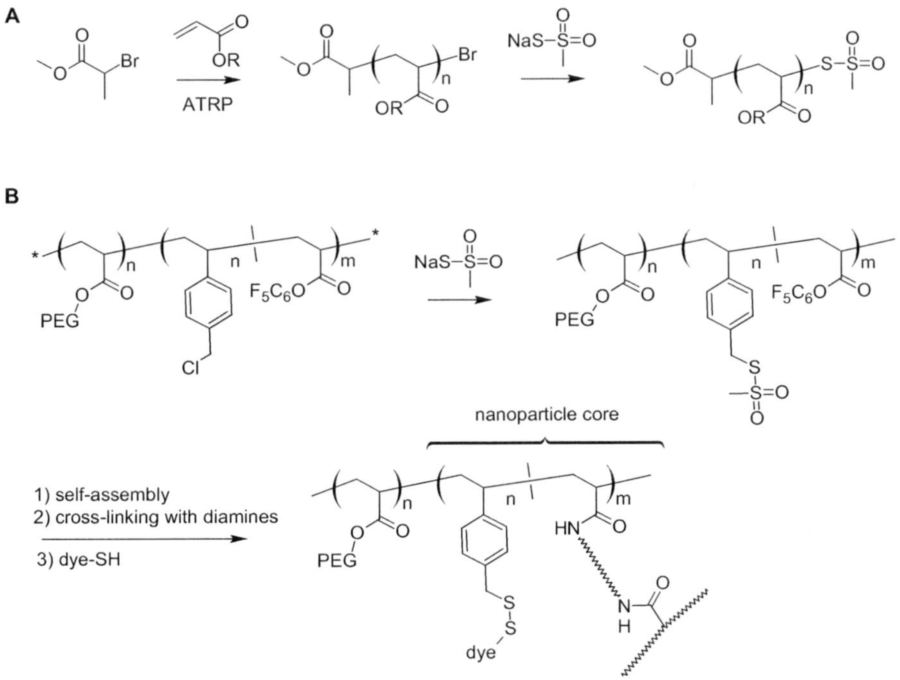


**Scheme 4.5** Introduction of MTS functionality into polymers carrying displaceable halides. (A) End group modification on ATRP polymers;[66] (B) side group modification on *p*-chloromethylstyrene monomer units PEG = poly(ethylene glycol).[67]

the formation of disulfide linkages. After self-assembly into nanoparticles in water, the activated ester groups were used to introduce acid degradable crosslinks. A second type of encapsulated model drug was released from the vehicles by an acid stimulus, whereas the molecules bound *via* disulfides were released in a reductive environment, thus constituting a smart carrier for the co-delivery of different drug types.

### 4.3.3 Polymeric Leaving Group and Polymeric Functional Group ($R^1$, $R^2$ = polymer)

Polymers containing thiosulfonate linkages along the backbone were described by Mack *et al.*[61] The authors employed a step growth polymerization of difunctional alkane thiols with difunctional sulfonyl chlorides in benzene in the presence of aq. NaOH or in *N*-methylpyrrolidinone as solvent. Solubility and melting points were determined in this patent literature.

## 4.4 Modification of RAFT polymers (R-SH = Polymer)

Reversible addition-fragmentation chain transfer (RAFT) polymerization is arguably one of the most versatile reversible deactivation radical polymerization (RDRP) techniques, giving access to a large range of functional (co)polymers. Chain growth is mediated by thiocarbonylthio chain transfer agents (CTAs), such as dithioesters, trithiocarbonates, dithiocarbamates, and xanthates.[68–71] These groups are retained as chain end groups after polymerization. Polymer end group modification plays an important role in the preparation of well-defined polymeric architectures.[72] For RAFT polymers, the scope of possible end group reactions is largely dictated by the available thiocarbonylthio chemistry.[73] Possibly the most common modification is a nucleophilic acyl substitution of the thiocarbonylthio group with amines[68,74] during which a thiol terminated polymer is expelled from the thiocarbonyl moiety. Other techniques yielding thiol terminated polymers include reduction with borohydrides[75] or (for some polymers), the thermal Chugaev reaction.[76]

### 4.4.1 Introducing Methyl Disulfide End Groups

Polymeric thiols prepared by aminolysis from RAFT end groups may undergo side reactions during the basic aminolytic conditions. One side reaction is an oxidative thiol–thiol coupling giving double molecular weight side products.[77,78] A nucleophilic back biting side reaction can occur on poly(meth)acrylates, where the terminal thiol substitutes the alcohol on the penultimate ester group.[79] This reaction consumes the desired sulfhydryls and produces thiolactone end groups.

Roth *et al.*[80] proposed that an excess of methyl methanethiosulfonate (MMTS) added to a RAFT-prepared polymer solution before addition of amine can circumvent these side reactions. In this case the thiol reacts

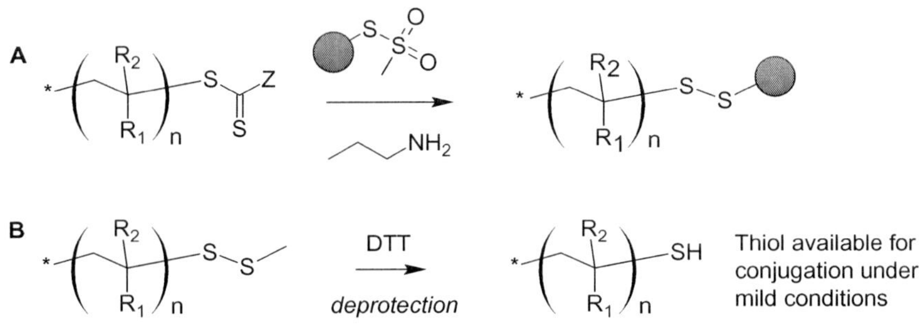


**Scheme 4.6** (A) Aminolysis of RAFT polymers in the presence of MTS reagents; (B) methyl disulfide as reversible thiol protecting group. DTT = dithiothreitol.

preferably with the MTS reagent, resulting quantitatively in well-defined methyl disulfide end groups (Scheme 4.6A).

A methyl disulfide end group can further be exploited as a protecting group for the thiol while under the harsh conditions required for transformation of the RAFT end group. The protecting group can then readily be cleaved using mild reducing agents avoiding typical side reactions that occur during aminolysis and making the thiol available for reaction with delicate (*e.g.* amine reactive) substrates (Scheme 4.6B). Carlson *et al.*[81] utilized this strategy to protect the thiol end group of a polystyrene-*block*-polyacrylonitrile copolymer during aminolysis. Treatment with lithium aluminium hydride transformed the pendant cyano groups of the second polyacrylonitrile block into amino groups. Subsequent removal of the methyl disulfide protecting group with dithiothreitol (DTT) produced cysteine like end groups with the thiol end group and the ultimate pendant amino group spaced by two carbon atoms. This combination of a thiol with an amine was used for a selective coupling to aldehyde moieties. Isophthalaldehyde was used as a difunctional linker to connect two polymer chains.

Methyl disulfide end groups have also been employed to produce self-assembled monolayers on spherical and planar metal and semiconductor surfaces,[80,82] *vide infra*.

## 4.4.2 Functional Disulfide End Groups

With MMTS successfully preventing unwanted side reactions and producing well-defined methyl disulfide end groups on RAFT-prepared polymers, the group of Theato investigated the use of functional MTS reagents to broaden the scope of RAFT end group modification *via* MTS chemistry.[57,60,82–85] Disulfides are typically stable in the absence of reducing agents, making MTS chemistry a powerful tool to introduce functional end groups into polymers. With a weak S–S bond strength, however, disulfides can easily be cleaved with a variety of easy to handle reducing agents, such as DTT (thiol

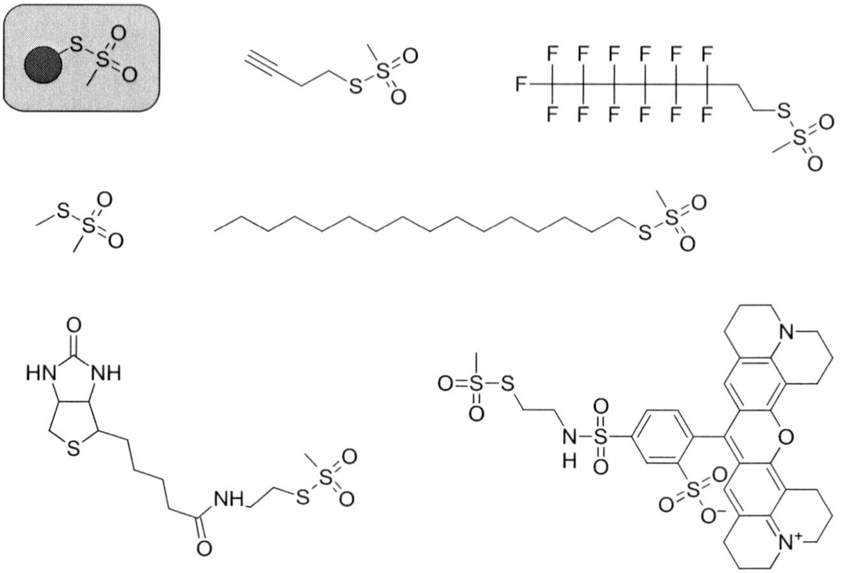

**Scheme 4.7** Structures of functional MTS reagents used for RAFT polymer end group modification.

containing), phosphines (requiring equimolar water), or sodium dithionite ($Na_2S_2O_4$, soluble in water).

In a proof-of-concept study, various methacrylate, acrylamide and styrenic polymers were prepared by the RAFT process and their end groups aminolyzed in the presence of *S*-but-3-ynyl methanethiosulfonate (Scheme 4.7, top middle). This reaction quantitatively equipped the polymers with alkyne end functionality. Polymers were grafted to an azide modified surface through copper catalyzed azide–alkyne cycloaddition (CuAAC). Reduction of the disulfide bonds successfully cleaved the polymers from the surface again as demonstrated by contact angle measurements. This chemistry provides an example of orthogonal attachment (CuAAC) and detachment (disulfide reduction) methods.[60]

The synthetic concept was expanded using a variety of other functional MTS reagents, including hexadecyl,[57] perfluorooctyl,[57] biotinylaminoethyl,[83,85] and Texas Red sulfonamidoethyl[84] MTS reagents (Scheme 4.7). Excellent conversions were achieved when the MTS reagent was employed in excess to RAFT end group, except for the Texas Red derivative, where steric hindrance was believed to reduce the end group functionality.

### 4.4.3 Simultaneous Use with Activated Esters

Within the paradigm of robust efficient orthogonal chemistries in materials science,[20] it is noteworthy to mention that thiol–thiosulfonate chemistry is compatible with other conjugation methods. In particular, the reaction of activated esters with amines can easily be performed simultaneously with the transformation of RAFT end groups into functional disulfides (Scheme 4.8).

Jochum *et al.*[83] prepared a diblock copolymer of poly[oligo(ethylene glycol) methylether methacrylate]-*block*-poly(pentafluorophenyl methacrylate)-dithiobenzoate (POEGMA-*b*-PPFPA-DTB) (Scheme 4.8A). Employing isopropyl amine, triethylamine and *N*-biotinylaminoethyl MTS, the activated ester block was converted into the thermally responsive poly(*N*-isopropyl methacrylamide) (PNIPMAM) species, while the end group was simultaneously converted into a biotin functional disulfide moiety. In this case the amine performed both the task of aminolyzing the activated ester side groups as well as cleavage of the dithioester end group. The modified diblock copolymer self-assembled into micelles by means of a thermal trigger, trapping the biotin target groups in the cores. Specific recognition of a streptavidin modified surface only took place at temperatures below the lower critical solution temperature (LCST) of the PNIPMAM at which the biotin target was exposed.

Exploiting similar chemistries, Roth *et al.* prepared various α,ω-hetero telechelic polymers synthesized by the RAFT process utilizing a pentafluorophenyl modified CTA (Scheme 4.8B).[57,82,84–86] In this case, the amine required to cleave the dithiobenzoate ω-end group simultaneously reacted with the PFP ester yielding an amide at the α-end group. Due to a higher reactivity of PFP esters towards amines, a selective modification of the PFP α-end groups was possible, leaving the dithiobenzoate ω-end groups intact. These were then

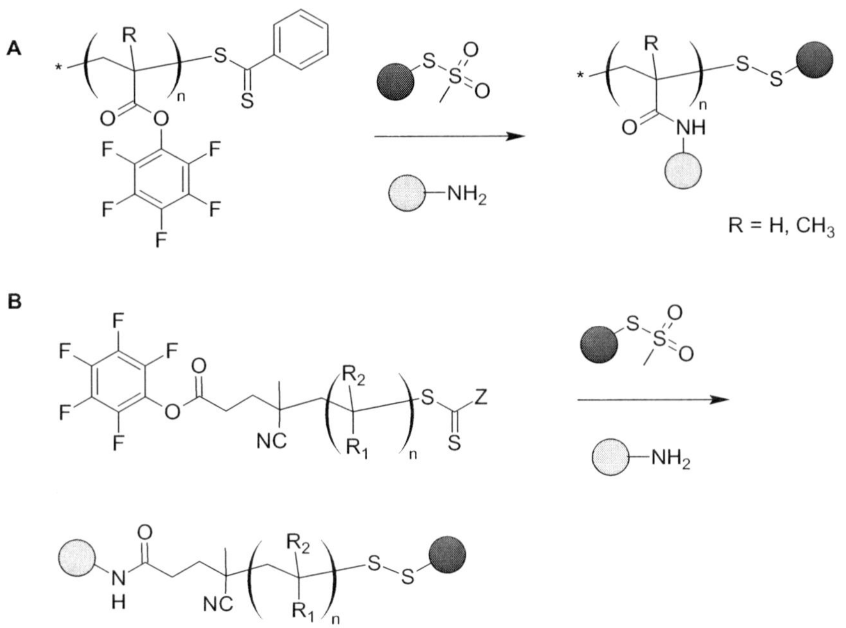


**Scheme 4.8** Simultaneous conversion of pentafluorophenyl activated esters and dithiobenzoate RAFT end groups into functional amides disulfides, respectively.[57,86]

converted into functional disulfides in the presence of a different (cheaper) amine.[84,85] In one example, the authors prepared a difunctional polymer carrying the two different bio target end groups biotin and thyroxine, each capable of recognizing and binding to a specific protein.[85] Similarly, a polymer carrying two different fluorescent dye end groups capable of communicating *via* Förster resonance energy transfer (FRET) was described.[84] A combination of amines of varying polarity with MTS reagents shown in Scheme 4.7 gave access to a library of hetero-telechelic POEGMA species. The influence of end group polarity and size with two functional end groups onto the thermal phase behaviour of POEGMA in water and alcohols could thus be assessed.[57,86]

## 4.5 Self-Assembled Monolayers of Polymeric Methyl Disulfides Prepared by MMTS

The formation of self-assembled monolayers (SAMs) is an important technique in current nanotechnology and nanoscience.[87,88] Polymer based SAMs, *i.e.* polymer brushes, attract significant attention in the field of biotechnology and sensing applications.[89] A prototypical example of surface self-assembly is the coordination of thiols and disulfides to gold surfaces, whereby symmetrical disulfides are known to produce the same SAMs on gold surfaces as the corresponding thiols.[90–92] Heister *et al.*[93] studied self-assembled monolayers from unsymmetrical disulfides having one hexadecyl chain and one butyl, octyl, or dodecyl chain. The authors found that monolayers consisting preferably of the larger chains were formed, suggesting that after dissociation of the S–S bond, lower molecular weight species were replaced on the surface and desorbed into solution. This deviation from an expected 1 : 1 composition was more pronounced with a bigger difference between the disulfide alkyl chains, with only an estimated 5–20% of butyl chains found on the surface after self-assembling butyl hexadecyl disulfide for 24 hours on a gold surface. These findings are promising for polymeric methyl disulfides which have an extreme length difference between the methyl and the polymer groups. The use of methyl disulfide (SSMe) functional/terminated polymers to produce essentially the same SAMs that the corresponding thiols would do, provides advantages wherever the isolation of polymeric thiols is hampered by side reactions.

SSMe end groups can be introduced into RAFT polymers as described above. Disulfide end groups have been shown to enable a stronger binding to planar gold surfaces than the corresponding dithiobenzoate RAFT agent end groups.[80] For example, in a surface plasmon resonance (SPR) experiment, PMMA ($M_n = 9.8$ kg mol$^{-1}$) carrying a dye and an SSMe end group was shown to produce a stable SAM on gold of 3.68 nm average thickness, whereas the same PMMA with RAFT agent end groups adsorbed onto the gold producing a layer of 0.47 nm average thickness, indicating a lower surface coverage and weaker binding.[82]

Polymeric methyl disulfide end groups have also been used to displace oleic amine ligands on gold nanoparticles (AuNP) and oleic acid ligands on CdSe/ZnS

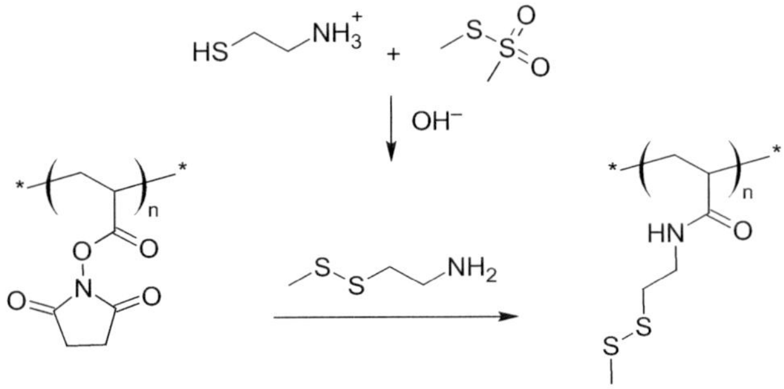

**Scheme 4.9** Synthesis of 2-aminoethyl methyl disulfide (AEMD) and use with NHS activated ester polymers to produce efficient metal anchoring groups.

semiconductor quantum dots (QD). Employing polymers carrying a fluorescent dye at the opposite end group, dye decorated nanoparticles were achieved which showed energy transfer from dye to AuNP or from QD to dye.[82]

Polymeric structures carrying multiple SSMe groups in monomeric units have been designed to bind to metal or semiconductor surfaces. In this case, a chelating effect can provide a stronger binding than a single end group. A convenient method is to introduce methyl disulfide groups into reactive polymers by way of polymer analogous reactions. 2-Aminoethyl methyl disulfide (AEMD), available from reaction of cysteamine with MMTS, can easily be coupled to polymers by way of activated esters (Scheme 4.9). Theato *et al.*[94–97] prepared copolymers of *N*-hydroxysuccinimide (NHS) acrylate and various acrylamide monomers. The NHS esters were converted using a combination of amines, including cationic amines, phospholipids, and AEMD. Cationic statistical[95,96] or block[94,97] copolymers containing dimyristylphosphatidyl ethanolamine (DMPE) and SSMe groups were immobilized onto a gold surface. The DMPE groups served to build up surface tethered lipid bilayers by way of fusion with phospholipid vesicles.

Thermally responsive copolymers containing *N*-isopropylacrylamide, immobilised onto a gold surface by way of a short SSMe-containing block and featuring a hydrophobic group at the far end of the chains, were used to adsorb lipid vesicles onto the surface. Above the LCST, when the chains were collapsed on the surface and the hydrophobic recognition sites were not available, no vesicle adsorption was observed.[94]

Nanosheets assembled from SSMe functionalized bis-acylureas have been used as templates to immobilize gold nanoparticles onto surfaces.[98]

## 4.6 Experimental Section

Following are selected experimental descriptions that should enable the reader to quickly get started in synthesis. First, general reactions yielding functional methanethiosulfonates are given. Next, the end group modification protocols of polymers prepared by ATRP and RAFT are described.

### 4.6.1 Synthesis of Sodium Methanethiosulfonate (NaMTS) from Sodium Methanesulfinate[55,56,66]

7.36 g (61.3 mmol) of sodium methanesulfinate (Sigma-Aldrich, technical grade, 85%), 1.96 g (61.3 mmol) of sulfur and 375 mL of methanol were combined in a round bottom flask and refluxed for 20 min, after which the mixture was cooled down. If traces of sulfur remained, the mixture was filtered. Removal of the solvent by rotary evaporation following by high vacuum afforded sodium methanethiosulfonate (9.0 g, 96%) as a slightly yellow solid. The product was sufficiently pure for conversion with alkyl halides. $^1$H NMR (300 MHz, $D_2O$), $\delta$/ppm = 3.39; $^{13}$C NMR (75 MHz, $D_2O$), $\delta$/ppm = 54.5.

### 4.6.2 Synthesis of *S*-But-3-ynyl Methanethiosulfonate[60]

Sodium methanethiosulfonate, prepared as above, (2.86 g, 21.3 mmol) was dissolved in dry DMF (28 mL) and a solution of 4-bromo-but-1-yne (Sigma-Aldrich, 97%, 1 mL, 1.42 g, 10.65 mmol) in dry DMF (2 mL) was added. The mixture was stirred overnight at 40 °C. Precipitated sodium bromide was removed by filtration. DMF was removed in vacuum and the residue was extracted with diethyl ether several times. Upon removal of diethyl ether, spectroscopically pure product (1.38 g, 79%) was obtained as a viscous liquid. $^1$H NMR (300 MHz, $CDCl_3$) $\delta$/ppm = 3.32 (s, 3 H, $CH_3$); 3.26 (t, 2 H, S–$CH_2$); 2.65 (2 H, td, S–$CH_2CH_2$), 2.06 (t, 1 H, CCH). $^{13}$C NMR (75 MHz, $CDCl_3$) $\delta$/ppm = 80.75 (*C*CH); 70.99 (*C*CH); 50.93 (S–$CH_2$); 34.95 ($CH_3$); 19.92 (S–$CH_2CH_2$).

### 4.6.3 Synthesis of *S*-1*H*,1*H*,2*H*,2*H*-Perfluorooctyl Methanethiosulfonate[57]

Sodium methanethiosulfonate (3.15 g, 23.5 mmol) was dissolved in 5 mL of dry DMF. 1*H*,1*H*,2*H*,2*H*-perfluorooctyl bromide (ABCR, Germany, 99%, 5 g, 11.7 mmol) was separately dissolved in 2 mL of dry DMF. Both solutions were combined and stirred at 40 °C. Upon mixing, some precipitation occurred. After 3 h, two liquid phases had formed. Stirring at 40 °C was continued overnight. After that time, there was one liquid phase. The solvent was removed at reduced pressure, and the light-yellow residue was extracted with diethyl ether. After removing the diethyl ether, 4.51 g (84%) of product was obtained which was used for RAFT polymer modification without further purification. $^1$H NMR (300 MHz, $CDCl_3$) $\delta$/ppm = 3.35 (m, 5 H, –$CH_3$ and –$SCH_2$–), 2.64 (m, 2 H, –$CF_2CH_2$–). $^{13}$C NMR (75 MHz, $CDCl_3$) $\delta$/ppm = 50.47 ($CH_3$–), 32.32 (t, $J$ = 16.5 Hz, –$CF_2CH_2$–), 27.03 (t, $J$ = 3.5 Hz, –$SCH_2$–). $^{19}$F NMR (376 MHz, $CDCl_3$) $\delta$/ppm = –81.3 (3 F), –114.7, –122.3, –123.2, –123.7, –126.5 (2 F each). MS (FD) $m/z$ (%): 457.59 (100.00), 458.60 (12.59), 459.60 (9.43).

### 4.6.4 Modification of ATRP End Groups with Sodium Methanthiosulfonate[66]

An atom transfer radical polymerisation using CuBr, *tert*-butyl acrylate, toluene, methyl 2-bromopropionate, and *N*,*N*,*N*′,*N*″,*N*″-pentamethyldiethylenetriamine (PMDETA) as catalyst, monomer, solvent, initiator and ligand, respectively, was terminated by the addition of $CuBr_2$ and cooling.

Bromide terminated poly(*tert*.-butyl acrylate) (2.70 g, 2 mmol terminal bromides) was dissolved in 5 mL of dry DMF. Sodium methanethiosufonate (0.810 g, 6 mmol) was added at room temperature. The reaction was carried out overnight at 30 °C. The solvent was removed by evaporation at low temperature and the polymer was dissolved in diethyl ether while excess sodium methanethiosulfonate and side product NaBr precipitated. The diethyl ether solution was dialyzed against methanol. The polymer was dried under vacuum to yield a transparent crystalline solid. GPC, mass spectroscopy and $^1$H NMR confirmed full reaction.

### 4.6.5 Reaction of (Polymeric) Methanethiosulfonates with Thiols

Reactions of MTS reagents and thiols usually proceed very rapidly in a wide range of solvents including water. Basic catalysis increases the reaction rate but is not necessary. Depending on material cost and purification techniques, either MTS reagent or thiol can be used in slight excess. Excess of MTS reagents can be extracted from an organic phase (*e.g.* dichloromethane) and hydrolyzed using 20% aqueous NaOH.[99]

A procedure for modification of polymeric MTS species prepared from ATRP polymers is given:[66] methanethiosulfonate terminated poly(*tert*-butyl acrylate) (0.135 g, 0.1 mmol of MTS end groups) was dissolved in 2 mL of THF at room temperature in the presence of 2-mercaptoethanol (23.0 mg, 0.3 mmol). The reaction was carried out overnight at 30 °C. The solvent was partially removed by evaporation at low temperature and the polymer was precipitated twice into a mixture of water/methanol (80/20 v/v). The polymer was dried under vacuum to yield a transparent crystalline solid. GPC, mass spectroscopy and $^1$H NMR confirmed the attachment of 2-mercaptoethanol. Similar conditions were used for benzyl mercaptan and 3-mercaptopropionic acid. 2-Aminoethanethiol hydrochloride was reacted in DMF as solvent.

### 4.6.6 Modification of RAFT End Groups with Functional Methanthiosulfonates[57,60]

Example for poly[oligo(ethylene glycol) methylether methacrylate] (POEGMA) with dithiobenzoate end groups and $M_n = 3800\ g\ mol^{-1}$: POEGMA (76 mg, 20 µmol) was dissolved in 1 mL of chloroform giving a deeply red solution. MTS reagent (*e.g.* methyl, but-3-ynyl, hexadecyl or 1*H*,1*H*,2*H*,2*H*-perfluorooctyl methanethiosulfonate) (400 µmol, 20 equiv.) was added and the mixture was stirred for 1 minute until the MTS reagent had dissolved. Then, amine (*e.g.* *n*-propylamine) (50 equiv.) was injected. The reaction mixture was stirred overnight at room temperature, turning first orange then yellow within a few hours. The resulting polymers can be purified either by dialysis (*e.g.* against methanol in 3500 $g\,mol^{-1}$ MWCO regenerated cellulose membranes) or by repeated precipitation.

Notes: for poorly soluble MTS reagents, the reaction can be performed in other solvents, *e.g.* DMF. The excess of MTS can be reduced for expensive reagents. After reaction, MMTS cannot be removed by precipitation into water or hexane due to limited solubility of the reagent. $^{1}H$ NMR (400 MHz, $CDCl_3$) POEGMA–SSC$H_3$ $\delta$/ppm = 2.36.

## 4.7 Conclusions

The reaction of thiosulfonates with thiols represents a unique but versatile class of thiol-X reactions. In particular, in combination with RAFT prepared polymers it provides a powerful and very selective conjugation chemistry, which is extremely easy to conduct due to the variety of commercially, or synthetically, available MTS reagents. It has been shown that the formation of asymmetric disulfides from the reaction of thiosulfonates with thiols can be applied to post-polymerization functionalization.

Even though the number of studies utilizing MTS chemistry so far is relatively small when compared to thiol-ene based modifications, we are convinced that it will attract the full attention of polymer and materials scientists in the future and deservedly so.

## References

1. R. M. Broyer, G. N. Grover and H. D. Maynard, *Chem. Commun.*, 2011, **47**, 2212–2226.
2. H.-A. Klok, *Macromolecules*, 2009, **42**, 7990–8000.
3. A. S. Hoffman and P. S. Stayton, *Prog. Polym. Sci.*, 2007, **32**, 922–932.
4. K. L. Heredia and H. D. Maynard, *Org. Biomol. Chem.*, 2007, **5**, 45–53.
5. J. Nicolas, G. Mantovani and D. M. Haddleton, *Macromol. Rapid Commun.*, 2007, **28**, 1083–1111.
6. P. Theato, *J. Polym. Sci., Part A: Polym. Chem.*, 2008, **46**, 6677–6687.
7. P. Theato and H.-A. Klok, *Functional Polymers by Post-Polymerization Modification – Concepts, Guidelines, and Applications*, Wiley-VCH, Weinheim, 2012.
8. A. Dondoni, *Angew. Chem., Int. Ed.*, 2008, **47**, 8995–8997.
9. A. Dondoni and A. Marra, *Chem. Soc. Rev.*, 2012, **41**, 573–586.
10. C. Hoyle and C. Bowman, *Angew. Chem., Int. Ed.*, 2010, **49**, 1540–1573.
11. C. E. Hoyle, T. Y. Lee and T. Roper, *J. Polym. Sci., Part A: Polym. Chem.*, 2004, **42**, 5301–5338.

12. C. E. Hoyle, A. B. Lowe and C. N. Bowman, *Chem. Soc. Rev.*, 2010, **39**, 1355–1387.
13. M. J. Kade, D. J. Burke and C. J. Hawker, *J. Polym. Sci., Part A: Polym. Chem.*, 2010, **48**, 743–750.
14. A. B. Lowe, *Polym. Chem.*, 2010, **1**, 17–36.
15. A. B. Lowe, C. E. Hoyle and C. N. Bowman, *J. Mater. Chem.*, 2010, **20**, 4745–4750.
16. B. S. Sumerlin and A. P. Vogt, *Macromolecules*, 2010, **43**, 1–13.
17. M. van Dijk, D. T. S. Rijkers, R. M. J. Liskamp, C. F. van Nostrum and W. E. Hennink, *Bioconjugate Chem.*, 2009, **20**, 2001–2016.
18. M. Gauthier, M. Gibson and H. A. Klok, *Angew. Chem., Int. Ed.*, 2009, **48**, 48–58.
19. K. Kempe, A. Krieg, C. R. Becer and U. S. Schubert, *Chem. Soc. Rev.*, 2012, **41**, 176–191.
20. R. K. Iha, K. L. Wooley, A. M. Nystrom, D. J. Burke, M. J. Kade and C. J. Hawker, *Chem. Rev.*, 2009, **109**, 5620–5686.
21. P. J. Roth, C. Boyer, A. B. Lowe and T. P. Davis, *Macromol. Rapid Commun.*, 2011, **32**, 1123–1143.
22. W. H. Binder and R. Sachsenhofer, *Macromol. Rapid Commun.*, 2007, **28**, 15–54.
23. P. L. Golas and K. Matyjaszewski, *Chem. Soc. Rev.*, 2010, **39**, 1338–1354.
24. A. N. Koo, H. J. Lee, S. E. Kim, J. H. Chang, C. Park, C. Kim, J. H. Park and S. C. Lee, *Chem. Commun.*, 2008, 6570–6572.
25. J. K. Oh, D. J. Siegwart, H.-i. Lee, G. Sherwood, L. Peteanu, J. O. Hollinger, K. Kataoka and K. Matyjaszewski, *J. Am. Chem. Soc.*, 2007, **129**, 5939–5945.
26. D. J. Smith, E. T. Miggio and G. L. Kenyon, *Biochemistry*, 1975, **14**, 766–771.
27. T. Bruice and G. Kenyon, *J. Protein Chem.*, 1982, **1**, 47–58.
28. L.-W. Guo, A. R. Hajipour, M. L. Gavala, M. Arbabian, K. A. Martemyanov, V. Y. Arshavsky and A. E. Ruoho, *Bioconjugate Chem.*, 2005, **16**, 685–693.
29. L. M. Mannuzzu, M. M. Moronne and E. Y. Isacoff, *Science*, 1996, **271**, 213–216.
30. B. G. Davis, M. A. T. Maughan, M. P. Green, A. Ullman and J. B. Jones, *Tetrahedron: Asymm.*, 2000, **11**, 245–262.
31. D. P. Gamblin, P. Garnier, S. J. Ward, N. J. Oldham, A. J. Fairbanks and B. G. Davis, *Org. Biomol. Chem.*, 2003, **1**, 3642–3644.
32. S. I. van Kasteren, H. B. Kramer, D. P. Gamblin and B. G. Davis, *Nat. Protocols*, 2007, **2**, 3185–3194.
33. A. Spura, R. U. Riel, N. D. Freedman, S. Agrawal, C. Seto and E. Hawrot, *J. Biol. Chem.*, 2000, **275**, 22452–22460.
34. J. G. Newell and C. Czajkowski, in *Handbook of Neurochemistry and Molecular Neurobiology: Practical Neurochemistry Methods*, ed. A. Lajtha, 2007, 3rd edn, pp. 439–454.
35. M. Akabas, D. Stauffer, M. Xu and A. Karlin, *Science*, 1992, **258**, 307–310.
36. M. H. Akabas and A. Karlin, *Biochemistry*, 1995, **34**, 12496–12500.

37. N. Yang, A. L. George Jr and R. Horn, *Neuron*, 1996, **16**, 113–122.
38. N. Yang and R. Horn, *Neuron*, 1995, **15**, 213–218.
39. M. Holmgren, Y. Liu, Y. Xu and G. Yellen, *Neuropharmacology*, 1996, **35**, 797–804.
40. Y. Liu, M. E. Jurman and G. Yellen, *Neuron*, 1996, **16**, 859–867.
41. H. P. Larsson, O. S. Baker, D. S. Dhillon and E. Y. Isacoff, *Neuron*, 1996, **16**, 387–397.
42. L. Y. Foong, S. You, D. C. J. Jaikaran, Z. Zhang, V. Zunic and G. A. Woolley, *Biochemistry*, 1997, **36**, 1343–1348.
43. L. J. Berliner, J. Grunwald, H. O. Hankovszky and K. Hideg, *Anal. Biochem.*, 1982, **119**, 450–455.
44. H. S. McHaourab, M. A. Lietzow, K. Hideg and W. L. Hubbell, *Biochemistry*, 1996, **35**, 7692–7704.
45. H. S. McHaourab, K. J. Oh, C. J. Fang and W. L. Hubbell, *Biochemistry*, 1997, **36**, 307–316.
46. D. A. Stauffer and A. Karlin, *Biochemistry*, 1994, **33**, 6840–6849.
47. J. L. Kice and T. E. Rogers, *J. Am. Chem. Soc.*, 1974, **96**, 8015–8019.
48. N. S. Zefirov, N. V. Zyk, E. K. Beloglazkina and A. G. Kutateladze, *Sulfur Rep.*, 1993, **14**, 223–240.
49. J. P. Weidner and S. S. Block, *J. Med. Chem.*, 1972, **15**, 564–567.
50. J. E. Dunbar and J. H. Rogers, *J. Org. Chem.*, 1966, **31**, 2842–2846.
51. K. Khodakhah, A. Melishchuk and C. M. Armstrong, *Biophys. J.*, 1998, **75**, 2821–2829.
52. S. Hallén, J. Fryklund and G. Sachs, *Biochemistry*, 2000, **39**, 6743–6750.
53. A. Yoshida, S. Oida and E. Ohki, *Chem. Pharm. Bull.*, 1975, **23**, 2507–2517.
54. J. H. Verheijen and H. Kloosterziel, *Synthesis*, 1975, 451–452.
55. J. D. Macke and L. Field, *J. Org. Chem.*, 1988, **53**, 396–402.
56. E. J. Grayson, S. J. Ward, A. L. Hall, P. M. Rendle, D. P. Gamblin, A. S. Batsanov and B. G. Davis, *J. Org. Chem.*, 2005, **70**, 9740–9754.
57. P. J. Roth, F. D. Jochum, F. R. Forst, R. Zentel and P. Theato, *Macromolecules*, 2010, **43**, 4638–4645.
58. M. D. Bentley and A. Fang, *PCT Int. Appl.*, 2004, WO 2004060967 A1 20040722.
59. Z. E. Shaked, R. P. Szajewski and G. M. Whitesides, *Biochemistry*, 1980, **19**, 4156–4166.
60. P. J. Roth, D. Kessler, R. Zentel and P. Theato, *J. Polym. Sci., Part A: Polym. Chem.*, 2009, **47**, 3118–3130.
61. E. Mack and H. R. Thoennessen, *Ger. Offen.*, Germany, 1972, DE 2051888 A 19720427.
62. A. Rajca and M. Wiessler, *Tetrahedron Lett.*, 1990, **31**, 6075–6076.
63. J. B. Arterburn, M. C. Perry, S. L. Nelson, B. R. Dible and M. S. Holguin, *J. Am. Chem. Soc.*, 1997, **119**, 9309–9310.
64. M.-T. Cai, G.-S. Lv, J.-X. Chen, W.-X. Gao, J.-C. Ding and H.-Y. Wu, *Chem. Lett.*, 2010, **39**, 368–369.
65. Y.-J. Zhao, Y.-q. Zhai, Z.-G. Su and G.-H. Ma, *Polym. Adv. Technol.*, 2010, **21**, 867–873.

66. C. Boyer, A. H. Soeriyadi, P. J. Roth, M. R. Whittaker and T. P. Davis, *Chem. Commun.*, 2011, **47**, 1318–1320.
67. H. T. T. Duong, C. P. Marquis, M. Whittaker, T. P. Davis and C. Boyer, *Macromolecules*, 2011, **44**, 8008–8019.
68. G. Moad, E. Rizzardo and S. H. Thang, *Polym. Int.*, 2011, **60**, 9–25.
69. G. Moad, E. Rizzardo and S. H. Thang, *Aust. J. Chem.*, 2009, **62**, 1402–1472.
70. G. Moad, E. Rizzardo and S. H. Thang, *Aust. J. Chem.*, 2005, **58**, 379–410.
71. G. Moad, E. Rizzardo and S. H. Thang, *Aust. J. Chem.*, 2006, **59**, 669–692.
72. C. Boyer, M. H. Stenzel and T. P. Davis, *J. Polym. Sci., Part A: Polym. Chem.*, 2011, **49**, 551–595.
73. M. A. Harvison, P. J. Roth, T. P. Davis and A. B. Lowe, *Aust. J. Chem*, 2011, **64**, 992–1006.
74. R. T. A. Mayadunne, E. Rizzardo, J. Chiefari, J. Krstina, G. Moad, A. Postma and S. H. Thang, *Macromolecules*, 2000, **33**, 243–245.
75. A. B. Lowe, B. S. Sumerlin, M. S. Donovan and C. L. McCormick, *J. Am. Chem. Soc.*, 2002, **124**, 11562–11563.
76. A. Postma, T. P. Davis, G. Moad and M. S. O'Shea, *Macromolecules*, 2005, **38**, 5371–5374.
77. G. Moad, Y. K. Chong, A. Postma, E. Rizzardo and S. H. Thang, *Polymer*, 2005, **46**, 8458–8468.
78. V. Lima, X. Jiang, J. Brokken-Zijp, P. J. Schoenmakers, B. Klumperman and R. Van Der Linde, *J. Polym. Sci., Part A: Polym. Chem.*, 2005, **43**, 959–973.
79. J. Xu, J. He, D. Fan, X. Wang and Y. Yang, *Macromolecules*, 2006, **39**, 8616–8624.
80. P. J. Roth, D. Kessler, R. Zentel and P. Theato, *Macromolecules*, 2008, **41**, 8316–8319.
81. J. S. Carlson, M. R. Hill, T. Young and P. J. Costanzo, *Polym. Chem.*, 2010, **1**, 1423–1426.
82. P. J. Roth, K. S. Kim, S. H. Bae, B. H. Sohn, P. Theato and R. Zentel, *Macromol. Rapid Commun.*, 2009, **30**, 1274–1278.
83. F. D. Jochum, P. J. Roth, D. Kessler and P. Theato, *Biomacromolecules*, 2010, **11**, 2432–2439.
84. P. J. Roth, M. Haase, T. Basche, P. Theato and R. Zentel, *Macromolecules*, 2010, **43**, 895–902.
85. P. J. Roth, F. D. Jochum, R. Zentel and P. Theato, *Biomacromolecules*, 2010, **11**, 238–244.
86. P. J. Roth, F. D. Jochum and P. Theato, *Soft Matter*, 2011, **7**, 2484–2492.
87. C. Miller, P. Cuendet and M. Graetzel, *J. Phys. Chem.*, 1991, **95**, 877–886.
88. J. C. Love, L. A. Estroff, J. K. Kriebel, R. G. Nuzzo and G. M. Whitesides, *Chem. Rev.*, 2005, **105**, 1103–1170.
89. W. Senaratne, L. Andruzzi and C. K. Ober, *Biomacromolecules*, 2005, **6**, 2427–2448.

90. H. A. Biebuyck, C. D. Bain and G. M. Whitesides, *Langmuir*, 1994, **10**, 1825–1831.
91. H. A. Biebuyck and G. M. Whitesides, *Langmuir*, 1993, **9**, 1766–1770.
92. C. Jung, O. Dannenberger, Y. Xu, M. Buck and M. Grunze, *Langmuir*, 1998, **14**, 1103–1107.
93. K. Heister, D. L. Allara, K. Bahnck, S. Frey, M. Zharnikov and M. Grunze, *Langmuir*, 1999, **15**, 5440–5443.
94. P. Théato and R. Zentel, *J. Colloid Interface Sci.*, 2003, **268**, 258–262.
95. P. Théato and R. Zentel, *Langmuir*, 2000, **16**, 1801–1805.
96. P. Théato and R. Zentel, *J. Macromol. Sci., Part A: Pure Appl. Chem.*, 1999, **36**, 1001–1015.
97. P. Théato, R. Zentel and S. Schwarz, *Macromol. Biosci.*, 2002, **2**, 387–394.
98. J.-U. Kim, K.-H. Kim, N. Haberkorn, P. J. Roth, J.-C. Lee, P. Theato and R. Zentel, *Chem. Commun.*, 2010, **46**, 5343–5345.
99. M. D. Conner, V. Janout, I. Kudelka, P. Dedek, J. Zhu and S. L. Regen, *Langmuir*, 1993, **9**, 2389–2397.

CHAPTER 5

# *Nucleophilic Thiol-alkene Michael Addition for the Functionalization of Polymers and for Bioconjugation*

M W JONES AND D M HADDLETON*

Department of Chemistry, University of Warwick, Coventry, CV4 7AL, UK
*Email: d.m.haddleton@warwick.ac.uk

## 5.1 Introduction

The use of thiols as nucleophiles in conjugation reactions is an increasingly popular reaction. In some ways thiols have similar properties to alcohols, however, sulfur is a third row element and thus the third level orbitals are larger than the valence orbitals for oxygen, which results in a weaker carbon–heteroatom bond. In addition, sulfur is significantly less electronegative than oxygen and bonding electrons have access to 3d orbitals. The result of this is a much more rich, complex and diverse chemistry for thiols as compared to alcohols. In addition the bond dissociation energy for an S–H bond is significantly less than for O–H. This weakening of the bond allows thiols to undergo facile oxidative coupling to disulfides. Thiols are also stronger acids than their alcohol analogues, and thus thiols are much more powerful nucleophiles, as sulfur is larger and more polarisable. In addition, the availability of the 3d orbitals assists in electron delocalisation. This rich chemistry has resulted in sulfur having an important role in technological applications

RSC Polymer Chemistry Series No. 6
Thiol-X Chemistries in Polymer and Materials Science
Edited by Andrew B. Lowe and Christopher N. Bowman

Published by the Royal Society of Chemistry, www.rsc.org

*e.g.* vulcanisation, whereby natural rubber is cross-linked with elemental sulphur. Due to this combination of availability of d-orbitals and a high electron density associated with sulfur, thiols exhibit high reactivity with substrates prone to reaction with strong nucleophiles. In terms of "click" reactions, several thiol-based reactions have been suggested as candidates under this class of chemical transformation. Epoxides ($S_N2$ ring opening), electron rich enes (radical), alkynes (radical), electron poor alkenes (Michael addition), isocyanates (carbonyl addition) and halogens ($S_N2$ nucleophilic substitution) have all been shown to react readily with thiols under appropriate conditions with high efficiency. Nucleophilic thiol-reactions with substrates such as epoxides, isocyanates and activated alkenes are traditionally initiated by strong bases that may be added as a catalyst or produced by photo-latent bases that serve as photoinitiators.[1]

## 5.2 Cysteine Conjugation

Conjugation of both small molecules and polymers to the thiol-containing amino acid cysteine is an attractive alternative to lysine targeting for conjugation of peptides, proteins and other biological reagents. The high nucleophilicity of cysteine often enables selective targeting over other functional amino acids that may be present. Often, free cysteines are found partially buried within a protein structure and therefore are often less accessible to conjugating agents under typical reaction conditions.[2] Although free cysteine residues are relatively rare in native proteins due to disulfide formation, they can be introduced by genetic modification or by the reduction of disulfide bridges. This reduction can be readily accomplished by treatment of proteins or peptides with mild reducing agents such as dithiothreitol (DTT)[3] or with the water soluble phosphine tris(2-carboxyethyl)phosphine (TCEP).[4]

This highly nucleophilic nature of thiols has resulted in the development of a range of substrates capable of exploitation for the covalent conjugation of synthetic polymers to free cysteines. Although a great deal of work has been conducted on the development of polymers bearing thiol-reactive chain-ends, such as pyridyl disulfide[5,6] and haloacetamides,[7] this chapter concentrates on the development of ene-containing reagents for peptide and protein conjugation. The propensity of thiols to undergo addition to alkenes is one that has been exploited for the development of a variety of reagents for cysteine conjugation. Thiol-ene chemistry[8,9] has recently enjoyed a resurgence, with the reaction gaining popularity as an alternative to the traditionally employed copper catalyzed azide-alkyne cycloaddition (CuAAC)[10] for the synthesis of complex macromolecules.

## 5.3 *N*-Substituted Maleimides Substrates

*N*-Substituted maleimides have been studied as thiol-conjugating agents for the introduction of functionality and pegylation to proteins. The fast reaction rates and high chemo-selectivity observed for the thiol addition to maleimides makes them ideal functional groups for the conjugation of synthetic polymers to proteins.[11] The double bond of maleimides undergoes addition reactions with sulfhydryl groups to form relatively stable thioether linkages. Maleimide addition reactions are specific to thiols in the pH range 6.5–7.5.[12,13] At pH 7.0, for example, the reaction of maleimides with thiols proceeds approximately 1000 times faster than with amines, allowing for excellent selectivity over lysine residues and terminal amines. At higher pH values, some cross-selectivity with amino groups can take place as the difference in rates of reaction decreases.[14] One of the carbons adjacent to the maleimide double bond undergoes nucleophilic attack by the thiolate anion to generate the addition product. It is noted that maleimides are also susceptible to hydrolysis resulting in the ring opened maleamic acid form that is unreactive towards sulfhydryls.

Several groups have reported the use of maleimide-functional polymers synthesised by living and controlled radical polymerisation techniques for conjugation and reaction with thiols. α-Maleimide functional polymers are readily synthesised by copper(I) mediated living radical polymerisation (ATRP),[15] although one of the drawbacks associated with this approach is the reactivity and propensity for copolymerisation of the maleimide group towards propagating radicals.[16] The vinyl group can copolymerise, albeit with a low reactivity, and it has been demonstrated that the maleimide can be incorporated into a propagating polymer chain, which inevitably leads to loss of chain-end fidelity and an effective lowering of initiator efficiency. To overcome this, furan-protected maleimide initiators, and monomers, have been used successfully to synthesise a range of polymers, followed by subsequent retro Diels–Alder deprotection of the furan group to yield the free maleimide. For example, Haddleton *et al.* have demonstrated the utility of a furan-protected maleimide initiator to synthesise a range of polymeric structures such as PEG- and sugar-based polymers.[15,17] α-Maleimide initiators have also been successfully utilised in the synthesis of giant amphiphiles[18] following polymerisation of protected monomers, followed by deprotection, functionalisation and conjugation to the free cysteine residue in bovine serum albumin (BSA). This has been further utilised in polyester functionalisation by Dove *et al.* who used furan protected maleimide initiators for ring opening polymerisation of lactides prior to some coupling reactions.[19] This has been covered by an excellent recent mini review.[20]

The emergence of the maleimide group as a popular target for free cysteine conjugation has resulted in a number of research groups describing the synthesis of polymers bearing this functionality prepared from CRP techniques for direct protein conjugation. ATRP has been successfully employed by a number of research groups in order to synthesise target maleimide-functional polymers.[15] A number of methods have been described whereby the maleimide functionality has been introduced for the covalent attachment of synthetic polymers to free cysteine residues in proteins and peptides (Scheme 5.1).

An initial report by Mantovani *et al.* described the synthesis of furan-protected maleimide initiators which were used to polymerise both PEG- and glycerol-containing monomers. Subsequently, the furan end-group was

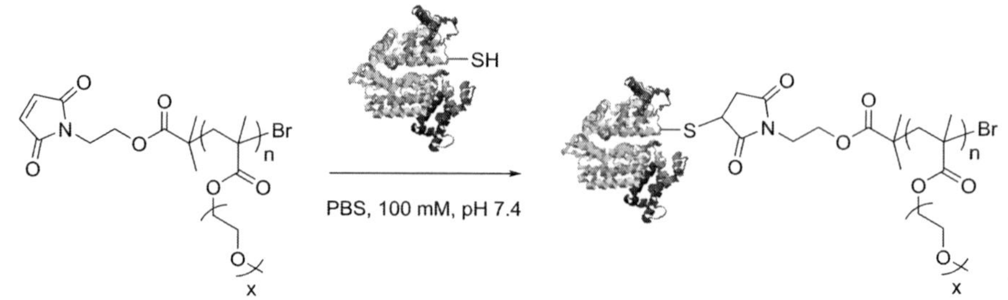


**Scheme 5.1** Conjugation of maleimide-functional poly(PEGMA) synthesised by ATRP to the free cysteine of BSA.[15]

removed to yield the desired maleimide functionality. The obtained polymers were coupled to the free cysteine residue in BSA to yield a polymer–protein conjugate (Scheme 5.1).[15] This approach has also been successfully utilised by Pieve *et al.* for the conjugation of a similar branched-PEG chain to a 25 base DNA aptamer.[21] Furthermore, maleimide-based initiators have been conjugated directly to proteins to yield macroinitiators in a "grafting from" approach.[22] This method has also been successfully employed for the synthesis of giant amphiphiles[23] that can be used as nanoreactors, whereby a catalytically active enzyme can be incorporated within the macromolecule.[24] Protected maleimide-initiators have since been used for the synthesis of synthetic glycopolymers and again conjugated to the free cysteine residue of BSA following deprotection.[17]

Due to the sensitivity of the maleimide functionality towards copolymerisation during radical polymerisations, the maleimide group has also been incorporated into polymeric chains by post-polymerisation modification. In an attempt to eliminate the necessity for protecting groups, Nolte *et al.* reported the direct functionalisation of polystyrene chains with maleimide.[25] The functionalised polymer was coupled to a free cysteine in the enzyme Cal B to yield a giant amphiphile comprising of the hydrophobic polymer and hydrophilic enzyme.

Sumerlin and coworkers have described the exploitation of the maleimide-thiol coupling reaction for the synthesis of a range of macromolecular species. The maleimide group was successfully utilised to introduce terminal-alkyne functionality to BSA for subsequent conjugation *via* CuAAC.[26] In a separate report, the same group installed a RAFT agent onto BSA using a hetero-Michael-addition reaction.[27] "Grafting from" this modified protein yielded well-defined thermoresponsive polymer–protein conjugates. Bis-maleimides have also been employed to synthesise well-defined conjugates with thiol-functional polymers. These were synthesised by cleavage of the RAFT end-group and immediately quenched with a large excess of a bis-maleimide (Scheme 5.2).[28] Following purification, a maleimide-functional polymer was isolated and conjugated to free cysteine residues in BSA and ovalbumin.

The strategy of transformation of a polymer prepared by RAFT to a thiol for Michael addition was also exploited by Fontaine and Pascual who showed the

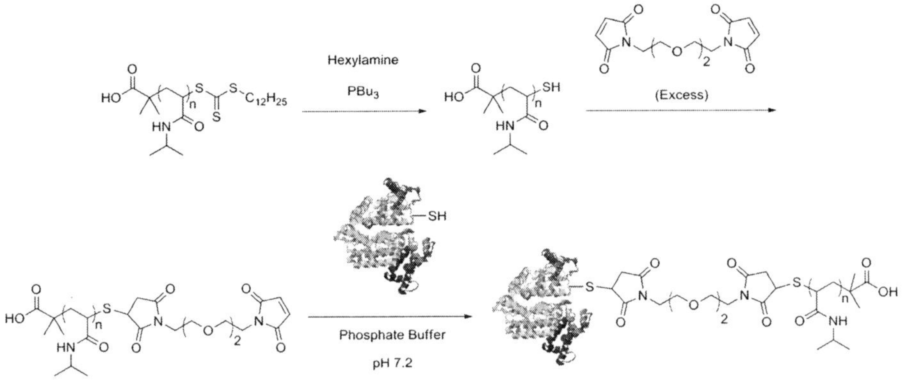


**Scheme 5.2** Introduction of maleimide-functionality to poly(*N*-isopropylacrylamide) (PNIPAM) *via* a post-polymerisation step, followed by conjugation to the free cysteine residue of BSA.[29]

use of triethylphosphite ($P(OEt)_3$) as an efficient reducing agent for the aminolysis of trithiocarbonate chain-end functionalized PNIPAM obtained by RAFT polymerization. The use of $P(OEt)_3$ allowed for a one-pot aminolysis/thiol-ene reaction of trithiocarbonate chain-end functionalized PNIPAM.[29] Lowe and Davis used a similar approach for the preparation of doubly thermoresponsive polymers consisting of a poly[oligo(ethylene glycol) methyl ether methacrylate] (POEGMA) block displaying UCST behaviour in alcohols and a block of PNIPAM or poly(*N*,*N*-diethylacrylamide) (PDEAM), each of which has an LCST in water. These were synthesized using RAFT polymerization followed by simultaneous activated ester/amine and nucleophilic thiol-ene post-polymerization reactions.[30]

Maynard *et al.* reported a novel method for the introduction of maleimide functionality into polymer chains *via* a modified approach first described by Perrier *et al.* In the initial report, RAFT polymer end-groups were removed by heating polymers in the presence of a large excess of azo-initiators.[31] The same group have utilised this approach to synthesise protected maleimide-functional azo-initiators in order to functionalise PNIPAM in a post-polymerisation modification step.[32] Deprotection of the furan terminal-group yielded a hetero-telechelic polymer, which was subsequently conjugated to two different proteins. This approach was further exemplified by the same group in order to synthesise hetero-telechelic polymer protein conjugates to reversibly modify surfaces.[33] In a separate report, they described the synthesis of semi-telechelic polymers bearing maleimide end-groups prepared by RAFT from *N*-maleimido functional trithiocarbonates.[34]

## 5.4 Vinyl Sulfone Substrates

Vinyl sulfone groups have also been popular groups for selectively targeting cysteine residues within proteins.[35] Relative to maleimides, vinyl sulfones are

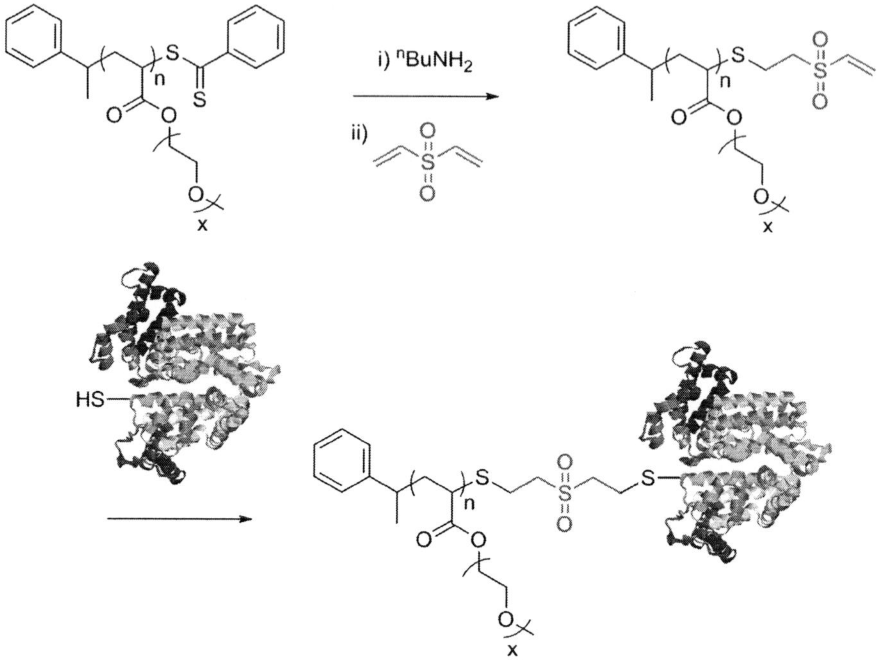


**Scheme 5.3** Conjugation of vinyl-sulfone functional poly(PEG-A) to the free cysteine residue of BSA.

not as electrophilic but do couple efficiently with cysteine residues at slightly alkaline pH to give stable β-thiosulfonyl linkages.[36] Conjugations to hydroxyl- and amino-groups can occur at higher pH values but selectivity towards cysteine residues can be can be maintained with careful choice of conditions.[37] A number of reports have described the synthesis of thermo-responsive PNIPAM–protein conjugates *via* vinyl sulfone conjugation.[38,39]

Maynard *et al.* reported the synthesis of vinyl sulfone-terminated poly(PEGA) for targeting cysteine residues (Scheme 5.3). Poly(PEGA) was synthesised *via* RAFT and the ω-thiocarbonylthio end group cleaved with butyl amine and converted to the corresponding ω-vinyl sulfone polymer following reaction with divinylsulfone in the presence of TCEP. The resultant vinyl sulfone-function poly(PEGA) was conjugated to reduced BSA containing three free thiols and coupling confirmed by SDS-PAGE.

## 5.5 Disulfide Bridging Reactions

Brocchini *et al.* reported an extremely elegant approach to the PEGylation of interferon (IFN) using a novel linking agent. Following reduction of the disulfide bridge in IFN's, the two free cysteines were re-joined using a three-carbon linked functional PEG (Scheme 5.4).[40] Consequently, following conjugation, biological activity as well as the tertiary structure of the protein

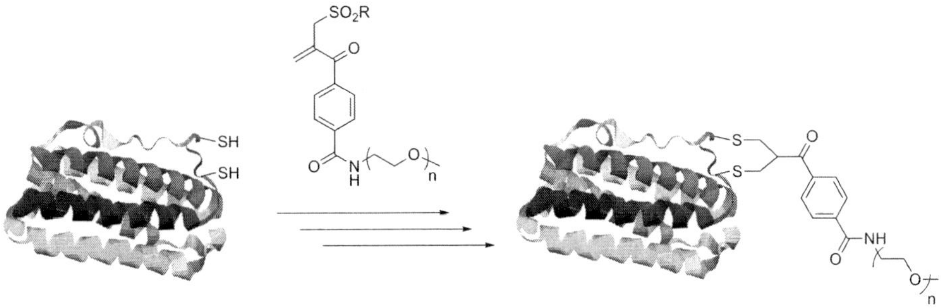


**Scheme 5.4** PEGylation of interferon using an activated three-carbon linker.[41]

was maintained. Mechanistically, the three-carbon bridge is formed following cysteine-addition to the unsaturated carbon–carbon bond, resulting in elimination of sulfinic acid to generate a second double bond. The second cysteine adds to the newly generated double bond, thus forming the stable three-carbon linker product.[41,42] This is a two-step approach to PEGylation and during this process, irreversible denaturation of the protein is shown not to occur. Conjugation is conducted by a sequential, interactive bis-alkylation using α,β-unsaturated β′-monosulfone functionalized PEG reagents, resulting in a combination of (a) maintaining the protein's tertiary structure after disulfide reduction, (b) the mechanism for bis-thiol selectivity of the PEG reagent, and (c) the steric shielding of PEG ensuring that only one PEG molecule is conjugated at each disulfide bond. The authors state that the stoichiometric efficiency of this approach also enables recycling of any unreacted protein. It therefore offers the potential to make PEGylated biopharmaceuticals as cost-effective medicines. Instead of chemically functionalising each cysteine sulfur atom with a separate molecule to block the reformation of the disulfide bond – as many modifications do – a covalent connectivity with this PEGylation reagent is always maintained between the two cysteines of the original disulfide bond through the three-carbon bridge. The approach has been developed commercially by Polytherics Ltd under the trademark "TheraPEG" and has been demonstrated in the literature with Interferon α-2 and leptin.

## 5.6 Nucleophilic Thiol-ene "Click"

Lowe *et al.* have demonstrated the potency of nucleophilic catalysts in thiol-Michael addition reactions.[8] Using amine catalysts (hexylamine and diisopropylamine) with similar p$K_a$ values, a striking difference was observed in the kinetic profiles for the reaction between hexanethiol and hexyl acrylate. The observation was rationalised if the amines are thought of as nucleophiles rather than bases. Several phosphines were also screened as potential catalysts for the thiol-Michael addition and were shown to be effective catalysts for the thiol-Michael addition of hexyl acrylate with hexanethiol. Trialkyl phosphines were the most active with decreasing activity observed with increasing aryl substitution.

Several groups have reported the successful utilisation of the phosphine-mediated thiol-Michael addition to activated alkenes. The Lowe group in particular have published several elegant reports on the utilisation of organophosphines as catalysts for thiol-ene "click" reactions of macromolecules. Macromolecular stars were synthesised when the dithioester end-group of polymers synthesised by RAFT were cleaved, yielding a thiol functionality for conjugation with triacrylates.[43,44] Nucleophilic thiol-ene chemistry has since been exploited in a number of reports to functionalise polymers synthesised by RAFT, whereby the masked thiol end-group is cleaved and modified with appropriate Michael-acceptors.[45–47]

This chemistry is useful for the functionalisation of polymers and in the case of this Michael addition reaction, this is a facile reaction between nucleophilic species and activated olefins under basic conditions.[48] This Michael reaction benefits from mild reaction conditions, minimal by-product formation, high functional group tolerance and high conversions when optimised, which are requirements for efficient polymer functionalisation. Various amines, enolates, thiols and phosphines can react with (meth)acrylates, (meth)acrylamides, maleimides, acrylonitriles and cyanoacrylates, which are Michael donors and Michael acceptors, respectively.[49,50] There are a few parameters that need to be taken into account when optimising the synthetic protocol, such as the base strength, solvent and substrate, all of which need to be systematically investigated.[51]

The synthesis of end functional polymers with sufficiently high fidelity has been challenging, even though several controlled radical polymerisation techniques have been developed.[52] Metal mediated radical polymerisation techniques have been widely employed to obtain bromide or chloride end-functionalized polymers that can be subsequently converted into azide groups to facilitate Cu-catalysed azide-alkyne click reactions. Similarly, reversible addition-fragmentation chain transfer (RAFT) polymerization has been used to give thiol terminated polymers following the cleavage of the chain transfer agent.[53,54] In general practice, amine based catalysts have been used to cleave the chain transfer agent from the chain end, however, the bulky structure of the polymer may reduce the cleaving efficiency. Additionally, the presence of free thiol end-groups can result in oxidation and disulfide formation. To avoid this side reaction, a strategy was developed proposing the *in-situ* aminolysis of RAFT end-groups in the presence of pyridyl disulfide to yield thiol protected polymers.[55] In addition, RAFT polymerization requires dedicated chain transfer agents (CTA) designed according to the monomer structure and also optimized reactions conditions such as initiator/CTA ratio, solvent type and reaction temperature.[56–61] Nevertheless, thiol terminated polymers have been successfully used in thiol-ene, thiol-yne, thio-bromo, thio-isocyanate and thiol-maleimide click reactions.[62–70]

Alternatively, vinyl terminated polymers have been prepared to react with thiol compounds.[71–73] One of the most successful techniques to obtain vinyl-terminated polymers is catalytic chain transfer polymerization (CCTP) which proceeds *via* a free-radical mechanism in the presence of certain low spin Co(II) catalysts.[74–78] In particular, this is an ideal technique to synthesize very short oligomers, *i.e.* dimer, trimer and tetramers of most methacrylates, with very high fidelity. This technique has been generally employed for the preparation of macromonomers and has been used successfully, and it is noted that this is exclusively for methacrylates. It is also noted that, advantageously, CCTP requires a very low amount of catalyst (in low ppm levels), yielding quasi-pure polymers with a very low amount of residual catalyst.[79]

Haddleton investigated the effect of different catalysts for the base-catalyzed thiol-ene reaction of monomers and oligomers as obtained by CCTP.[51] Different catalysts, including amine catalysts, pentyl amine and hexyl amine (primary amines), triethyl amine (tertiary amine), and two different phosphine catalysts, dimethylphenyl phosphine (DMPP) and tris(2-carboxyethyl) phosphine (TCEP) were investigated in the presence of different non- and functional thiols. This study determined the optimum reaction conditions for base-catalyzed thiol-ene click reactions. It was shown reproducibly that triethylamine as a catalyst, even in excess amounts, does not give any noticeable side reactions; however, the reaction takes a relatively longer period of time to reach high conversions. Moreover, primary amines, pentyl amine and hexylamine, were tested as appropriate catalysts. However, it is noted that these can react with the vinyl group and form stable species, as monitored by ESI-MS. Phosphine catalysts provide the fastest reaction, but this also brings the highest amount of side reactions. Nevertheless, using the catalyst in catalytic amounts provides relatively fast reactions rates for the desired compounds. It should also be noted that the effect of solvent, type of thiol and the basicity of the medium have a great influence on this reaction and should be selected with care based on these optimization reactions. These model reactions were performed on monomers, dimers and oligomers.

Highly functionalized thiol end functionalized polyethylenes (PE-SHs), which are compatible with the solubilization of PE-SH, have been functionalised with acrylates and methacrylates. This was further exploited to give both an acrylate and a methacrylate moiety to obtain a methacrylate macromonomer based on PE. This was further exploited for the synthesis of block copolymers incorporating a PE segment and a poly(ethylene glycol) (PEG) segment *via* thia-Michael addition chemistry. There was a lack of reactivity towards methacrylate groups, which was further confirmed and indeed exploited in order to successfully synthesize macromonomers based on PE using 2-(acryloyloxy) ethyl methacrylate, a molecule containing both an acrylate and a methacrylate group. An efficient alternative to the synthesis of the methacrylate-type macromonomer based on PE was the quantitative reaction of 2-isocyanatoethyl methacrylate with PE-SH. The efficiency of thia-Michael addition onto acrylates was eventually used to produce PE-*b*-PEG block copolymers after reaction of a commercially available PEG acrylate macromonomer with PE-SH.[80]

A combination of CCTP with protected alkyne functional monomers led to oligomers with two complementary highly reactive sites. The alkyne was used to attach a range of sugars to glycopolymers with different carbohydrate epitopes, leaving the terminal vinyl group available for nucleophilic thio-ene reaction. The polymer terminal group was reacted using Michael addition with different thiols to obtain end-funtionalised polymers, and glycopolymers have been produced by clicking sugar moieties to the side chain of these polymers by CuAAC. The ability of the resulting end-functionalised glycopolymers to be recognised by lectins was demonstrated and the polymers were able to function as multivalent ligands for the recognition of the lectins con A and RCA I, respectively.[81]

The reverse reaction, reacting a thiol functional polymer with an acrylate, was shown by the use of a disulphide containing difunctional initiator to give well-defined polymers by SET-LRP.[82,83] The resulting polymers were reduced by dimethyl phenylphosphine in the presence of a range of acrylates and methacrylates.[84] Two different functional acrylates were chosen to exemplify the versatility of the reaction; hydroxyethyl acrylate and a hostasol functional acrylate to give a hydroxyl-functional and a UV/fluorescently tagged polymer, respectively, with $^{1}$H NMR indicating complete conjugation. This allows for facile introduction of functionality into polymers for further modification (HEA and propargyl acrylate) or to introduce a specific functionality *e.g.* TFEMA and EGMAP. This can be applied to any disulfide-containing polymer, followed by subsequent modification with any (meth)acrylate or acrylamide, allowing for efficient post-polymerisation modification.

## 5.7 Thiol-ene Chemistry for Protein/Peptide Modification

Heise *et al.* reported on the synthesis of well-defined cysteine-containing polypeptides *via* the ring-opening polymerisation of protected-thiol *N*-carboxyanhydrides (Scheme 6.6).[85] Following deprotection, these polymers were subjected to post-polymerisation modification with a range of activated alkenes in order to incorporate further functionality to the backbone. The authors report the coupling of PEG-acrylate chains to the polypeptide *via* two separate thiol-ene methods, a base-catalysed route, as well as a radical-induced reaction. The base-catalysed strategy proved relatively unsuccessful, with low coupling efficiencies obtained, and a second approach, initiated by a radical source (AIBN) gave the desired shift to a higher molecular weight species upon heating at 70 °C for 1 hour. Side reactions were reported, with the observation of low-molecular weight species attributed to free-radical polymerisation of PEG-acrylate (Schemes 5.5 and 5.6).

In a similar vein, Schlaad *et al.* have reported the synthesis of thiol-ene clickable polypeptides, this time describing the synthesis and polymerisation of allyl-functional NCAs (Scheme 5.7).[86] This approach requires no protecting groups and eliminates crosslinking of polymer chains *via* thio-residues. The synthesised poly(DL-allylglycine) homopolymers and PEG-block-copolymer were functionalised by thiol-ene "click" with a range of simple thiols including thio-sugars to yield glycopolypeptides.

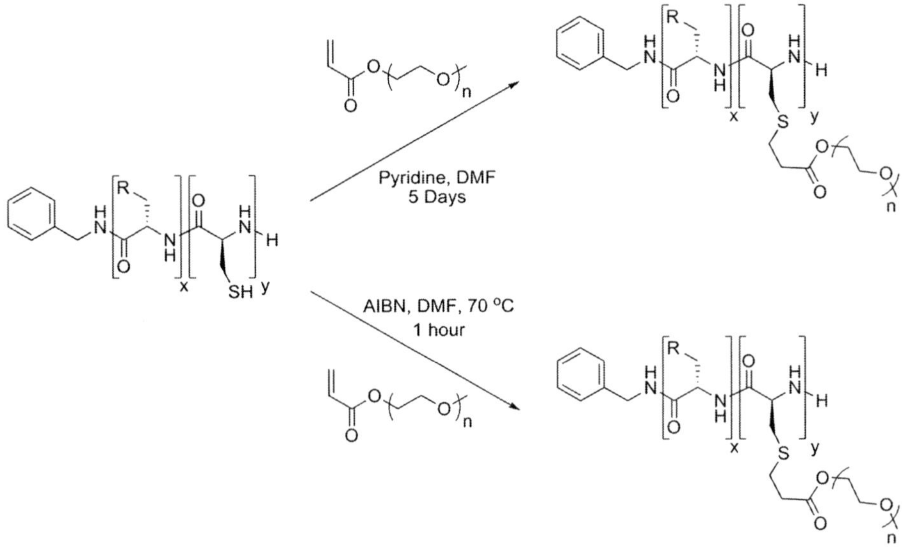


**Scheme 5.5** Functionalisation of cysteine-containing polypeptides with base-catalysed and radical-mediated thiol-ene chemistry.

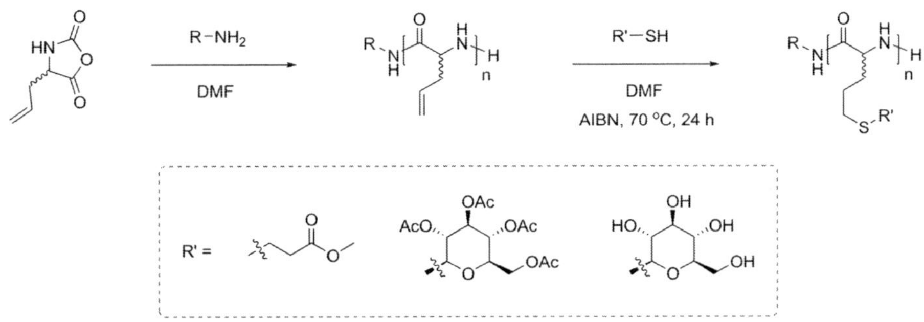


**Scheme 5.6** Synthesis of alkene-functional polypeptides and subsequent functionalisation by radical-mediated thiol-ene chemistry.[87]

In a report by Waldmann *et al.*, *S*-alkylation of cysteine was reported using radical-initiated thiol-ene.[87] Although the authors reported successful functionalisation of the amino acid, relatively low yields were observed, even at prolonged reaction times and high temperatures. Kunz *et al.* reported a novel route to protein–glycopeptide conjugates using thiol-ene chemistry.[88] Alkene and thiol-functionalities were successfully introduced to the solvent-accessible lysine residues in BSA *via* carbodiimide coupling. Glycopeptides bearing complementary thiol and alkene functionalities were synthesised by solid-phase synthesis and coupled to the functionalised protein by radical thiol-ene reactions. In the case of functionalisation of BSA with an alkene group, 24 spacer molecules were introduced on average to the 30 accessible lysine residues. Following thiol-ene functionalisation, only weak signals attributed to

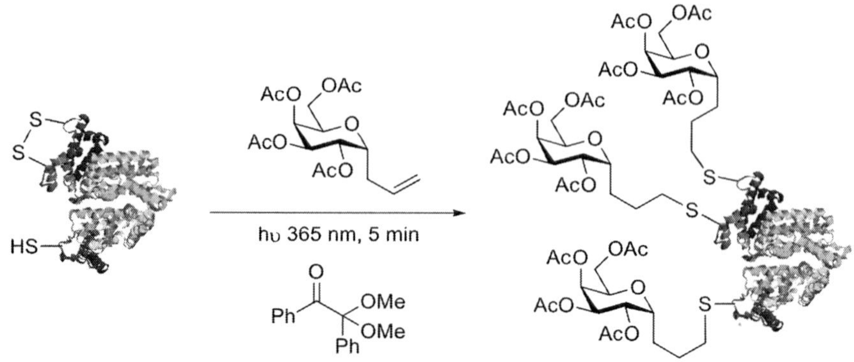


**Scheme 5.7** Radical-mediated glycosylation of BSA using alkene-functional saccharides.[90]

the product were observed by MALDI-ToF-MS. Upon further characterisation, relatively low coupling efficiencies were observed, with as little as nine glycopeptides conjugated to the modified protein. Dondoni *et al.* have reported the utilisation of radical-mediated thiol-ene coupling as a means to introduce glycosides to proteins.[89] Initial studies performed by the group involved coupling reactions between an allyl-functional galactose and *N*-Fmoc cysteine using AIBN as the radical source in a refluxing solution of dichloroethane. Under these conditions, very low yields were observed, and therefore photochemical reactions were investigated. Using DMPA as the radical source, far greater yields were observed, with cystine as the only observable by-product.

Following optimisation of the reaction conditions, larger polypeptides were studied and successfully conjugated with the allyl-functional sugar. The free cysteine residue of BSA was also targeted, although when analysed by MALDI-ToF-MS, three sugars were shown to be conjugated to the protein, despite the lack of a disulfide reducing agent. Irradiation for longer periods resulted in further degrees of functionalisation and attempts to solely functionalise Cys-34 by modifying the reaction conditions proved unsuccessful.

## 5.8 Polymeric Acrylates for Peptide Conjugation

Poly(monomethoxy ethylene glycol)(meth)acrylates (p(mPEG)(M)A) were employed with water-soluble organic phosphines (TCEP) in a one-pot synthetic protocol where a (poly)peptide disulfide bridge was first reduced, followed by subsequent reaction of the two thiols *in situ*. This approach used commercially available macromolecular conjugates which are normally employed as monomers, or synthesised in one step from mPEGs and (meth)acryloyl chloride (Scheme 5.8).[90]

Haddleton used salmon calcitonin (sCT) as a disulfide bridge-containing peptide which is a 32 aminoacid calcitropic hormone currently administered for the treatment of a number of hypercalcemia-related diseases, including

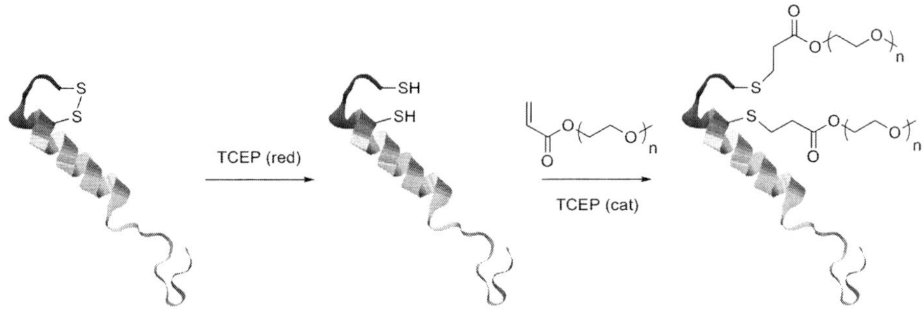


**Scheme 5.8** Synthetic protocol for one-pot phosphine-mediated approach to polymer-protein conjugates.

post-menopausal osteoporosis and Paget's disease. sCT contains a disulfide bridge – $Cys^1$–$Cys^7$ – that can be reduced to give two sulphydryl units available for thiol functionalisation. It has been previously reported that this $Cys^1$–$Cys^7$ bridge can be reduced and functionalised with no loss of bioactivity.[91]

Tris(2-carboxyethyl)phosphine (TCEP),[92] a commercially available water-soluble phosphine was used as both the reducing agent, for the opening of the $Cys^1$–$Cys^7$ bridge, and as catalyst for the subsequent thiol-ene PEGylation of the two resulting sulphydryl units (Scheme 5.9). Bioactivity of the conjugate was maintained, illustrating that this phosphine-mediated one-pot thiol-ene click is a very general reaction that should find application for the site-selective conjugation of peptides and proteins where the disulfide bridge is not essential for retention of bioactivity. In addition, it allows for quantitative substrate modification in a relatively short reaction time, a factor that can be very important for peptides with limited stability in solution. We have since described the introduction of a simple inimer to the reduced diulfide of salmon calcitonin using thiol-ene chemistry in order to synthesise a well defined polypeptide macroinitiator. A number of oligo(ethylene glycol) methacrylates were polymerised directly from the peptide macroinitiator to synthesise a range of thermoresponsive conjugates.[93,94] It was noted that this versatile approach will find application as general method for the functionalisation of macromolecular entities (*i.e.* polymers, dendrimer and naturally occurring scaffolds) and natural and synthetic materials.[90]

## 5.9 Radical Thiol-ene

The radical-mediated thiol-ene reaction is an alternative to the nucleophile-mediated process and proceeds in a similar manner to the conventional chain transfer polymerisation. A recent report by Junkers and co-workers[95] described a study on the radical-mediated thiol-ene reaction for polymer–polymer conjugation. Both thermal and photo-chemically induced reactions were investigated, although both systems were shown to be beset with considerable difficulties in reaching quantitative conversion. Based on their findings, they recommend that radical-mediated thiol-ene should not be referred to as a "click" reaction due to the lack of criteria fulfilled by the system, such as low reaction efficiency and a number of undesirable side-reactions. Radical-initiated thiol-ene has been successfully employed to functionalise the pendant vinyl groups of poly(butadiene) (PB) with aryl,[96] bio-inspired thiols,[97] as well as with oligopeptides, to yield self-assembling amphiphiles.[98] Alkene-functional poly(oxazoline)s have also been successfully functionalised with a number of functional thiols under radical conditions.[99] However, to date little success has been achieved in the application of the radical initiated thiol-ene for protein modification.

## 5.10 Development of Novel Maleimide Derivatives for Cysteine Targeting

Baker *et al.* have reported the development of a new variation on the traditional maleimide for cysteine modification.[100] Although not strictly a Michael addition, bromomaleimides have been reported as reversible cysteine conjugating agents following cleavage with TCEP In contrast to unmodified maleimides, bromomaleimide was also shown to display an interesting secondary characteristic in its ability to facilitate the conversion of cysteine to dehydroalanine (Scheme 5.9).

Dibromomaleimides, developed by the same group, offer the same advantages as the singly modified derivative in terms of selectivity for cysteine residues, as well as a third point for functionalisation at the second vinylic carbon using a second thiol functionality (Scheme 5.10).[101]

One of the most interesting aspects of the work demonstrated by Baker was the ability of the dibromomaleimides to bridge free cysteines derived from the reduction of disulfide bridges. Somatostatin was utilised as a model for the study and, following reduction of the disulfide bridge with TCEP, dibromomaleimide was shown to successfully re-bridge the two generated cysteine residues. Again, the bridging maleimide was shown to be cleavable under reducing conditions to yield the reduced native polypeptide. The synthesis of dibromomaleimide functional polymers for disulfide bridging applications has since been explored by Haddleton *et al.*[102] Two independent approaches to dibromomaleimide functional polymers were described and employed for the disulfide re-bridging of sCT upon reduction with TCEP. Rapid re-bridging was

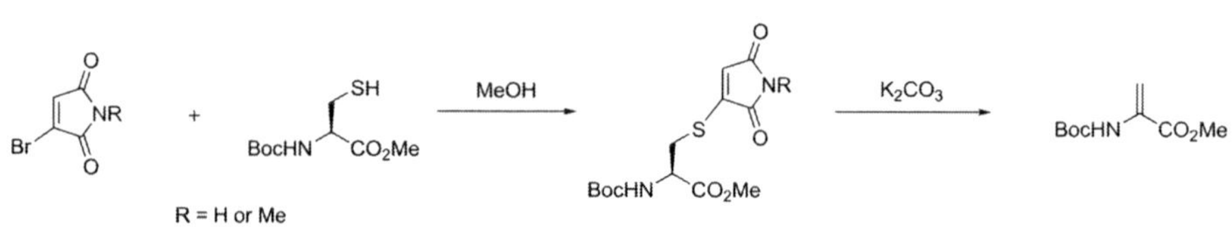


**Scheme 5.9** Modification of cysteine residues with bromomaleimide, followed by subsequent cleavage to yield dehydroalanine.[101]

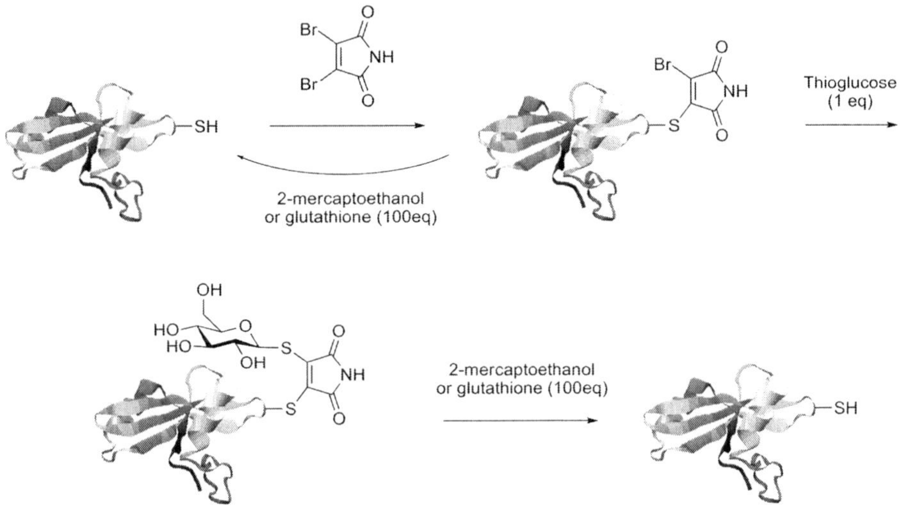


**Scheme 5.10** Reversible modification of the free cysteine residue of the Grb2 SH2 domain with dibromomaleimide.[102]

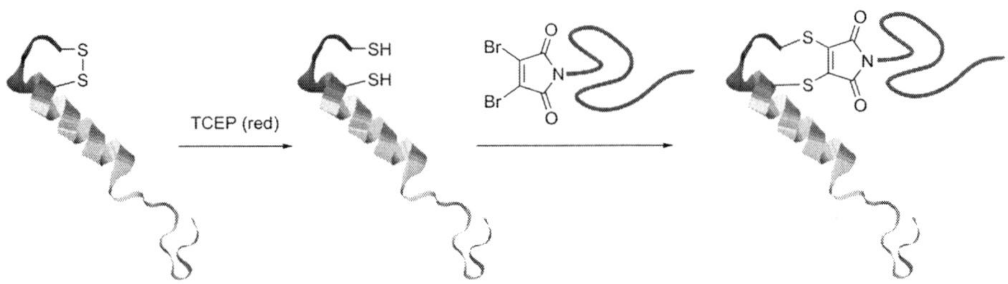


**Scheme 5.11** Disulfide re-bridging of sCT using dibromomaleimide functional polymers synthesised by ATRP.[103]

observed, with complete consumption of the polypeptide observed in minutes with 1.1 eq. of the conjugating polymers (Scheme 5.11).

Baker *et al.* have more recently described the development a second maleimide variant capable of polypeptide disulfide re-bridging. Dithiomaleimides possess the same rapid disulfide bridging capability as their dibromomaleimide siblings, although the thio-derivatives showed a greater tolerance to TCEP and were capable of *in situ* bridging protocols. This greater stability led us to investigate these substrates for the direct synthesis of dithiophenol maleimide functional polymers for polypeptide conjugation without the need for protecting groups.[103] A number of poly(oligo ethylene glycol) methacrylates were synthesised from a dithiophenol maleimide functional initiator and directly employed for the disulfide re-bridging of sCT following reduction with TCEP (Scheme 5.12).

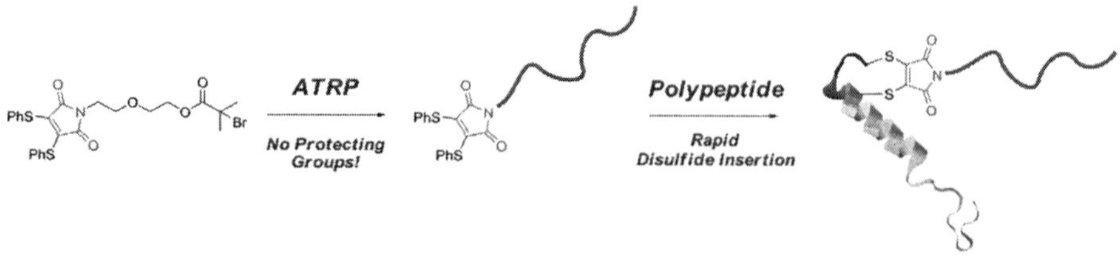


**Scheme 5.12** Synthesis of dithiophenol-maleimide functional polymers by ATRP, followed by disulphide bridging of salmon calcitonin.[104]

## 5.11 siRNA and Aptamer Conjugation

Targeted delivery of nucleic acid -based therapeutics is an active areas of research in the field of nanomedicine. The combination of both ATRP and RAFT polymerisation with maleimide groups to synthetic (co)polymers has potential for therapeutic delivery of small interfering RNA (siRNA) and subsequent gene "knockdown".[70,104]

An alternative approach is to employ a RAFT CTA containing a pyridyl disulfide R-group; this allows the polymer to be directly conjugated to siRNA containing a thiol group *via* a disulphide link.[105] This has been extended to look at the effects of RAFT-synthesized comb-type PEG conjugation on siRNA's *in vitro* stability and gene inhibition activity. The results support the comb-like PEG (meth)acrylate polymers synthesized by RAFT polymerization as an effective component of delivery systems for siRNA.[106] A similar approach was used for conjugation of a 3′-thiol-modified (disulfide protected) 25 base DNA aptamer, selected against the protein core of MUC1 glycoprotein (AptD),[107] which was successfully conjugated *via* a maleimide functional group to a range of PEG-based polymers of varying topology and molecular weight. Reduced binding affinity was only apparent for the AptD-LinPEG20 and AptD-BrPEG40 conjugates.[108] Maynard and Davis demonstrated the use of linear poly(ethylene glycol) (PEG) conjugates of Dicer-substrate small interfering RNA (DsiRNA) to investigate an siRNA release strategy. 3′-Sense or 5′-antisense thiol-modified, blunt-ended DsiRNAs, inhibiting enhanced green fluorescent protein (eGFP) expression, were covalently conjugated to PEG with varying molecular weights (2, 10, and 20 $kg\,mol^{-1}$) through a stable thioether bond using a Michael addition reaction.[30]

## 5.12 Antibody Drug Conjugates (ADCs)

The use of antibody drug conjugates (ADCs) is emerging as a very promising drug delivery route. These ADCs must have a number of key attributes including specificity, safety and stability to give effective payload delivery. Conjugation is required and, as with peptide and protein conjugates, both cysteine or lysine residue conjugation are popular.[109,110] Thiol-maleimide chemistry is a commonly used method, whereby a cysteine residue, either native or engineered, is used for conjugation.[111] Antibody conjugates for therapeutic applications must satisfy demanding criteria concerning specificity, safety and stability to achieve effective delivery of the payload to the target. Antibodies are generated with engineered cysteine that provides reactive thiols. A range of cytotoxic drugs have been used for conjugation, including auristatin and maytansine.[111,112] These ADCs were shown to display reduced liver and bone marrow toxicity. The extension of this to the substituted bromomaleimides maleimides allow for facile conjugation of thiolated biomolecules and the potential for reversibility. It is noted that this is not strictly a thiol-ene Michael addition reaction but a substitution reaction at the maleimide. Caddick *et al.* have shown that bromomaleimide-linked GFP–rhodamine FRET pairs cleave in the cytoplasm of mammalian cells. This gives the potential for bromo-maleimide scaffolds to provide a potential core structure for prodrugs designed to release bioactive cargo following cell internalisation.[113]

## 5.13 Summary

The realisation that thiol-ene chemistry can be efficiently catalysed by simple nucleophiles leading to high yields of functionalised polymers is rapidly expanding. The use of maleimides for this purpose has been quite widespread in protein bioconjugation with efficient reactions under mild conditions. The extension to non-biological polymers is quickly being exploited in many areas. In turn, this has led to further elegant polymer functionalisation reactions which is helping the use of existing polymers, from CCTP for example. This increase in activity has again, in turn, led to the development of more reactive/efficient vinyl reagents and disubstituted maleimides. The absence of metal catalysts, and often auto-catalysis and tandem reactions, make this approach very attractive and it is anticipated it will continue to grow rapidly.

## References

1. C. E. Hoyle, A. B. Lowe and C. N. Bowman, *Chem. Soc. Rev.*, 2010, **39**, 1355–1387.
2. P. Thordarson, B. Droumaguet and K. Velonia, *Appl. Microbiol. Biotechnol.*, 2006, **73**, 243–254.
3. W. W. Cleland, *Biochemistry*, 1964, **3**, 480–482.
4. J. A. Burns, J. C. Butler, J. Moran and G. M. Whitesides, *J. Org. Chem.*, 1991, **56**, 2648–2650.
5. J. Carlsson, H. Drevin and R. Axen, *Biochem. J.*, 1978, **173**, 723–737.
6. D. Bontempo, K. L. Heredia, B. A. Fish and H. D. Maynard, *J. Am. Chem. Soc.*, 2004, **126**, 15372–15373.
7. F. R. N. Gurd, *Methods Enzymol.*, 1967, **11**, 532–541.
8. J. W. Chan, C. E. Hoyle, A. B. Lowe and M. Bowman, *Macromolecules*, 2010, **43**, 6381–6388.
9. A. Dondoni, *Angew. Chem., Int. Edit.*, 2008, **47**, 8995–8997.

10. H. C. Kolb, M. G. Finn and K. B. Sharpless, *Angew. Chem., Int. Edit.*, 2001, **40**, 2004–2021.
11. D. G. Smyth, O. O. Blumenfeld and W. Konigsberg, *Biochem. J.*, 1964, **91**, 589–595.
12. G. Gorin, P. A. Martic and G. Doughty, *Arch. Biochem. Biophys.*, 1966, **115**, 593–597.
13. J. R. Heitz, C. D. Anderson and B. M. Anderson, *Arch. Biochem. Biophys.*, 1968, **127**, 627–636.
14. C. F. Brewer and J. P. Riehm, *Anal. Biochem.*, 1967, **18**, 248–255.
15. G. Mantovani, F. Lecolley, L. Tao, D. M. Haddleton, J. Clerx, J. J. L. M. Cornelissen and K. Velonia, *J. Am. Chem. Soc.*, 2005, **127**, 2966–2973.
16. G. H. Deng and Y. M. Chen, *Macromolecules*, 2004, **37**, 18–26.
17. J. Geng, G. Mantovani, L. Tao, J. Nicolas, G. J. Chen, R. Wallis, D. A. Mitchell, B. R. G. Johnson, S. D. Evans and D. M. Haddleton, *J. Am. Chem. Soc.*, 2007, **129**, 15156–15163.
18. B. Le Droumaguet, G. Mantovani, D. M. Haddleton and K. Velonia, *J. Mater. Chem.*, 2007, **17**, 1916–1922.
19. R. J. Pounder, M. J. Stanford, P. Brooks, S. P. Richards and A. P. Dove, *Chem. Commun.*, 2008, 5158–5160.
20. D. J. Hall, H. M. Van den Berghe and A. P. Dove, *Polym. Int.*, 2011, **60**, 1149–1157.
21. C. Da Pieve, P. Williams, D. M. Haddleton, R. M. J. Palmer and S. Missailidis, *Bioconjugate Chem.*, 2010, **21**, 169–174.
22. J. Nicolas, V. San Miguel, G. Mantovani and D. M. Haddleton, *Chem. Commun.*, 2006, 4697–4699.
23. B. Le Droumaguet, G. Mantovani, D. M. Haddleton and K. Velonia, *J. Mater. Chem.*, 2007, **17**, 1916–1922.
24. B. Le Droumaguet and K. Velonia, *Angew. Chem., Int. Edit.*, 2008, **47**, 6263–6266.
25. K. Velonia, A. E. Rowan and R. J. M. Nolte, *J. Am. Chem. Soc.*, 2002, **124**, 4224–4225.
26. M. Li, P. De, S. R. Gondi and B. S. Sumerlin, *Macromol. Rapid Commun.*, 2008, **29**, 1172–1176.
27. P. De, M. Li, S. R. Gondi and B. S. Sumerlin, *J. Am. Chem. Soc.*, 2008, **130**, 11288–11289.
28. M. Li, P. De, H. M. Li and B. S. Sumerlin, *Polym. Chem.*, 2010, **1**, 854–859.
29. H. T. Ho, M. E. Levere, S. Pascual, V. Montembault, J. C. Soutif and L. Fontaine, *J. Polym. Sci. Polym. Chem.*, 2012, **50**, 1657–1661.
30. P. J. Roth, T. P. Davis and A. B. Lowe, *Macromolecules*, 2012, **45**, 3221–3230.
31. S. Perrier, P. Takolpuckdee and C. A. Mars, *Macromolecules*, 2005, **38**, 2033–2036.
32. K. L. Heredia, G. N. Grover, L. Tao and H. D. Maynard, *Macromolecules*, 2009, **42**, 2360–2367.

33. K. L. Heredia, L. Tao, G. N. Grover and H. D. Maynard, *Polym. Chem.*, 2010, **1**, 168–170.
34. E. Bays, L. Tao, C. W. Chang and H. D. Maynard, *Biomacromolecules*, 2009, **10**, 1777–1781.
35. M. S. Masri and M. Friedman, *J. Protein Chem.*, 1988, **7**, 49–54.
36. G. N. Grover, S. N. S. Alconcel, N. M. Matsumoto and H. D. Maynard, *Macromolecules*, 2009, **42**, 7657–7663.
37. M. Morpurgo, F. M. Veronese, D. Kachensky and J. M. Harris, *Bioconjugate Chem.*, 1996, **7**, 363–368.
38. Z. Ding, C. J. Long, Y. Hayashi, E. V. Bulmus, A. S. Hoffman and P. S. Stayton, *Bioconjugate Chem.*, 1999, **10**, 395–400.
39. V. Bulmus, Z. Ding, C. J. Long, P. S. Stayton and A. S. Hoffman, *Bioconjugate Chem.*, 2000, **11**, 78–83.
40. S. Shaunak, A. Godwin, J.-W. Choi, S. Balan, E. Pedone, D. Vijayarangam, S. Heidelberger, I. Teo, M. Zloh and S. Brocchini, *Nat. Chem. Biol.*, 2006, **2**, 312–313.
41. S. Brocchini, A. Godwin, S. Balan, J.-W. Choi, M. Zloh and S. Shaunak, *Adv. Drug Delivery Rev.*, 2008, **60**, 3–12.
42. S. Balan, J.-W. Choi, A. Godwin, I. Teo, C. M. Laborde, S. Heidelberger, M. Zloh, S. Shaunak and S. Brocchini, *Bioconjugate Chem.*, 2007, **18**, 61–76.
43. J. W. Chan, B. Yu, C. E. Hoyle and A. B. Lowe, *Chem. Commun.*, 2008, 4959–4961.
44. J. W. Chan, B. Yu, C. E. Hoyle and A. B. Lowe, *Polymer*, 2009, **50**, 3158–3168.
45. J. M. Spruell, B. A. Levy, A. Sutherland, W. R. Dichtel, J. Y. Cheng, J. F. Stoddart and A. Nelson, *J. Polym. Sci. Polym. Chem.*, 2009, **47**, 346–356.
46. C. Boyer, A. Granville, T. P. Davis and V. Bulmus, *J. Polym. Sci. Polym. Chem.*, 2009, **47**, 3773–3794.
47. P. J. Roth, C. Boyer, A. B. Lowe and T. P. Davis, *Macromol Rapid Comm*, 2011, **32**, 1123–1143.
48. B. D. Mather, K. Viswanathan, K. M. Miller and T. E. Long, *Prog. Polym. Sci.*, 2006, **31**, 487–531.
49. C. Gimbert, M. Lumbierres, C. Marchi, M. Moreno-Manas, R. M. Sebastian and A. Vallribera, *Tetrahedron*, 2005, **61**, 8598–8605.
50. C. Gimbert, M. Moreno-Manas, E. Perez and A. Vallribera, *Tetrahedron*, 2007, **63**, 8305–8310.
51. G. Z. Li, R. K. Randev, A. H. Soeriyadi, G. Rees, C. Boyer, Z. Tong, T. P. Davis, C. R. Becer and D. M. Haddleton, *Polym. Chem.*, 2010, **1**, 1196–1204.
52. Y. Yagci and M. A. Tasdelen, *Prog. Polym. Sci.*, 2006, **31**, 1133–1170.
53. C. Boyer, V. Bulmus, T. P. Davis, V. Ladmiral, J. Liu and S. Perrier, *Chem. Rev.*, 2009, **109**, 5402–5436.
54. J. T. Xu, C. Boyer, V. Bulmus and T. P. Davis, *J. Polym. Sci., Part A: Polym. Chem.*, 2009, **47**, 4302–4313.

55. C. Boyer, V. Bulmus and T. P. Davis, *Macromol. Rapid Commun.*, 2009, **30**, 493–497.
56. G. Moad, E. Rizzardo and S. H. Thang, *Aust. J. Chem.*, 2006, **59**, 669–692.
57. G. Moad, E. Rizzardo and S. H. Thang, *Aust. J. Chem.*, 2009, **62**, 1402–1472.
58. G. Moad and S. H. Thang, *Austr. J. Chem.*, 2009, **62**, 1379–1381.
59. G. Moad, E. Rizzardo and S. H. Thang, *Acc. Chem. Res.*, 2008, **41**, 1133–1142.
60. G. Moad, E. Rizzardo and S. H. Thang, *Polymer*, 2008, **49**, 1079–1131.
61. M. Benaglia, J. Chiefari, Y. K. Chong, G. Moad, E. Rizzardo and S. H. Thang, *J. Am. Chem. Soc.*, 2009, **131**, 6914.
62. J. T. Xu, L. Tao, C. Boyer, A. B. Lowe and T. P. Davis, *Macromolecules*, 2010, **43**, 20–24.
63. M. Semsarilar, V. Ladmiral and S. Perrier, *Macromolecules*, 2010, **43**, 1438–1443.
64. B. M. Rosen, G. Lligadas, C. Hahn and V. Percec, *J. Polym. Sci., Part A: Polym. Chem.*, 2009, **47**, 3931–3939.
65. M. Li, P. De, S. R. Gondi and B. S. Sumerlin, *J. Polym. Sci., Part A: Polym. Chem.*, 2008, **46**, 5093–5100.
66. H. Li, B. Yu, H. Matsushima, C. E. Hoyle and A. B. Lowe, *Macromolecules*, 2009, **42**, 6537–6542.
67. R. M. Hensarling, V. A. Doughty, J. W. Chan and D. L. Patton, *J. Am. Chem. Soc.*, 2009, **131**, 14673–14675.
68. A. B. Lowe, C. E. Hoyle and C. N. Bowman, *J. Mat. Chem.*, 2010, **20**, 4745–4750.
69. C. E. Hoyle, A. B. Lowe and C. N. Bowman, *Chem. Soc. Rev.*, 2010, **39**, 1355–1387.
70. C. Boyer, V. Bulmus, T. P. Davis, V. Ladmiral, J. Q. Liu and S. Perrier, *Chem. Rev.*, 2009, **109**, 5402–5436.
71. T. P. Davis, D. M. Haddleton and S. N. Richards, *J. Macromol. Sci.*, 1994, **C34**, 243–324.
72. T. P. Davis, D. Kukulj, D. M. Haddleton and D. R. Maloney, *Trends Polym. Sci.*, 1995, **3**, 365–373.
73. D. Kukulj, T. P. Davis, K. G. Suddaby, D. M. Haddleton and R. G. Gilbert, *J. Polym. Sci., Part A: Polym. Chem.*, 1997, **35**, 859–878.
74. K. G. Suddaby, D. R. Maloney and D. M. Haddleton, *Macromolecules*, 1997, **30**, 702–713.
75. R. A. Sanayei and K. F. Odriscoll, *J. Macromol. Sci.*, 1989, **A26**, 1137–1149.
76. J. P. A. Heuts, D. Kukulj, D. J. Forster and T. P. Davis, *Macromolecules*, 1998, **31**, 2894–2905.
77. A. Gridnev, *J. Polym. Sci., Part A: Polym. Chem.*, 2000, **38**, 1753–1766.
78. J. P. A. Heuts, G. E. Roberts and J. D. Biasutti, *Aust. J. Chem.*, 2002, **55**, 381–398.
79. J. P. A. Heuts and N. M. B. Smeets, *Polym. Chem.*, 2011, **2**, 2407–2423.

80. J. Mazzolini, O. Boyron, V. Monteil, F. D'Agosto, C. Boisson, G. C. Sanders, J. P. A. Heuts, R. Duchateau, s. D. Gigme and D. Bertin, *Polym. Chem.*, 2012.
81. L. Nurmi, J. Lindqvist, R. Randev, J. Syrett and D. M. Haddleton, *Chem. Commun.*, 2009, 2727–2729.
82. B. M. Rosen and V. Percec, *Chem. Rev.*, 2009, **109**, 5069–5119.
83. M. E. Levere, I. Willoughby, S. O'Donohue, A. de Cuendias, A. J. Grice, C. Fidge, C. R. Becer and D. M. Haddleton, *Polym. Chem.*, 2010, **1**, 1086–1094.
84. J. A. Syrett, M. W. Jones and D. M. Haddleton, *Chem. Commun.*, 2010, **46**, 7181–7183.
85. G. J. M. Habraken, C. E. Koning, J. P. A. Heuts and A. Heise, *Chem. Commun.*, 2009, 3612–3614.
86. J. Sun and H. Schlaad, *Macromolecules*, 2010, **43**, 4445–4448.
87. G. Triola, L. Brunsveld and H. Waldmann, *J. Org. Chem.*, 2008, **73**, 3646–3649.
88. S. Wittrock, T. Becker and H. Kunz, *Angew. Chem., Int. Edit.*, 2007, **46**, 5226–5230.
89. A. Dondoni, A. Massi, P. Nanni and A. Roda, *Chem.–Eur. J.*, 2009, **15**, 11444–11449.
90. M. W. Jones, G. Mantovani, S. M. Ryan, X. X. Wang, D. J. Brayden and D. M. Haddleton, *Chem. Commun.*, 2009, 5272–5274.
91. J. Wang, D. Chow, H. Heiati and W.-C. Shen, *J. Controlled Release*, 2003, **88**, 369–380.
92. J. A. Burns, J. C. Butler, J. Moran and G. M. Whitesides, *J. Org. Chem.*, 1991, **56**, 2648–2650.
93. M. W. Jones, M. I. Gibson, G. Mantovani and D. M. Haddleton, *Polym. Chem.*, 2011, **2**, 572–574.
94. K. Bebis, M. W. Jones, D. M. Haddleton and M. I. Gibson, *Polym. Chem.*, 2011, **2**, 975–982.
95. S. P. S. Koo, M. M. Stamenovic, R. A. Prasath, A. J. Inglis, F. E. Du Prez, C. Barner-Kowollik, W. Van Camp and T. Junkers, *J. Polym. Sci. Polym. Chem.*, 2010, **48**, 1699–1713.
96. R. L. A. David and J. A. Kornfield, *Macromolecules*, 2008, **41**, 1151–1161.
97. Z. Hordyjewicz-Baran, L. C. You, B. Smarsly, R. Sigel and H. Schlaad, *Macromolecules*, 2007, **40**, 3901–3903.
98. Y. Geng, D. E. Discher, J. Justynska and H. Schlaad, *Angew. Chem., Int. Edit.*, 2006, **45**, 7578–7581.
99. A. Gress, A. Volkel and H. Schlaad, *Macromolecules*, 2007, **40**, 7928–7933.
100. L. M. Tedaldi, M. E. B. Smith, R. I. Nathani and J. R. Baker, *Chem. Commun.*, 2009, 6583–6585.
101. M. E. B. Smith, F. F. Schumacher, C. P. Ryan, L. M. Tedaldi, D. Papaioannou, G. Waksman, S. Caddick and J. R. Baker, *J. Am. Chem. Soc.*, 2010, **132**, 1960–1965.

102. M. W. Jones, R. A. Strickland, F. F. Schumacher, S. Caddick, J. R. Baker, M. I. Gibson and D. M. Haddleton, *J. Am. Chem. Soc.*, 2011, **134**, 1847–1852.
103. M. W. Jones, R. A. Strickland, F. F. Schumacher, S. Caddick, J. R. Baker, M. I. Gibson and D. M. Haddleton, *Chem. Commun.*, 2012, **48**, 4064–4066.
104. D. Smith, A. C. Holley and C. L. McCormick, *Polym. Chem.*, 2011, **2**, 1428–1441.
105. K. L. Heredia, T. H. Nguyen, C. W. Chang, V. Bulmus, T. P. Davis and H. D. Maynard, *Chem. Commun.*, 2008, 3245–3247.
106. K. Gunasekaran, T. H. Nguyen, H. D. Maynard, T. P. Davis and V. Bulmus, *Macromol. Rapid Commun.*, 2011, **32**, 654–659.
107. C. S. M. Ferreira, C. S. Matthews and S. Missailidis, *Tumor Biol.*, 2006, **27**, 289–301.
108. C. Da Pieve, P. Williams, D. M. Haddleton, R. M. J. Palmer and S. Missailidis, *Bioconjugate Chem.*, 2010, **21**, 169–174.
109. R. J. Kreitman and I. Pastan, *Adv. Drug Delivery Rev.*, 1998, **31**, 53–88.
110. P. J. Carter and P. D. Senter, *Cancer J*, 2008, **14**, 154–169.
111. J. R. Junutula, H. Raab, S. Clark, S. Bhakta, D. D. Leipold, S. Weir, Y. Chen, M. Simpson, S. P. Tsai, M. S. Dennis, Y. M. Lu, Y. G. Meng, C. Ng, J. H. Yang, C. C. Lee, E. Duenas, J. Gorrell, V. Katta, A. Kim, K. McDorman, K. Flagella, R. Venook, S. Ross, S. D. Spencer, W. L. Wong, H. B. Lowman, R. Vandlen, M. X. Sliwkowski, R. H. Scheller, P. Polakis and W. Mallet, *Nat. Biotechnol.*, 2008, **26**, 925–932.
112. J. R. Junutula, K. M. Flagella, R. A. Graham, K. L. Parsons, E. Ha, H. Raab, S. Bhakta, T. Nguyen, D. L. Dugger, G. M. Li, E. Mai, G. D. L. Phillips, H. Hiraragi, R. N. Fuji, J. Tibbitts, R. Vandlen, S. D. Spencer, R. H. Scheller, P. Polakis and M. X. Sliwkowski, *Clin. Cancer Res.*, 2010, **16**, 4769–4778.
113. P. Moody, M. E. B. Smith, C. P. Ryan, V. Chudasama, J. R. Baker, J. Molloy and S. Caddick, *ChemBioChem.*, 2012, **13**, 39–41.

CHAPTER 6

# *Thiol-ene/yne Chemistry for the Synthesis of Branched and Star-shaped Macromolecules*

ABBY R. JENNINGS AND DAVID Y. SON*

Department of Chemistry, Southern Methodist University, Dallas, TX 75275-0314, United States
*Email: dson@smu.edu

## 6.1 Introduction

Since their initial synthesis and discovery, dendrimers and other hyperbranched compounds continue to be valuable materials in areas such as drug delivery, imaging, UV-cured films and many more.[1–3] Some of the difficulties facing researchers when fabricating these hyperbranched materials include protecting and deprotecting functional groups, using excess reagents, harsh reaction conditions and tedious purification methods.[4–6] To overcome these hurdles, researchers have turned to employing "click" reactions in order to obtain more efficient syntheses. More recently, the thiol-ene and thiol-yne reactions have been exploited for the synthesis of dendritic and hyperbranched materials because of their "click"-like attributes.[7,8] Herein, we describe the recent strides and discoveries that have been made by employing the thiol-Michael reaction, also known as the ionic thiol-ene reaction, the thiol-ene reaction, which proceeds *via* a free-radical mechanism, and the thiol-yne reaction, for the synthesis of dendrimers and other hyperbranched materials. The thiol-yne examples presented in this chapter also proceed by a free-radical mechanism.

RSC Polymer Chemistry Series No. 6
Thiol-X Chemistries in Polymer and Materials Science
Edited by Andrew B. Lowe and Christopher N. Bowman

Published by the Royal Society of Chemistry, www.rsc.org

## 6.2 The Thiol-Michael Reaction

One of the first examples of employing the ionic thiol-ene, or thiol-Michael reaction, for the synthesis of branched molecules was reported in 2006 by Uhrich *et al.*[9] In this instance, the authors used an amine-initiated thiol-Michael reaction to build a four-armed core for amphiphilic star-like macromolecules (ASMs). To synthesize the core, **1**, pentaerythritiol tetraacrylate (PETA) was reacted with a slight excess of 4-aminothiophenol in the presence of triethylamine (TEA) (Scheme 6.1).

Once compound **1** had been purified by silica gel chromatography, it was used to divergently construct methoxy, carboxy, and amine-terminated ASMs by reaction with mucic acid derivatives followed by coupling with methoxy poly(ethylene glycol) (mPEG), heterobifunctional PEG, and mPEG bis-amine, respectively.

In 2008 and 2009, Lowe, Hoyle and co-workers demonstrated that the thiol-Michael reaction was also a viable choice in the convergent synthesis of star polymers.[10,11] To illustrate the versatility of the thiol-Michael reaction, two different homopolymers, **2** and **3**, were synthesized using standard reversible addition-fragmentation chain transfer (RAFT) conditions. Subsequent to obtaining **2** and **3**, the polymers were reacted in a one-pot synthesis with a slight excess of trimethylolpropane triacrylate (TMPTA) in the presence of dimethylphenylphosphine (DMPP) and an excess of hexylamine (HexAM), yielding the 3-arm star polymers **4** and **5**, respectively (Scheme 6.2). The reactions were performed at room temperature (RT). Polymers **4** and **5** were purified by precipitation into hexanes.

To verify the roles of HexAm and DMPP in Scheme 6.2, polymer **3** was reacted with HexAm in the absence of DMPP. Using matrix-assisted laser desorption/ionization time-of-flight mass spectrometry (MALDI-TOF MS),

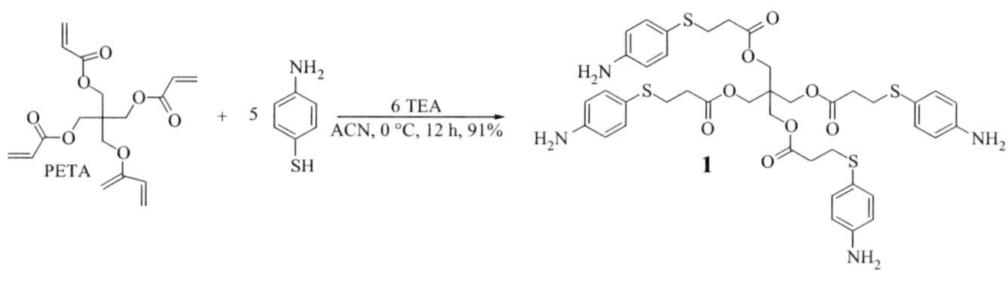


**Scheme 6.1** Synthesis of ASMs core.

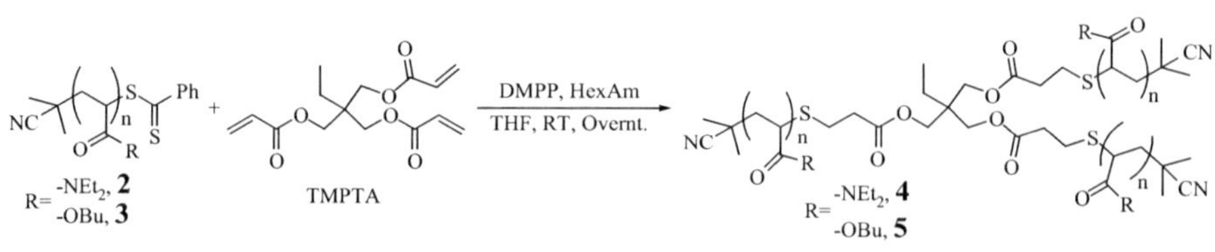


**Scheme 6.2** Convergent synthesis of 3-armed star polymers.

it was found the HexAm was responsible for cleaving the dithioester end group, resulting in the formation of a thiol terminated homopolymer as well as the undesirable disulfide (Figure 6.1).

When DMPP was added, only the thiol terminated homopolymer was observed in the MALDI-TOF MS spectrum. Thus, DMPP was employed not only to initiate the thiol-Michael reaction in Scheme 6.2, but to prevent disulfide formation as well.

In order to establish the speed of the thiol-Michael reaction, the authors used Fourier transform infrared spectroscopy (FT-IR), and $^{1}H$ and $^{13}C$ NMR spectroscopy to monitor the disappearance of the vinylic functional groups of TMPTA during the course of the reaction. In all cases, the vinyl bond appeared to be consumed after only 5 minutes. No yields were reported for **4** and **5**.

The thiol-Michael reaction has also been successfully employed for the divergent synthesis of dendritic molecules.[12] Using a two-step reaction sequence, also known as an alternate multi-addition (AMA) process, Sedaghat-Haerati *et al.* were able to synthesize a third generation hydroxyl-terminated dendrimer, **8** (Scheme 6.3).

In the first step, mPEG is reacted with acryloyl chloride (Ac-Cl) to yield the acrylate terminated polymer, **6**. After work-up, **6** readily undergoes the thiol-Michael reaction with 1-thioglycerol (TG), yielding the first generation

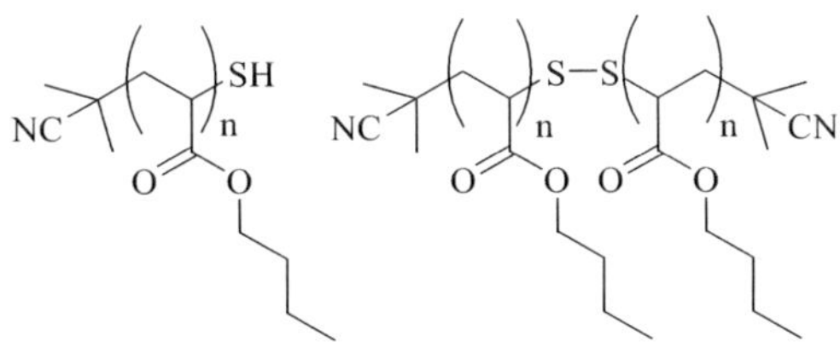


**Figure 6.1** Thiol terminated homopolymer and disulfide product.

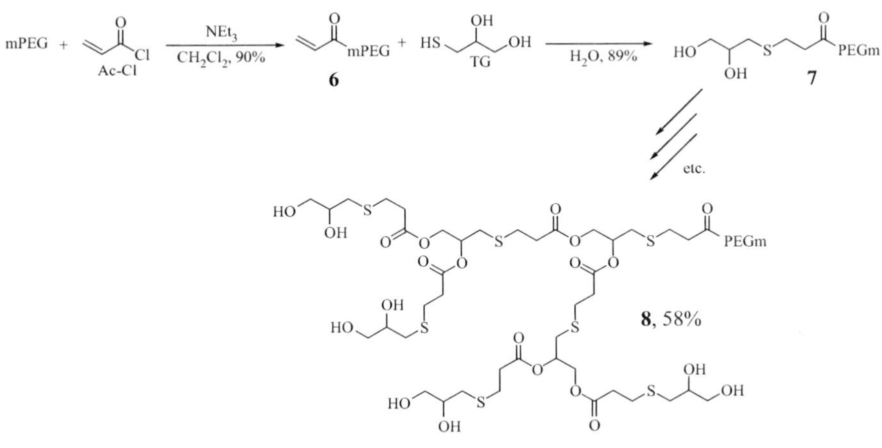


**Scheme 6.3** The divergent synthesis of dendrimer **8** using Ac-Cl and TG.

hydroxyl-terminated dendrimer, **7**. Repeating these steps with a slight excess of acryloyl chloride, TEA, and TG, generated the third generation hydroxyl-terminated dendrimer, **8**, which was isolated by precipitation. Each step of the dendrimer synthesis was monitored by $^{1}H$ and $^{13}C$ NMR spectroscopy. Analysis of **8** *via* MALDI-TOF MS showed that the dendritic molecule had few imperfections and a molecular mass that was within good agreement of the calculated value.

Later that same year, Rim and Son utilized the thiol-Michael reaction for the efficient synthesis of star-shaped oligomers.[13] Initially, 1,3,5,-triacryloylhexahydro-1,3,5-triazine (TAT) was reacted with three equivalents 2-mercaptoethanol (ME) *via* the thiol-Michael reaction employing *n*-propylamine as an initiator, producing a first generation hydroxyl-terminated star oligomer, **9**. Oligomer **9** was then reacted with three equivalents of Ac-Cl, with respect to each hydroxyl-group, to produce an alkene terminated star oligomer, **10**. Repeating these steps in an AMA process produced a hydroxyl-terminated star polymer, **11** (Scheme 6.4).

The compounds synthesized in Scheme 6.4 were characterized by $^{1}H$ and $^{13}C$, as well as COSY, HETCOR, and HMBC NMR spectroscopy. Further analysis using electrospray ionization (ESI) MS found that the experimental molecular weights of the star oligomers were within good agreement of the theoretical values.

Another significant example of utilizing the thiol-Michael reaction for the divergent synthesis of dendritic molecules was illustrated by Shen and co-workers in 2009.[14] In their first example, the authors employed 2-[(methacryloyl)oxy]ethyl acrylate (MAEA) and cysteamine (CS) for the highly efficient syntheses of various polyaminoester-based dendrimers.

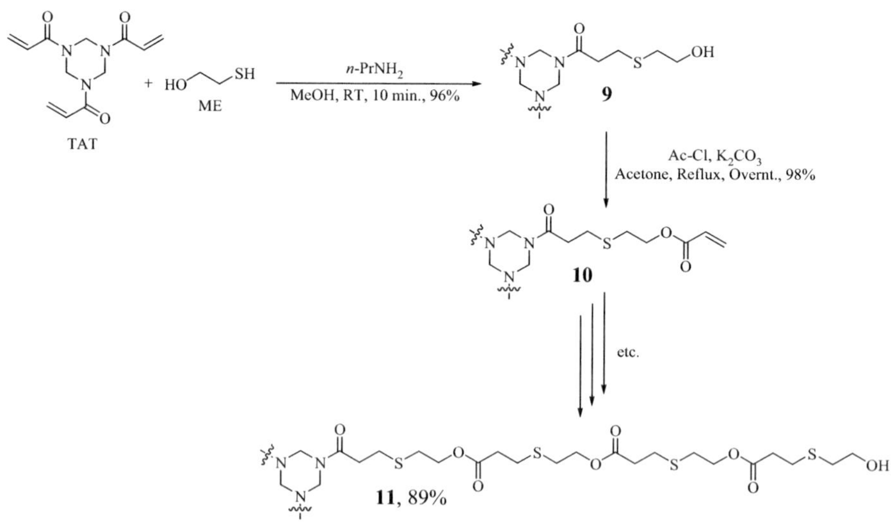


**Scheme 6.4** The divergent synthesis of star-shaped oligomers.

To begin with, the authors synthesized four different alkene terminated first generation dendrimers, compounds **12–15**, by reacting 1,4-diaminobutane, ethylenediamine, 1,6-diaminohexane, and tris(2-aminoethyl)amine with a slight excess of MAEA. Scheme 6.5 shows an example using 1,4-diaminobutane.

It was discovered that the reaction shown in Scheme 6.5 could be catalyzed using magnesium perchlorate or heat. To avoid removal of the catalyst and thus obtain more efficient dendrimer syntheses, heat was used for constructing the dendrimers.

Compounds **12–15** readily underwent the thiol-Michael reaction with CS. Although a stoichiometric amount of CS could be used, it was found that using a slight excess of CS increased the rate of the reaction. Employing an AMA process with MAEA and CS produced dendritic molecules in high yield. Scheme 6.6 shows a representative example using compound **12**.

Purification after each step of the dendrimer synthesis shown in Scheme 6.6 could be performed with ease by removing excess MAEA or CS by washing the crude product with either hexane or brine. Analysis of the dendrimers by MALDI-TOF MS showed that the experimental molecular weights were in agreement with the theoretical values. Characterization of the dendrimers by gel-permeation chromatography (GPC) confirmed that there was a steady increase in the molecular weight as the dendrimer growth progressed. Table 6.1 summarizes the results reported for each respective final generation of the polyaminoester dendrimers synthesized using the methods in Scheme 6.6.

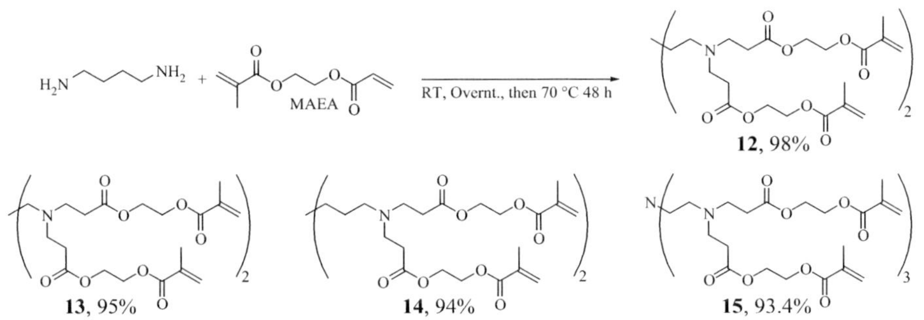


**Scheme 6.5** Synthesis of first generation alkene-terminated dendrimers, **12–15**.

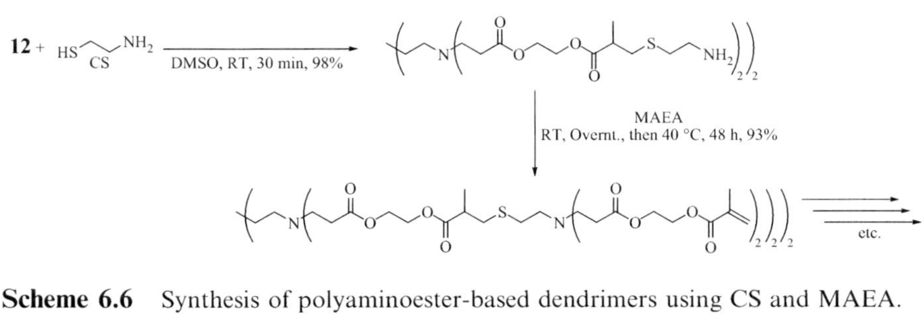


**Scheme 6.6** Synthesis of polyaminoester-based dendrimers using CS and MAEA.

**Table 6.1** Summary of results for the polyester dendrimers synthesized by Shen *et al.*

| *Amine* | *Generation* | *Terminus* | *Yield (%)* |
|---|---|---|---|
| 1,4-Diaminobutane | 3.5 | Amine | 93 |
| Ethylenediamine | 3.0 | methacrylate | 90 |
| 1,6-Diaminohexane | 3.0 | methacrylate | 91 |
| Tris(2-aminoethyl)amine | 3.0 | methacrylate | 94 |

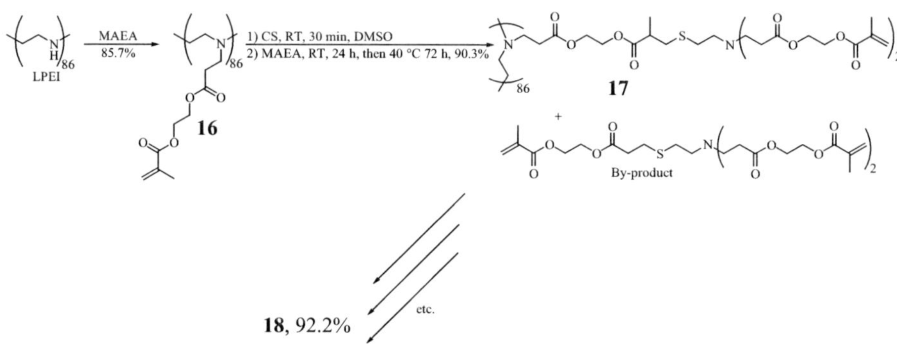


**Scheme 6.7** One pot synthesis of dendronized LPEI.

In the next example presented, the authors made use of the MAEA-CS pair and the thiol-Michael reaction to dendronize the linear polymer, polyethylenimine (LPEI), using a one-pot per generation method. Initially, the linear polymer was functionalized with methacrylate groups by reacting LPEI with MAEA, producing **16**. Subsequently, compound **16** was reacted with an excess of CS under ambient conditions for 30 minutes. To this solution was added an excess of MAEA, yielding **17** and a small molecule by-product that was easily removed by precipitation (Scheme 6.7).

Repeating steps one and two in Scheme 6.7 gave a fourth generation methacrylate-terminated dendronized linear polymer, **18**, in high yield. GPC and $^1$H NMR spectroscopic analyses of the various generations showed that the experimental molecular weights agreed well with the theoretical values and that there were very few defects in the dendrimer structure. Furthermore, compound **18** had a polydispersity of 1.2, which was close to the original linear polymer.

To expand further, Shen and co-workers employed TG and 2-isocyanatoethyl methacrylate (IEMA) to efficiently synthesize dendritic molecules *via* the thiol-Michael reaction. To build the first generation methacrylate-terminated dendrimer, **19**, a slight excess of IEMA was reacted with glycerol. Dendrimer **19** was then reacted with a slight excess of TG to give the hydroxyl-terminated dendrimer, **20**. Subsequently, a fourth generation methacrylate terminated dendrimer, **21**, was easily synthesized by employing an AMA process with the IEMA and TG (Scheme 6.8).

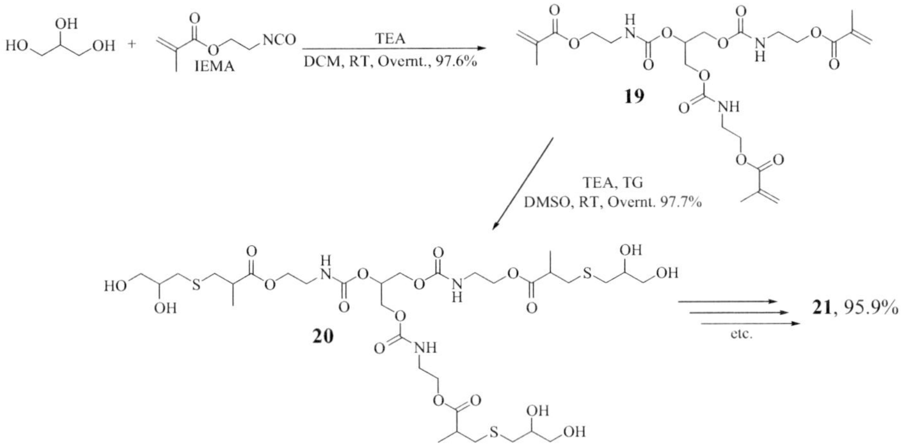


**Scheme 6.8** Synthesis of a fourth generation methacrylate-terminated dendrimer, **21**, using IEMA and TG.

Each generation was easily obtained and purified by simple precipitation and recovered in high yield, all greater than 90%. Analysis of the dendrimers by MALDI-TOF MS showed that the experimental molecular weights were within the expected mass range.

An extension of the this work was published later in 2011 by Shen *et al.*[15] Second, third, and fourth generation primary amine-terminated dendrimers were synthesized from compound **15** using the methods shown in Scheme 6.6. These dendrimers were also reacted with 2-(*N*,*N*-dimethylamino)ethyl acrylate to produce tertiary amine-terminated dendrimers. Furthermore, a fourth generation dendronized LPEI compound from Scheme 6.7 was also synthesized and converted to the corresponding tertiary amine compound by reaction with 2-(*N*,*N*-dimethylamino)ethyl acrylate. These dendritic compounds were analyzed for their pH and temperature responsiveness as well as their photoluminescent properties.

It was observed that the lower critical solution temperature (LCST) for the dendritic molecules increased as the pH decreased. Both the tertiary and primary amine-terminated dendrimers exhibited similar temperature and pH sensitivities; however, the tertiary amine dendrimers had sharper transitions and LCSTs that were not as sensitive to pH. The dendronized LPEI exhibited a higher LCST than the globular dendrimers at the same pH. Furthermore, the globular dendrimers were luminescent without the presence of a chromophore, and the fluorescence intensity increased as the generation increased and was stronger for the tertiary amine-terminated dendrimers.

Additionally, Aoki and Ichimura exploited the thiol-Michael addition and an AMA process for the divergent synthesis of dendritic molecules.[16,17] In their first example, the authors used PETA to build the core of the dendrimer and then reacted it with TG *via* the thiol-Michael reaction to yield compound **22**. Subsequently, **22** was reacted with an excess of 1,1-bis(acryloyloxymethyl)ethyl isocyanate (BEI), using cupferron as a polymerization inhibitor and dibutyltin dilaurate (DBTDL) as a catalyst. Treating the solution with *N*,*N*-dimethylaminoethanol (DMAE) and washing with acid removed any unreacted BEI, yielding dendrimer **23** (Scheme 6.9).

Analysis of **23** *via* $^1$H NMR spectroscopy and MALDI-TOF MS showed that the product was a complicated mixture, implying that imperfections and unreacted hydroxyl groups were present. To further evaluate **23** using UV spectroscopy, it was functionalized with 9-fluorenylmethoxycarbony (Fmoc-SH) *via* the thiol-Michael reaction, using dicyclohexylamine (DCHA) as an initiator (Scheme 6.10).

The data obtained from the UV spectroscopic experiments showed that only 5% of the secondary alcohols of **22** reacted. The low conversion was believed to be due to the sterically demanding tertiary isocyanate structure of BEI. To validate this, BEI in Scheme 6.9 was replaced with 2-isocyanatoethyl acrylate (AOI) to yield compound **24**. After functionalizing with Fmoc-SH *via* the thiol-Michael reaction, the same UV spectroscopic study was performed on compound **24** to determine the amount of terminal hydroxyl groups that were reacted (Scheme 6.11). Analysis of compound **24** showed that the reactions of

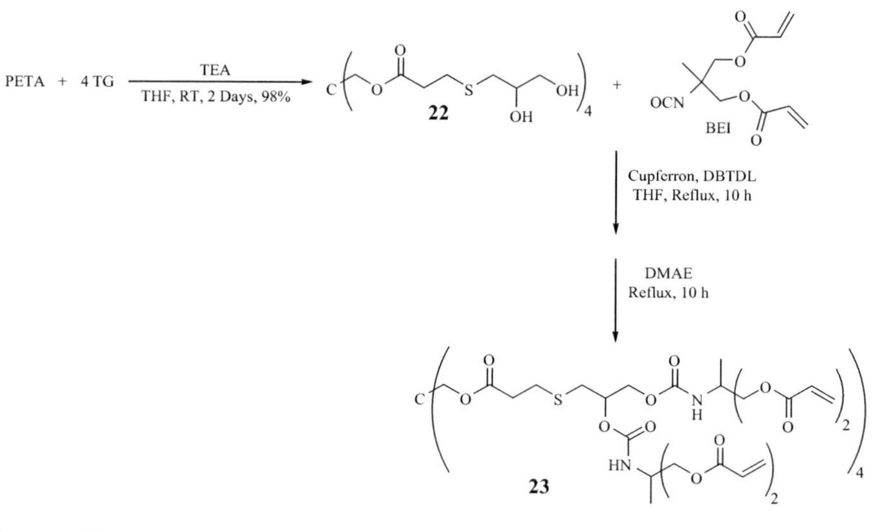


**Scheme 6.9** Synthesis of a second generation acrylate-terminated dendrimer, **23**, using TG and BEI.

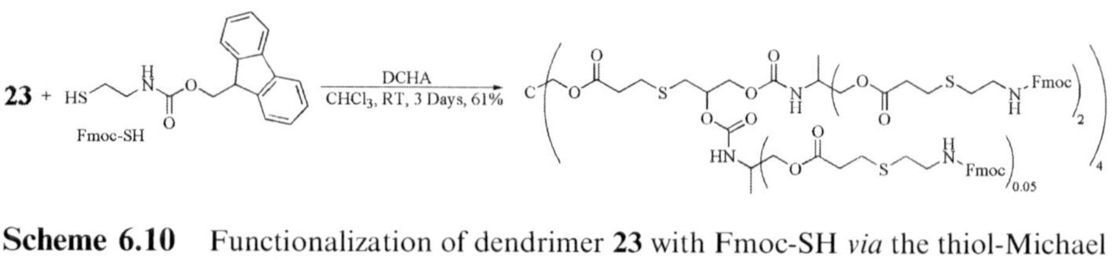


**Scheme 6.10** Functionalization of dendrimer **23** with Fmoc-SH *via* the thiol-Michael reaction.

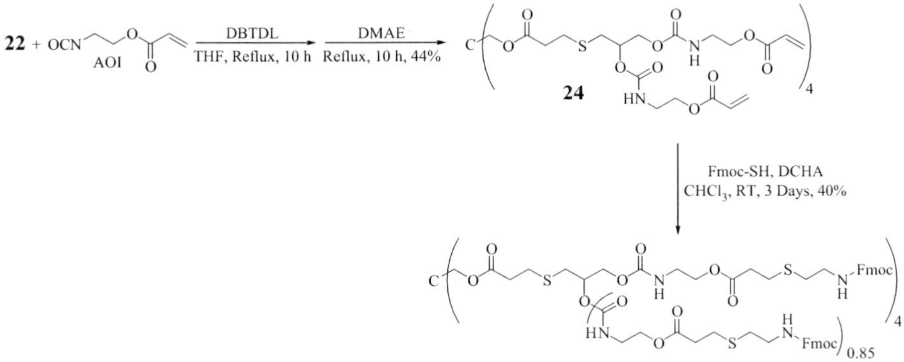

**Scheme 6.11** Synthesis and functionalizing of compound **24**.

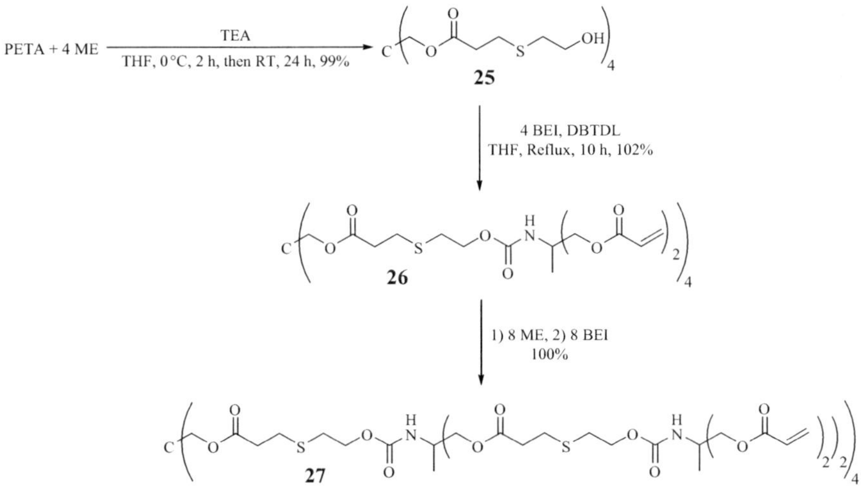

**Scheme 6.12** Synthesis of dendrimers **26** and **27** using ME and BEI.

the secondary terminal hydroxyl groups proceeded with 85% conversion when AOI was used.

The results from the analysis of **23** and **24** showed that using TG was not a practical choice for synthesizing dendrimers using the methods shown in Scheme 6.10 and Scheme 6.11 because of the unreactive nature of the secondary hydroxyl groups. As a result, a similar synthesis was developed using ME instead of TG (Scheme 6.12).

For the reactions shown in Scheme 6.12, stoichiometric amounts of BEI and ME could be used, thus there was no need to use DMAE to remove the unreacted BEI. After dendrimers **26** and **27** were obtained, they were reacted with a slight excess of 2-naphthalenethiol (NT-SH) using TEA as an initiator. This allowed for the analysis of the dendrimers *via* GPC and UV spectroscopy.

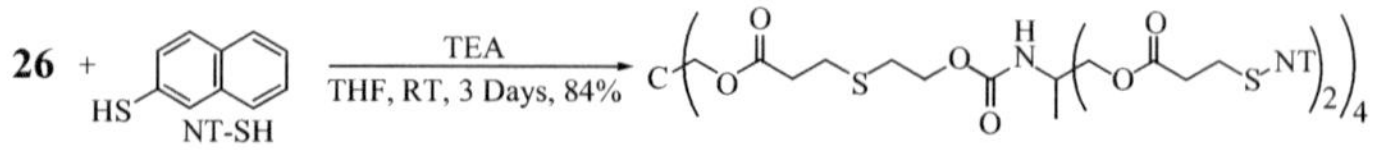


**Scheme 6.13** Functionalization of dendrimer **26** with NT-SH *via* the thiol-Michael reaction.

Scheme 6.13 shows an example using dendrimer **26**. When performing the reaction in Scheme 6.13 with dendrimer **27**, a 66% yield of the NT-SH functionalized material was obtained.

The UV data showed that the purities of dendrimers **26** and **27** were 98 and 88%, and the number of terminal naphthylthio-moieties was determined to be 7.8 and 14, respectively. Analysis of the functionalized dendrimers by GPC was used to determine the PDI and found to be 1.04 and 1.06. Employing dynamic light scattering (DLS), the diameter of **26** was found to be 2.1 +/− 0.2 nm and the diameter of **27** was determined to be 3.4 +/− 0.3 nm.

To further investigate the advantages of these dendritic molecules, specifically **26** and **27**, an extension of this work was published by Aoki *et al.*[18] In this study, the authors investigated the UV curing properties of **26** and **27** with respect to a conventional bi-functional monomer, tripropyleneglycol diacrylate (TPGDA). The authors determined that their dendritic molecules could function as polymerization enhancers, had a higher photoactivity then TPGDA, and exhibited low polymerization shrinkage.

Another example of utilizing the thiol-Michael reaction for the efficient divergent synthesis of dendritic molecules was demonstrated by Lupton *et al.* in 2010.[19] Initially, the first generation alkene terminated dendrimer, **28**, was synthesized by reacting benzene-1,3,5-triyltrimethanol and a slight excess of 1-(iodomethyl)-2-oxo-cyclohexanecarboxylic acid *via* a Grob/Eschenmoser fragmentation. Dendrimer **28** was then reacted with (5-(mercaptomethyl)-1,3-phenylene)dimethanol *via* the thiol-Michael reaction using DMPP as an initiator to produce compound **29**. The authors could perform the thiol-Michael reaction with or without solvent. By using an AMA process with the fragmentation and thiol-Michael steps, the authors were able to obtain a fourth generation alkene terminated dendrimer, **30**, in high yield (Scheme 6.14).

By monitoring the dendrimer synthesis shown in Scheme 6.14 by $^1$H NMR spectroscopy it was determined that each step reached greater than 95% conversion. MALDI-TOF MS analysis of the hydroxyl terminated generations showed that the experimental molecular masses were within the scope of the calculated values. GPC analysis showed that the alkene terminated generations had narrow PDI values (PDI < 1.2).

In late 2010, Li and co-workers also took advantage of the thiol-Michael reaction to synthesize various monomers for the construction of hyperbranched poly(ester-amide)s.[20] Initially, various amino acids were converted into their corresponding vinylic monomers by reaction with methanol and thionyl chloride, followed by reaction with acryloyl chloride. These vinylic monomers readily underwent the thiol-Michael reaction with TG using TEA as an initiator (Scheme 6.15).

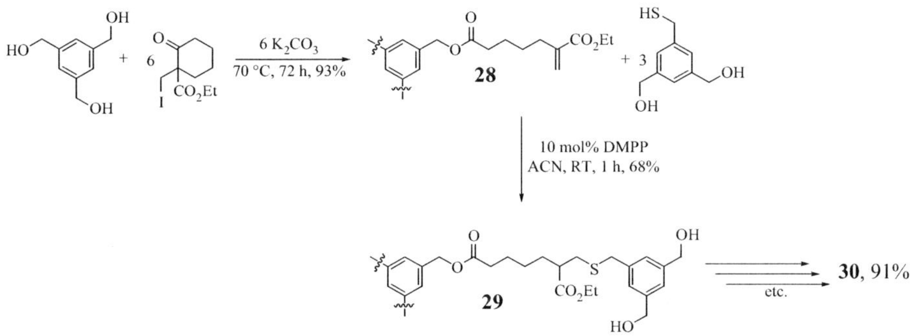


**Scheme 6.14** Divergent synthesis of dendrimer **30** using Grob/Eschenmoser fragmentations and thiol-Michael reactions.

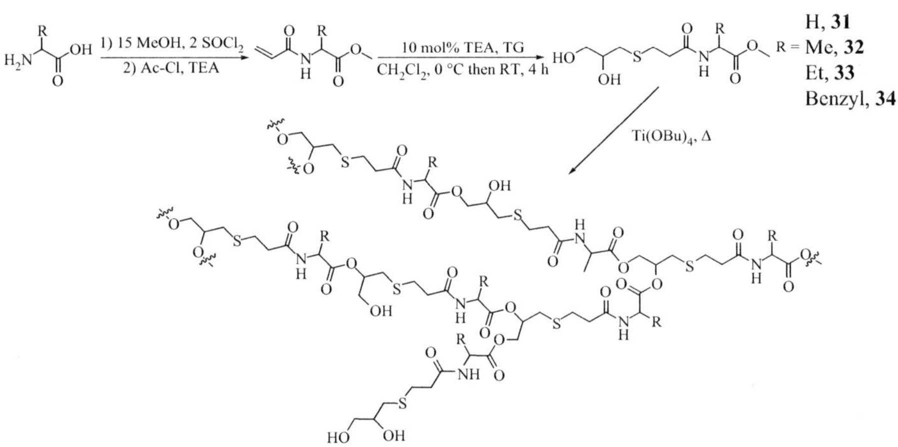


**Scheme 6.15** Synthesis of hyper-branched polymers from various vinylic monomers synthesized *via* the thiol-Michael reaction.

The thiol-Michael reactions were monitored by $^1H$ NMR spectroscopy and found to be 90% complete after one hour. The authors reported that near quantitative results were obtained for the synthesis of compounds **31–34**. Analysis by HPLC showed that the percent purities of **31–34** were greater than 95%. The products were also analyzed by $^{13}C$ NMR spectroscopy and ESI MS. After compounds **31–34** were obtained, they were used to make hyperbranched poly(ester-amide)s *via* self-condensation reactions under reduced pressure using heat and $Ti(OBu)_4$.

## 6.3 The Thiol-ene Reaction

The first example of employing the thiol-ene reaction for the synthesis of dendritic molecules was presented in 2007 by Heidecke and Lindhorst.[21]

In this example, the authors reacted 3,6-di-*O*-allyl-2,4,-di-*O*-benzyoly-α-D-mannopyranoside with an excess of ME to produce the first generation hydroxyl-terminated glycodendron, **35**. In this instance, AIBN was used as an initiator. Compound **35** was then reacted with *O*-(3,6-di-*O*-allyl-2,4,-di-*O*-benzyoly-α-D-mannopyranoside)trichloracetimidate to yield the second generation alkene-terminated glycodendron, **36**. Repeating the ME step followed by reaction with perbenzoylated mannosyl trichloroacetimidate, yielded a third generation benzoyl-terminated glycodendron, **37** (Scheme 6.16).

Each generation of the glycodendron was characterized by $^1$H and $^{13}$C NMR spectroscopy. Analysis by ESI and MALDI-TOF MS showed that the experimental molecular weights were in agreement with the theoretical values. The biological activity of compound **37** was investigated using an ELISA (enzyme-linked immunosorbent assay) formate, and found to inhibit the adhesion of *Escherichia coli*.

Another example of employing the thiol-ene reaction for the divergent synthesis of dendritic molecules was illustrated in 2008 by Hawker *et al.*[22] The first generation hydroxyl terminated dendrimer, **38**, was obtained by reacting TG with 2,4,6-triallyloxy-1,3,5-triazine. In this example, the authors used 0.02 equivalents (with respect to alkene) of 2,2-dimethoxy-2-phenylacetophenone (DMPA) as a photoinitiator and a hand held UV-lamp as the irradiation source ($\lambda_{ex} = 365$ nm). The reaction required no solvent, proceeded under ambient conditions, was complete in 30 minutes, and gave quantitative yields when 1.5 and 2 equivalents of TG were used with respect to the alkene. Compound **38** was then reacted with 4-pentenoic anhydride in THF to yield the first generation dendrimer, **39** (Scheme 6.17).

Repeating steps one and two using an AMA process yielded a fourth generation alkene-terminated dendrimer, **40**. The hydroxyl terminated

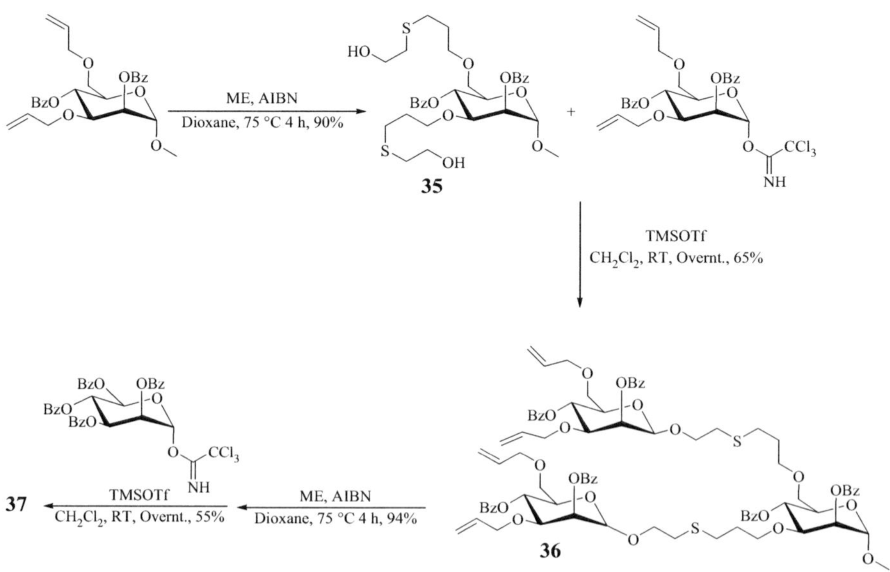


**Scheme 6.16** Divergent synthesis benzoyl-terminated glycodendrons.

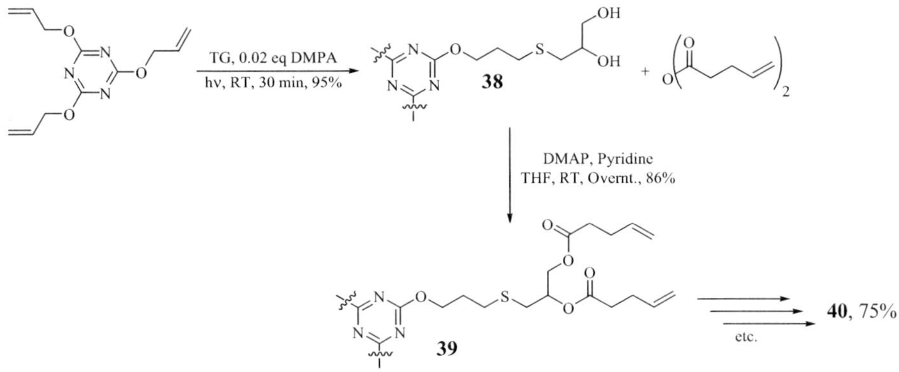


**Scheme 6.17** Divergent synthesis of a fourth generation hydroxyl-terminated dendrimer, **40**, using esterification and thiol-ene reactions.

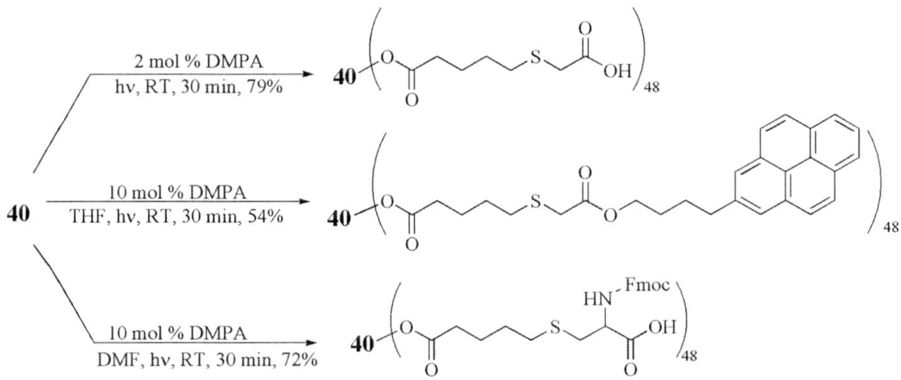


**Scheme 6.18** Functionalizing dendrimer **40** *via* the thiol-ene reaction.

dendrimers could be purified by simple precipitation. Analysis of the products *via* GPC showed that the dendrimers were monodisperse and that their size increased regularly with each respective generation. MALDI-TOF MS demonstrated that the experimental molecular weight agreed with the calculated values, indicating that the dendrimers lacked structural imperfections. These results were also validated with $^{1}H$ and $^{13}C$ NMR spectroscopy.

To further demonstrate the versatility of the thiol-ene reaction, Killops *et al.* utilized the thiol-ene reaction to functionalize dendrimer **40** with thioglycolic acid, 4-(pyren-1-yl)butyl 2-mercaptoacetate, and 9-fluorenylmethoxycarbonyl cysteine, respectively (Scheme 6.18). The products were characterized using $^{1}H$ and $^{13}C$ NMR spectroscopy.

Another significant example of using the thiol-ene reaction for the divergent synthesis of dendritic molecules was demonstrated by Rissing and Son in 2009.[23] Beginning with a tetravinylsilane core, the authors synthesized a fifth generation vinyl terminated dendrimer by employing an AMA process with the thiol-ene reaction and a nucleophilic substitution reaction (Scheme 6.19).

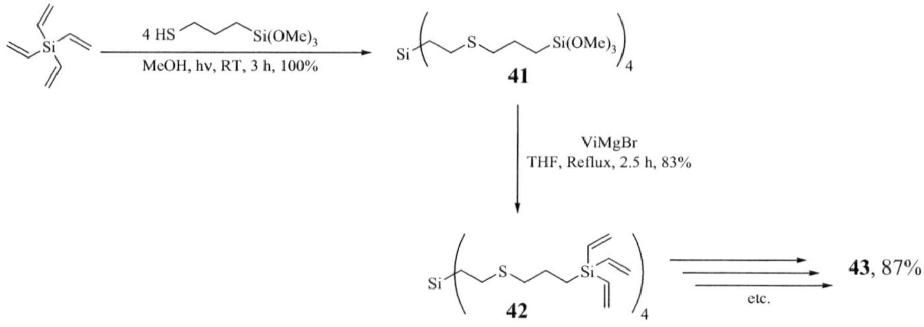


**Scheme 6.19** Synthesis of a fifth generation vinyl-terminated carbosilane-thioether dendrimer, **43**.

In the first step, 3-mercaptopropyltrimethoxysilane was reacted with tetravinylsilane in methanol using a General Electric 275 W sunlamp as a light source, yielding the first generation methoxy-terminated dendrimer, **41**. Refluxing compound **41** in THF with an excess of vinylmagnesium bromide yielded the first generation vinyl-terminated dendrimer, **42**. Repeating steps one and two yielded a fifth generation vinyl-terminated dendrimer, **43**.

For the synthesis of the third generation dendrimer, a 1 : 2 THF:MeOH solvent mixture was employed and 10% excess of the thiol was used. Furthermore, 1 mol% of benzophenone (with respect to the alkene) was added as a photoinitiator to augment reaction completeness. With the exception of the first generation, each methoxy-terminated generation could be purified by simple precipitation. The dendrimers were characterized by $^{1}$H, $^{13}$C, and $^{29}$Si NMR spectroscopy. GPC analysis of the second through fourth generation vinyl-terminated dendrimers showed that the dendrimers were monodisperse and the corresponding molecular weights increased as the generation increased. MALDI-TOF MS for the first three vinyl-terminated generations showed that the experimental molecular weights were in agreement with the theoretical values. The dendrimers synthesized in Scheme 6.19 were further utilized to stabilize various metal nanoparticles.[24]

The authors also demonstrated that the thiol-ene reaction could be applied to functionalize the third generation vinyl-terminated dendrimer, **44**, by reaction with ME. This reaction was performed under ambient conditions without the use of a photoinitiator (Scheme 6.20).

In early 2010, Hawker *et al.* demonstrated another significant example of employing the thiol-ene reaction for the efficient divergent synthesis of dendritic molecules.[25] The authors employed an AMA process using an 'amine-epoxy' step followed by a thiol-ene reaction. Initially, a diamine core, 1,4-bis(aminomethyl)benzene was used to ring open allyl glycidyl ether (AGE) to yield a first generation allyl-terminated dendrimer, **45**. Dendrimer **45** readily underwent the thiol-ene reaction with cysteamine hydrochloride (CS-HCl) in the presence of a small amount of DMPA, to produce **46**, a first generation amine-terminated dendrimer. Repetition of these two steps produced a third generation alkene-terminated dendrimer, **47** (Scheme 6.21).

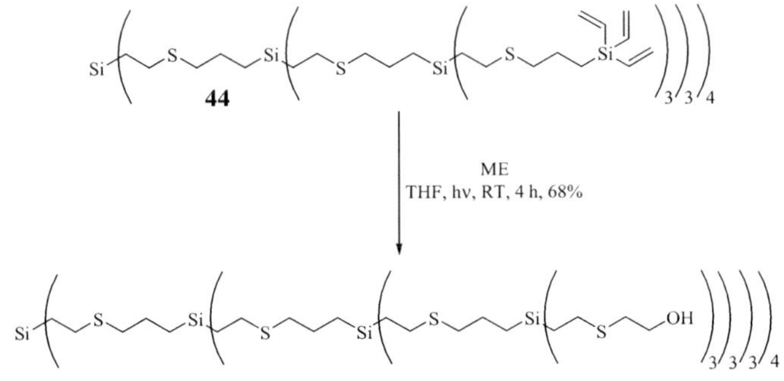


**Scheme 6.20** Functionalization of dendrimer **44** *via* the thiol-ene reaction.

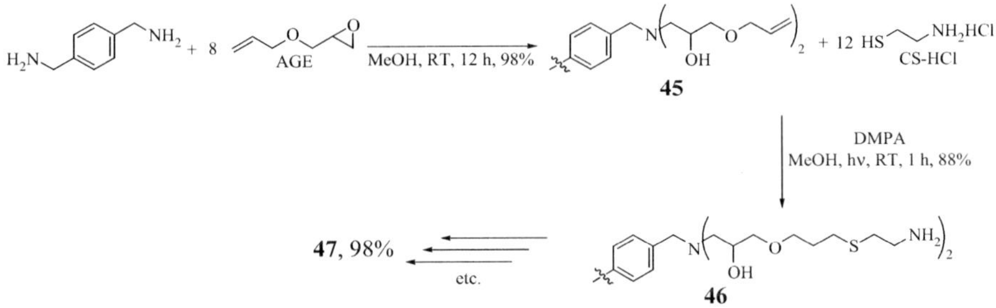


**Scheme 6.21** Divergent synthesis of a third generation alkene-terminated dendrimer using AGE and CS-HCl.

Each generation of the dendrimer was easily purified by precipitation into a suitable solvent. Analysis of the dendrimers by size exclusion chromatography (SEC) showed that they were monodisperse and that the size increased as the generation increased. The authors employed IR and NMR spectroscopy to confirm the presence of the internal hydroxyl groups. MALDI-TOF MS was also employed to verify the purity of each generation, and revealed only slight deviances from the expected spectra for the third generation.

Furthermore, Hawker *et al.* were able to functionalize the interior hydroxyl groups of dendrimer **47** using an esterification reaction with naphthalene-1-acetyl chloride. The exterior allyl groups were also functionalized by employing the thiol-ene reaction (Scheme 6.22).

Characterization of the functionalized dendrimer in Scheme 6.22 using $^1$H NMR spectroscopy showed that there was a high degree of terminal and internal functionalization.

Later that same year, Hawker, Malkock, and co-workers reported another novel synthesis of dendritic molecules using the thiol-ene reaction.[26] To begin with, the authors synthesized a first generation hydroxyl-terminated dendrimer, **48**, by reacting a triallyl core with a slight excess of 2,2-bis(hydroxymethyl)-2-mercaptoethylpropanoate using 0.02 equivalents of DMPA (per alkene unit) as an initiator. Dendrimer **48** then underwent an esterification reaction with an

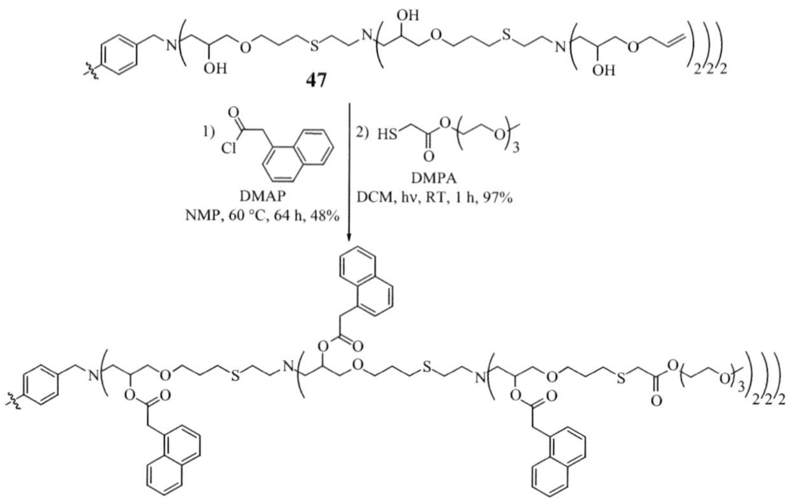


**Scheme 6.22** Internal and external functionalization of dendrimer **47**.

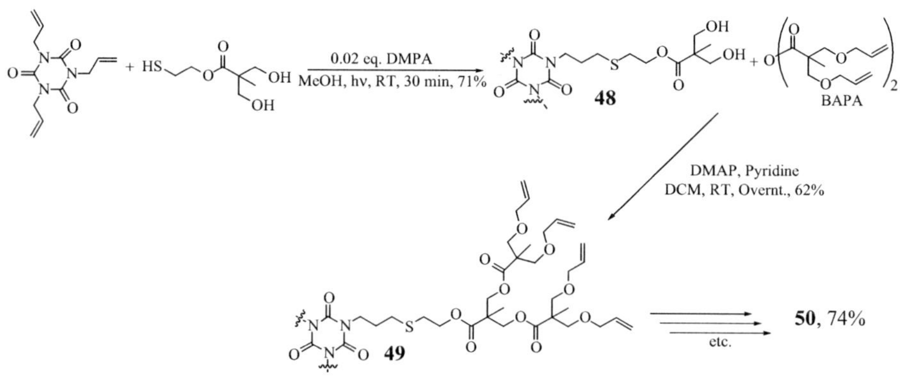


**Scheme 6.23** Divergent synthesis of a fifth generation hydroxyl-terminated dendrimer, **50**, using thiol-ene and esterification reactions.

excess of bis(allyl propionic acid anhydride) (BAPA), to produce **49**. Repeating these two steps using an AMA process yielded a fifth generation hydroxyl-terminated dendrimer, **50** (Scheme 6.23).

The dendrimers were characterized using $^{1}$H and $^{13}$C NMR spectroscopy and quantitative conversions were seen up to the fourth generation of the dendrimer. Analysis by SEC also showed that the dendrimers were mondisperse with PDIs between 1.02 and 1.04.

In a similar fashion, the authors synthesized a fourth generation allyl-terminated dendrimer by employing TG in place of 2,2-bis(hydroxymethyl)-2-mercaptoethylpropanoate. The first generation hydroxyl-terminated dendrimer was synthesized using the same procedures shown in Scheme 6.18 and was reacted with an excess of BAPA to give compound **51**. Subsequent to

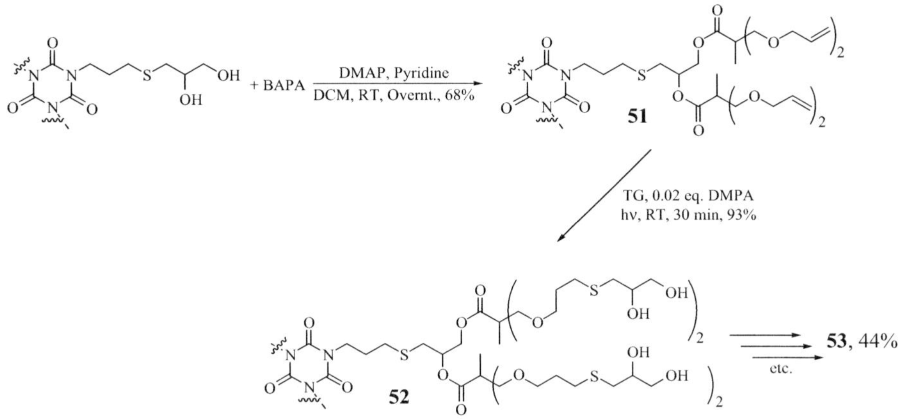


**Scheme 6.24** Divergent synthesis of a fourth generation alkene-terminated dendrimer using BAPA and TG.

obtaining **51**, it was reacted with TG, without the use of a solvent, to yield dendrimer **52**. Using an AMA process with the BAPA and TG steps, a fourth generation alkene-terminated dendrimer, **53**, was synthesized (Scheme 6.24). Using TG produced a dendrimer that had a lower molecular weight and higher density profile then compound **50**.

Each generation shown in Scheme 6.24 was characterized by $^{1}H$ and $^{13}C$ NMR spectroscopy. Analysis by SEC found that the generations were monodisperse and had PDIs ranging from 1.02 to 1.05.

Additionally, a fifth generation allyl-terminated dendrimer was synthesized by employing an AMA process. In this example, trimethylolpropane was reacted with a slight excess of trimethylolpropane diallyl-ether (TMPDE) anhydride using an esterification reaction to produce a first generation alkene-terminated dendrimer, **54**. Subsequently, dendrimer **54** was reacted with an excess of 2,2-bis(hydroxymethyl)-2-mercaptoethylpropanoate by the thiol-ene reaction to produce a second generation hydroxyl-terminated dendrimer, **55**. By repeating these steps, the authors were able to produce a fifth generation allyl-terminated dendrimer, **56** (Scheme 6.25).

By changing the sequence of the reactions, the authors were able to synthesize a fifth generation vinyl-terminated dendrimer which exhibited different solubility and thermal properties than the fifth generation hydroxyl-terminated dendrimer, **50**, from Scheme 6.23. Each generation synthesized was characterized by $^{1}H$ and $^{13}C$ NMR spectroscopy. Furthermore, it was found that the dendrimers had PDIs ranging from 1.02–1.04. Analysis of the dendrimers by differential scanning calorimetry (DSC) showed that the alkene-terminated dendrimers had lower $T_g$ values than the hydroxyl-terminated dendrimers.

To further demonstrate the versatility of using the thiol-ene reaction to tailor the thermal properties of dendrimers, dendrimer **49** was functionalized with 11-mercapto-1-undecanol and (monoethylether triethyleneglycol)-3-mercaptopropionate using the thiol-ene reaction (Scheme 6.26).

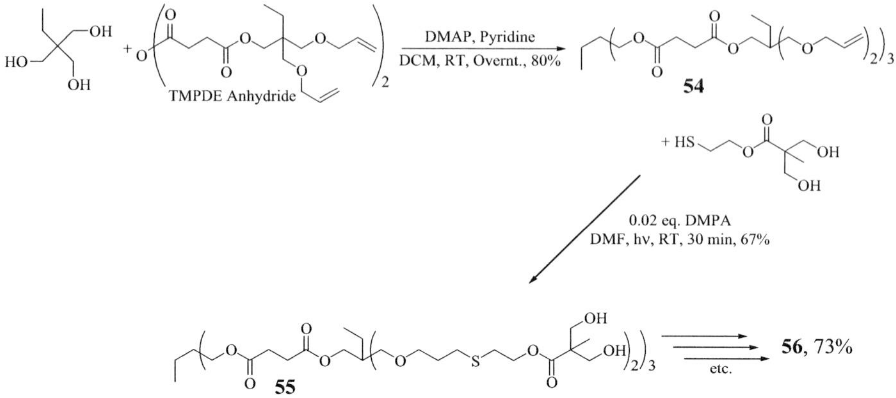


**Scheme 6.25** Synthesis of a fifth generation alkene-terminated dendrimer using TMPDE anhydride and 2,2-bis(hydroxymethyl)-2-mercapto-ethylpropanoate.

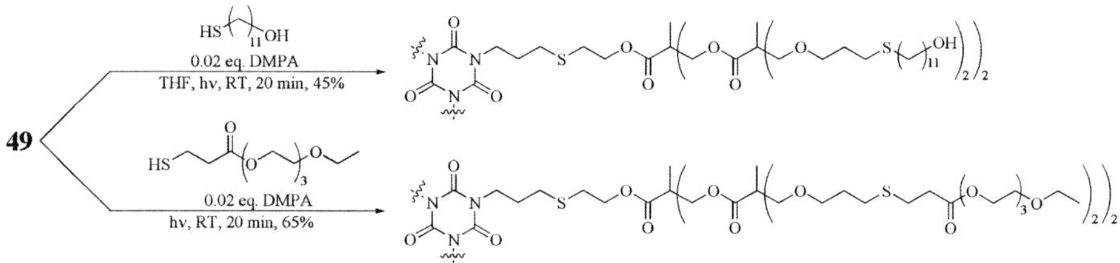


**Scheme 6.26** Functionalization of dendrimer **49** *via* the thiol-ene reaction.

The adduct formed by reaction with 11-mercapto-1-undecanol had a higher $T_g$ than the starting material, dendrimer **49**, while the adduct formed by reaction with monoethylether triethyleneglycol)-3-mercaptopropionate had a lower $T_g$ value.

This demonstration was expanded upon by reacting a fourth generation alkene-terminated dendrimer synthesized by the methods shown in Scheme 6.23 with 3-mercaptopropionic acid, benzyl mercaptan, and (9-fluorenylmethyloxycaronyl)-L-cysteine. The thermal properties of the resulting dendrimers were then analyzed by DSC (Scheme 6.27).

The adducts synthesized in Scheme 6.27 all had $T_g$ values that were higher than the unfunctionalized fourth generation alkene-terminated dendrimer.

As a demonstration of the potential applications of their dendritic materials, the authors used a third generation alkene-terminated dendrimer that was synthesized by the methods shown in Scheme 6.25 as a crosslinker with dithiol PEG in an effort to obtain a hydrogel. The authors were able to use equimolar amounts of the starting reagents and a small amount of DMPA as a photo-initiator. The authors obtained a white hydrogel that fully degraded at a pH of 11 within three days and was stable at a pH of 1 for 12 days.

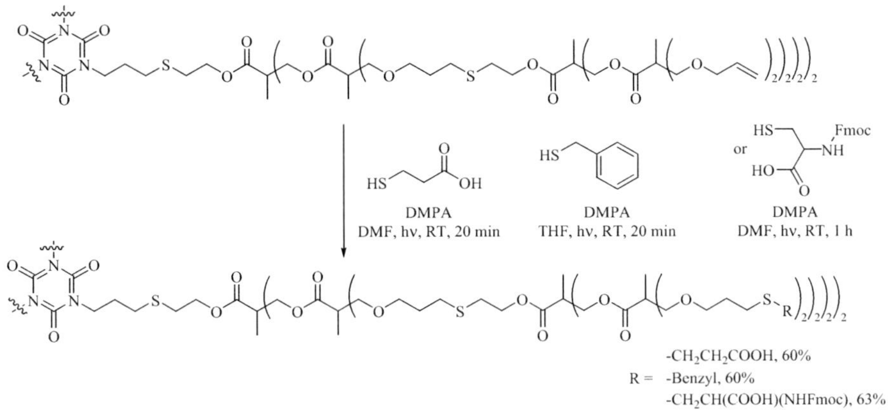


**Scheme 6.27** Functionalization of a fourth generation alkene-terminated dendrimer *via* the thiol-ene reaction.

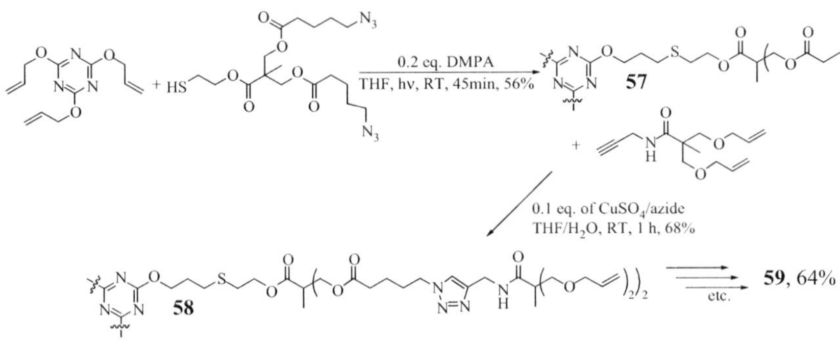


**Scheme 6.28** Synthesis of a sixth generation alkene-terminated dendrimer using the thiol-ene and CuAAC reactions.

Another example published in 2010 by Malkoch, Hawker, and coworkers employed the thiol-ene reaction followed by the copper catalyzed azide alkyne cycloaddition (CuAAC) in an AMA process to synthesize a sixth generation alkene-terminated dendrimer.[27] The first step in the synthesis involves the reaction of 2,2-bis(hydroxylmethyl)-propionic acid (bis-MPA) with an excess of 2-((2-mercaptoethoxy)carbonyl)-2-methylpropane-1,3-diyl-bis-(5-azidopentanoate) *via* the thiol-ene reaction using 0.2 equivalents of DMPA (per alkene) as an initiator. The first generation azide-terminated dendrimer, **57**, could then undergo the CuAAC reaction with a slight excess of 3-(allyloxy)-2-(allyloxymethyl)-2-methyl-*N*-(prop-2-ynyl)propanamide to yield compound **58**. Repeating these two steps yielded a sixth generation allyl-terminated dendrimer, **59** (Scheme 6.28).

The progression of the dendrimer growth shown in Scheme 6.28 was monitored by $^1$H NMR and FT-IR spectroscopy, and each purified generation could be obtained in a few hours. Characterization of the first four generations

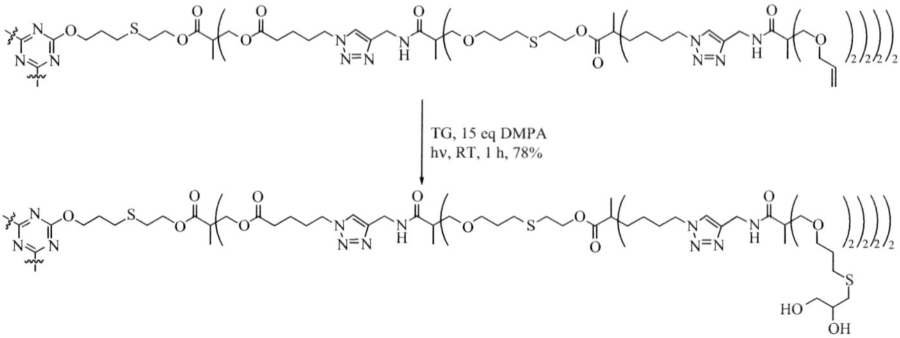


**Scheme 6.29** Functionalizing a fourth generation alkene-terminated dendrimer with TG.

of the dendrimers by MALDI-TOF MS showed that the experimental values were within agreement of the expected values. Deviations by the fifth and sixth generation could be attributed to the linear mode used. Moreover, analysis by GPC showed that the first four generations were monodisperse, and the fifth and sixth generations had larger PDI values as a result of dendrimer–dendrimer cross coupling. The thermal properties of the dendrimers synthesized in Scheme 6.28 were analyzed by thermogravimetric analysis (TGA) and DSC. The azide-terminated dendrimers exhibited low temperature degradation while the alkene-terminated dendrimers exhibited a higher onset of thermal degradation.

The authors further utilized the thiol-ene reaction to functionalize a fourth generation alkene-terminated dendrimer that was synthesized by the methods shown in Scheme 6.28 to prepare a hydrophilic dendrimer (Scheme 6.29).

An extension of this work was published by Hawker *et al.* in 2011.[28] The authors employed the thiol-ene reaction to functionalize a fourth generation alkene-terminated dendrimer that was synthesized by the methods shown in Scheme 6.28 (Scheme 6.30).

Furthermore, in 2010 Ortiz and co-workers made use of the thiol-ene reaction for the divergent synthesis of a second generation thiol-terminated dendrimer.[29] In the first step, the authors reacted pentaerythritol tetrakis-mercaptopropionate (PETKMP) with 3-allyloxy-1,2-propanediol (APD) using DMPA as a photoinitiator. The authors performed the thiol-ene reaction in a UV oven using a 300 W fusion lamp. Compound **60** was then reacted with an excess of mercaptopropionic acid (MPA) to yield a first generation thiol-terminated dendrimer, **61**. Repetition of these two steps yielded **62**, a second generation thiol-terminated dendrimer (Scheme 6.31).

With dendrimers **61** and **62** in hand, the authors investigated using these thiol-terminated dendrimers along with PETKMP for the photopolymerization of the following monomers: triallyl cyanurate, 1,6-hexanedioldiacrylate, and pentaerythritol triacrylate. The authors discovered that dendrimer **62** had a higher reactivity towards photopolymerization with respect to PETKMP and dendrimer **61**. Analysis of the resulting photopolymers by dynamic mechanical

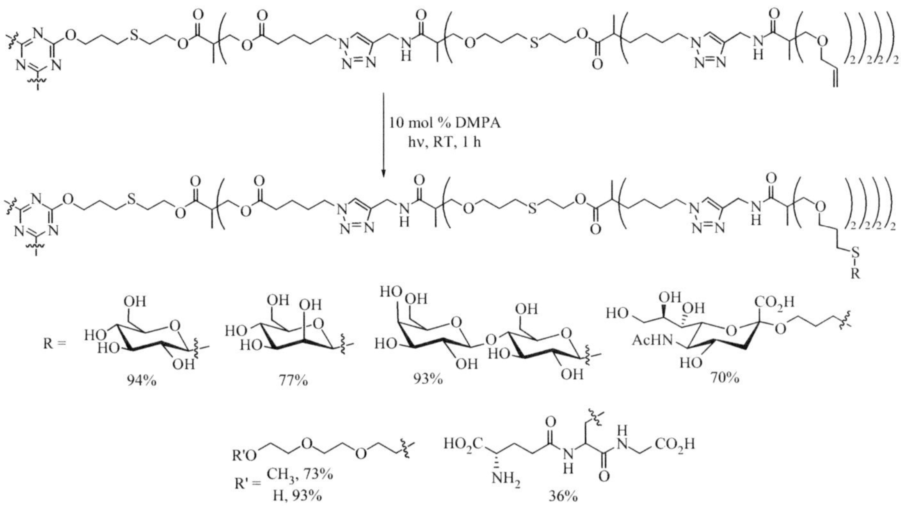

**Scheme 6.30** Functionalization a fourth generation alkene-terminated dendrimer with various thiol containing compounds.

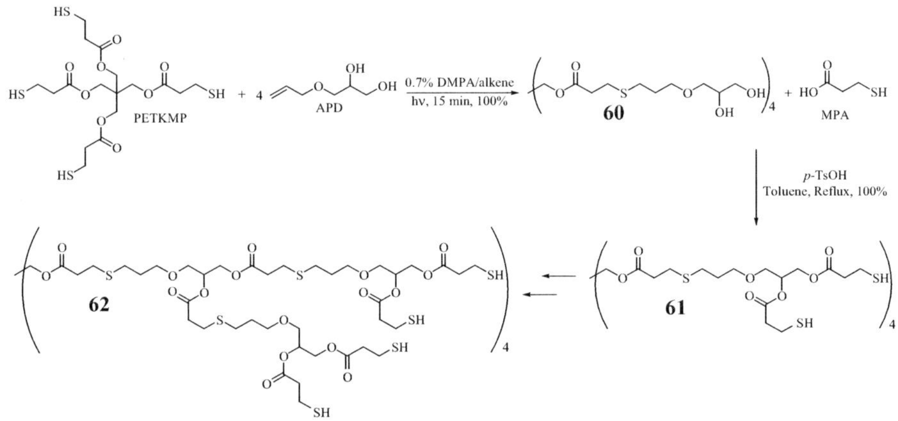

**Scheme 6.31** Synthesis of a second generation thiol-terminated dendrimer using APD and MPA.

analysis (DMA) revealed that the polymers synthesized using **62** were very flexible and had the lowest $T_g$ values in comparison to those synthesized using PETKMP and **61**. Analysis by TGA showed that the dendrimer-derived polymers also exhibited high thermal stability.

Sangermano and co-workers presented an example in 2011 in which the thiol-ene reaction was employed for the synthesis of dendritic molecules.[30] To synthesize the dendritic molecules, 1,1,1-tris(hydroxymethyl) propane (TMP) was reacted with 10-undecenoyl chloride via an esterification reaction to produce a first generation alkene-terminated dendrimer, **63**. Dendrimer **63** was then reacted with TG *via* the thiol-ene reaction to produce the first generation hydroxyl-terminated dendrimer, **64**. Repeating these two steps using an AMA process, a third generation hydroxyl-terminated dendrimer was synthesized, **65** (Scheme 6.32).

Dendrimer **65** was characterized by NMR spectroscopy, GPC, and DSC. The number-average molecular weights determined by $^{1}$H NMR spectroscopy and GPC were in agreement with each other but varied from the theoretical value. This was attributed to the lack of complete esterification during the dendrimer synthesis. The molecular weight distribution was found to be 1.35. Analysis by TGA showed a main process of degradation occurring around 300 °C. After synthesizing and characterizing compound **65**, it was used as a crosslinker to thermally cure epoxy resins. The resulting thermosets had a $T_g$ value that was higher than the cured epoxy material alone. Furthermore, it was found that the rigidity of the cured materials increased as the amount of **65** used increased.

Later that same year, Malkoch *et al.* used the thiol-ene reaction to synthesize a variety of symmetric and asymmetric dendritic molecules, as well as dendronized linear polymers.[31] In the first instance, the authors employed a divergent method to synthesize symmetrical dendrimers. A variety of branches were synthesized from disulfide-based dendrimers that were synthesized in house. Figure 6.2 shows the branches that were used, compounds **66–70**, and

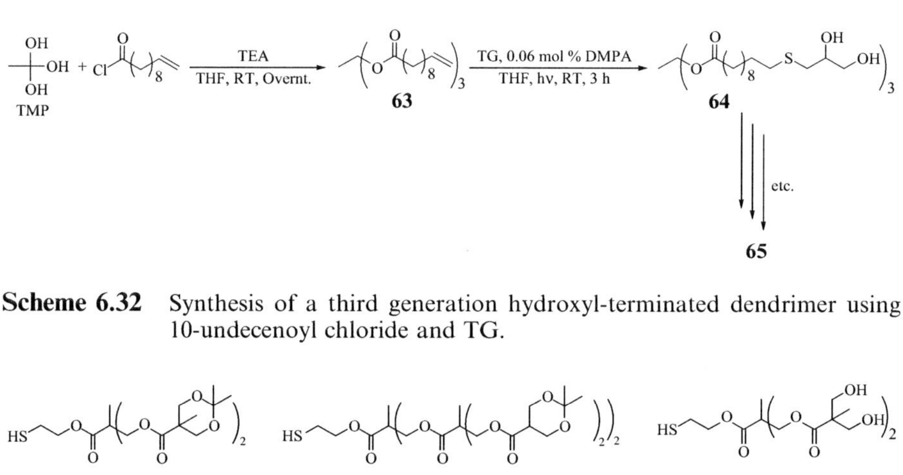


**Scheme 6.32** Synthesis of a third generation hydroxyl-terminated dendrimer using 10-undecenoyl chloride and TG.

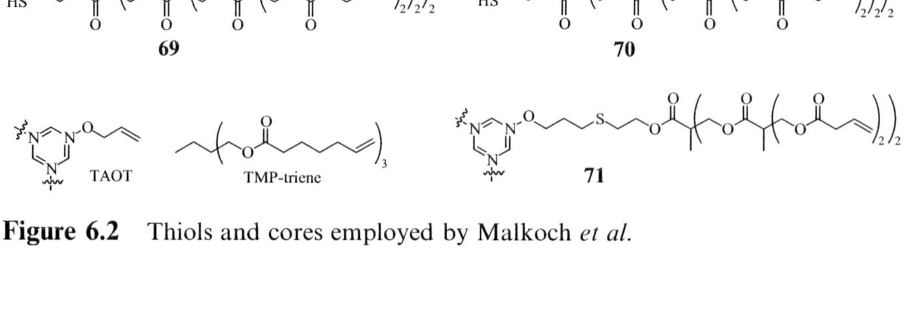


**Figure 6.2** Thiols and cores employed by Malkoch *et al.*

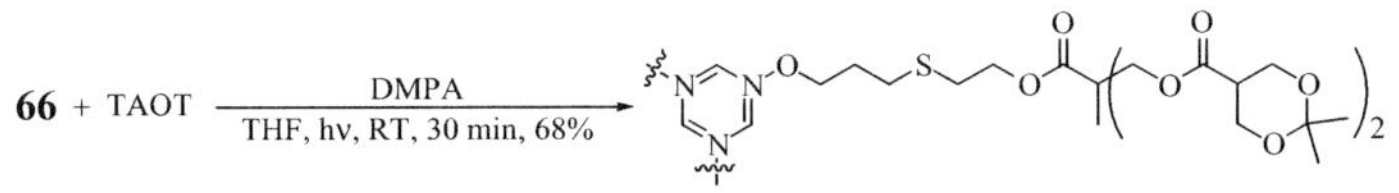

**Scheme 6.33** Free radical thiol-ene reaction between compound **66** and TAOT.

**Table 6.2** Summary of results for the symmetric dendrimers synthesized by Malkoch *et al.*

| *Core* | *Branch* | *Rxn Time* | *Solvent* | *Yield* | *PDI* |
|---|---|---|---|---|---|
| TAOT | **66** | 30 min | THF | 68% | 1.06[a] |
| TAOT | **67** | 30 min | THF | 62% | 1.05[a] |
| TAOT | **68** | 40 min | MeOH | 60% | |
| TAOT | **70** | 50 min | MeOH | 70% | |
| **71** | **66** | 30 min | THF | 74% | 1.05[a] |
| TMP-triene | **69** | 45 min | THF | 58% | |

[a]Value determined by SEC.

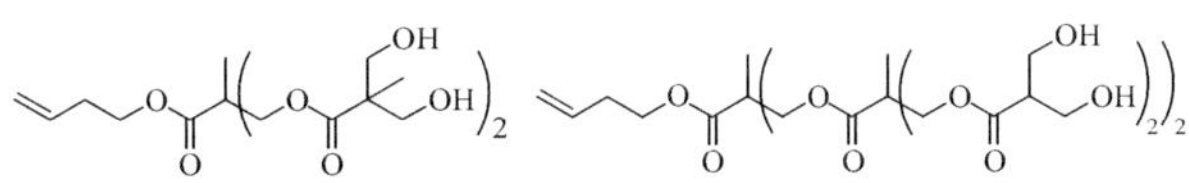

**72** **73**

**Figure 6.3** Branches used in the synthesis of asymmetric dendrimers.

the three core molecules that were employed, 2,4,6-tryallyloxy-1,3,5-triazine (TAOT), trimethylolpropane-triene (TMP-triene), **71**.

The cores, TAOT, TMP-triene, or **71**, were reacted with an excess of one of the thiol-containing reagents, using DMPA as an initiator to produce the symmetric dendrimers. The reactions were monitored and the products were characterized by NMR spectroscopy, MALDI-TOF MS and SEC. Scheme 6.33 shows an example of the dendrimer synthesis using TAOT and compound **66**.

Table 6.2 summarizes the results obtained from the symmetric dendrimers.

Similar methods were employed to synthesize the asymmetric dendritic molecules. In this study, the authors reacted compound **66** and **67** from Figure 6.2 with the two branches shown in Figure 6.3 using the thiol-ene reaction, producing asymmetric dendrimers.

For the synthesis of asymmetric dendrimers, similar methods were used to those shown in Scheme 6.33. An excess of the thiol containing reagent was used, and DMPA was used as an initiator. Table 6.3 summarizes the results from the synthesis of the asymmetric dendrimers. The resulting dendrimers were characterized by NMR spectroscopy, MALDI-TOF MS and SEC.

In their final example, the authors utilized the thiol-ene reaction to functionalize linear polymers to produce dendronized linear polymers. Malkoch and co-workers functionalized three linear polymers, allylated-poly(hydroxyethylmethylacrylate) (pHEMA), diallyl poly(ε-caprolactone)

**Table 6.3** Summary of results for the asymmetric dendrimers synthesized by Malkoch *et al.*

| *Branch 1* | *Branch 2* | *Rxn Time* | *Solvent* | *Yield* | *PDI* |
|---|---|---|---|---|---|
| **66** | **72** | 15 min | ACN | 73% | 1.08[a] |
| **67** | **73** | 1 h | DMF | 76% | 1.03[a] |

[a]Value determined by SEC.

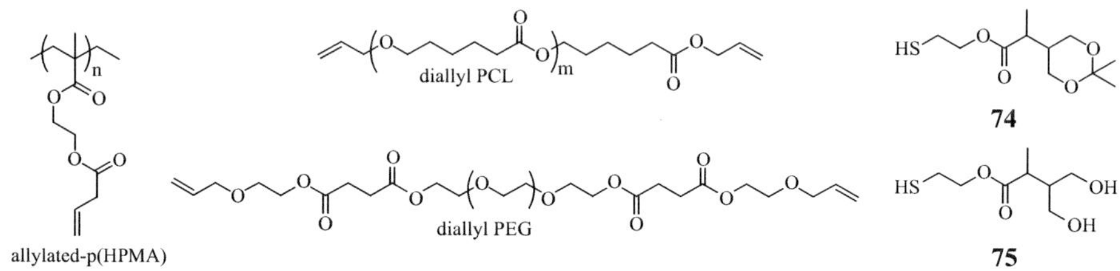


**Figure 6.4** Compounds used in the synthesis of dendronized linear polymers.

**Table 6.4** Summary of results for the dendronized linear polymers synthesized by Malkoch *et al.*

| *Polymer* | *Branch* | *Rxn Time* | *Solvent* | *Yield* | *PDI* |
|---|---|---|---|---|---|
| Allylated-p(HEMA) | **74** | 20 min | THF | 80% | 1.3[a] |
| Allylated-p(HEMA) | **66** | 1 h | THF | 68% | 1.2[a] |
| Allylated-p(HEMA) | **67** | 1.5 h | THF | 61% | |
| Allylated-p(HEMA) | **75** | 15 min | MeOH | 76% | |
| Diallyl PCL | **69** | 30 min | THF | 56% | 1.05[a] |
| Diallyl PEG | **68** | 30 min | $H_2O$ | 68% | 1.1[a] |

[a]Value determined by SEC.

(PCL) and diallyl PEG with compounds **66**, **67**, **68** and **69** compound **72** from Figure 6.3, and compounds **74** and **75** from Figure 6.4.

For the synthesis of dendronized linear polymer, similar methods were used to those shown in Scheme 6.33. In all cases, an excess of the thiol containing component was used and DMPA was employed as an initiator. Table 6.4 summarizes the results obtained. The dendronized polymers were characterized using NMR spectroscopy, MALDI-TOF MS and SEC.

## 6.4 The Thiol-yne Reaction

One of the first examples of using the thiol-yne reaction for the synthesis of hyperbranched molecules was presented in 2009 by Stenzel *et al.*[32] In the first step, the authors performed the thiol-yne reaction between excess TG and a tri-alkynyl core using DMPA as a photoinitator. The first generation hydroxyl-terminated dendrimer, **76**, was then reacted with an acetylenic anhydride *via* an esterification reaction to produce compound **77**. Repeating these steps *via* an

AMA process, the authors were able to obtain third generation hydroxyl-terminated dendrimer, **78** (Scheme 6.34).

The dendrimers were characterized using $^1$H NMR spectroscopy and GPC. The dendrimers were found to be monodisperse and symmetric. Analysis by MALDI-TOF MS revealed that the experimental molecular weights were in agreement with the expected values.

To expand further on the versatility and applications of the thiol-yne reaction, the authors synthesized a carboxylic acid-terminated dendrimer from dendrimer **77** using thioglycolic acid (Scheme 6.35).

The carboxylic acid dendrimer was then validated as a potential carrier for *cis*-dichlorodiammineplatinum(II) or CDDP. It was found that carboxylic acid dendrimer could be 86% loaded with CDDP through conjugation.

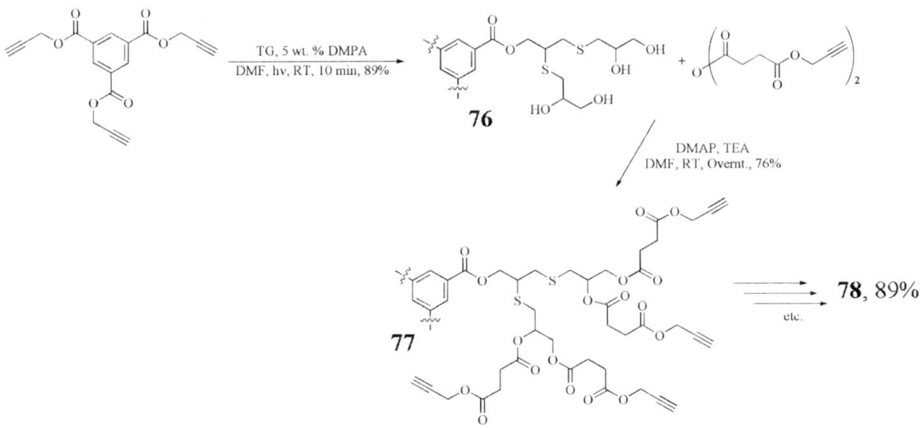


**Scheme 6.34** Synthesis of a third generation hydroxyl-terminated dendrimer using thiol-yne and esterification reactions.

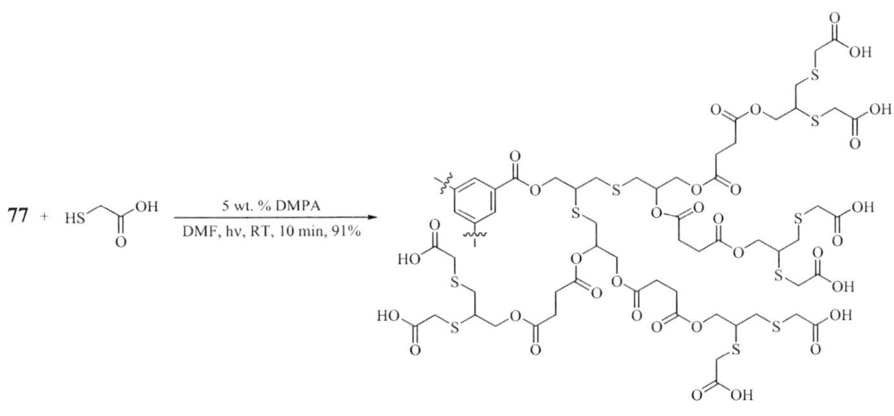


**Scheme 6.35** Functionalization of dendrimer **77** with carboxylic acid groups using the thiol-yne reaction.

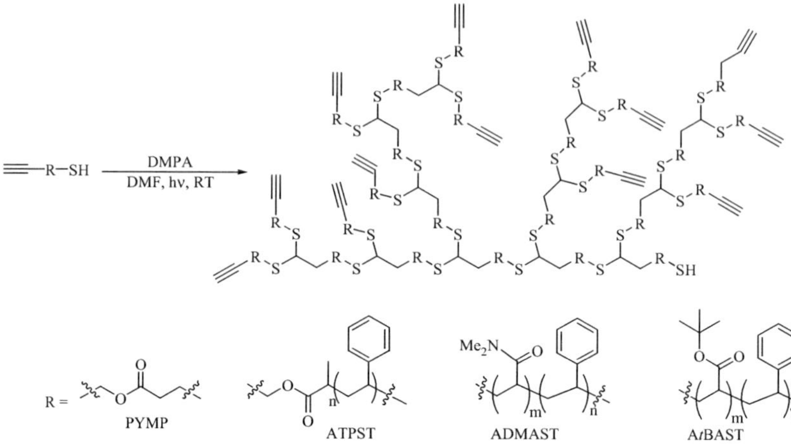


**Scheme 6.36** Synthesis of hyper-branched polymers from PYMP, ATPST, ADMAST, and A*t*BAST using the thiol-yne reaction.

In late 2009 and early 2011, Perrier and co-workers used the thiol-yne reaction to synthesize hyperbranched polymers.[33,34] In their first example, two different polymeric materials were synthesized, one using prop-2-ynyl-3-mercaptopropanoate (PYMP) as a monomer and one using an alkyne-terminated polystyrene-thiol (ATPST). In addition, the authors synthesized two diblock copolymers, alkyne-dimethyl acrylamide styrene thiol (ADMAST) and alkyne *tert*-butyl acrylate-styrene thiol (A*t*BAST). The authors employed the thiol-yne reaction, using DMPA as an initiator, to synthesize the hyperbranched polymers from PYMP, ATPST, ADMAST, and A*t*BAST (Scheme 6.36).

$^{1}$H NMR spectroscopy, universal calibration, triple detection, and differential refractive index (DRI) were used to characterize the hyperbranched polymers synthesized from PYMP and ATPST. For both polymers it was found that high molecular weight species were obtained as the reactions proceeded, the hyperbranched polymers had high polydispersity indices, and there were very few partially reacted groups present on the polymers.

SEC, viscometric and DRI-SEC were used to characterize the hyperbranched polymers that were synthesized from ADMAST and A*t*BAST and revealed that the formation of high molecular weight polymers was successful. The self-aggregation of the ADMAST and A*t*BAST hyperbranched polymers was also studied. The ADMAST hyperbranched polymers were found to form aggregates with an average size of 100 nm in a 1 : 4 acetone:water solution. After selectively hydrolyzing the *tert*-butyl ester groups of the A*t*BAST hyperbranched polymer, the products were found to form aggregates with an average size of 500 nm at a low pH and 10 nm at a high pH. Table 6.5 summarizes the results from the studies performed by Perrier and co-workers.

To demonstrate the applications of the hyperbranched polymers synthesized in Scheme 6.36, Perrier *et al.* reacted the hyperbranched polymer synthesized

**Table 6.5** Summary of results from Perrier *et al.*

| *Compound* | *DMPA* | *Time* | *Conversion* | *PDI* |
|---|---|---|---|---|
| PYMP | 3 wt. % | 1.5 h | 99 %[a] | 11.94[b] |
| ATPST | 5 wt. % | 3 h | 95 %[a] | 16.32[b] |
| ADMAST | 5 wt. % | 5 h | > 90 %[c] | 3.5[c] |
| A*t*BAST | 5 wt. % | 5 h | > 90 %[c] | 7.4[c] |

[a]Determine by[a] $^1$H NMR Spectroscopy.
[b]UC.
[c]SEC.

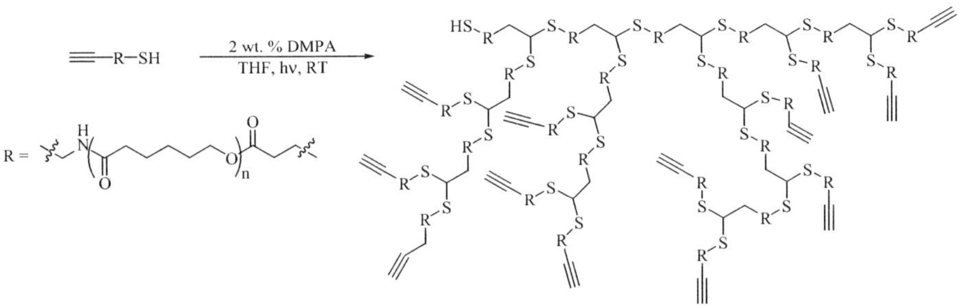


**Scheme 6.37** Synthesis of hyper-branched polymers from a α-thiol and ω-alkyne-terminated PCL polymer using the thiol-yne reaction.

from PYMP with [Au(IPr)(OH)] to produce gold functionalized polymers.[35] The luminescent properties of the resulting gold functionalized materials were investigated and found to have larger quantum yields than typical gold-alkynyl complexes.

Another example of using the thiol-yne reaction for the synthesis of hyper-branched polymers was demonstrated in 2010 by Liu and Dong.[36] In this example, the authors synthesized a PCL polymer with α-thiol and ω-alkyne terminal groups using a controlled ring-opening polymerization and used it to make hyperbranched polymers (Scheme 6.37).

The hyperbranched polymers synthesized in Scheme 6.37 were characterized by NMR spectroscopy, FT-IR spectroscopy, DSC and GPC. It was found that the hyperbranched polymers had a degree of branching close to 1 and the maximum melting and crystallization points decreased with increasing molecular weight. Furthermore, as the photopolymerization proceeded, the molecular weight increased, as did the PDI of the resulting hyperbranched polymers, with the exception of reactions that were irradiated for longer than 30 minutes. This deviation was attributed to cross-linking. As a result, a PCL polymer with α-thiol and ω-alkyne terminal groups was threaded with α-cylcodextrin, making the resulting polymer more rigid. The threaded polymer was used to make hyperbranched polymers using similar methods to those shown in Scheme 6.37. The hyperbranched polymers were again characterized using GPC. It was found that no cross-linking had occurred, even after three hours of irradiation.

In 2011, Albertazzi, Hawker, and co-workers synthesized dendritic molecules using a divergent method employing the thiol-yne reaction.[37] Bis-amine terminated PEG was reacted with an excess of propargyl glycidyl ether (PGE) to produce compound **79**, a first generation alkyne-terminated dendrimer. Compound **79** could then undergo the thiol-yne reaction with an excess of CS-HCl, to produce a second generation amine-terminated dendrimer, **80**. Repeating these steps in an AMA process produced a fourth generation amine-terminated dendrimer, **81** (Scheme 6.38).

Additionally, a fourth generation amine-terminated dendrimer with coumarin functionalized interior groups could be obtained through an esterification process from dendrimer **80** (Scheme 6.39).

The dendrimers shown in Scheme 6.38 and Scheme 6.39 were characterized by NMR spectroscopy, MALDI MS and GPC. The molecular weights found experimentally were within agreement of the theoretical values. The PDI values showed that the dendrimers were reasonably monodisperse.

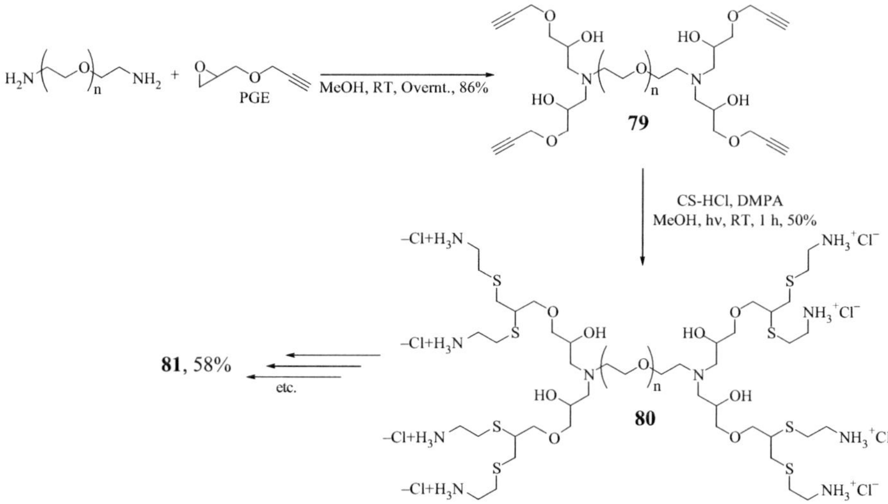


**Scheme 6.38** Synthesis of a fourth generation amine-terminated dendrimer, **81**, using AGE and CS-HCl.

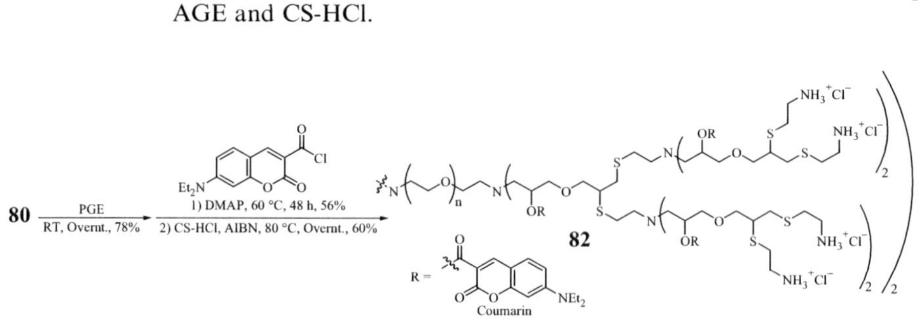


**Scheme 6.39** Functionalization of the interior hydroxyl groups of dendrimer **80** using an esterification reaction.

To demonstrate the ability of using their dendrimers as delivery vessels, fluorescence spectroscopy was used to monitor the functionalized dendrimer from Scheme 6.39 in the absence and in the presence of porcine liver esterase (PLE). In the absence of PLE, the functionalized dendrimer quenched fluorescence. In contrast, when PLE was present an enhancement was observed, as a result of the ester bonds being hydrolyzed. These results were also verified by HPLC experiments. Furthermore, compound **81** was labeled with fluorescein isothiocyanate (FITC) and Alexa647, while compound **82** was labeled with Alexa647, to study the uptake of the dendritic structures by B16 mouse melanoma cells. Analysis of the labeled dendrimers by confocal fluorescence microscopy revealed that compounds **81** and **82** had been internalized by the B16 cells, and that the coumarin payload of compound **82** could be released into the cytoplasm by intracellular enzymes.

## 6.5 Sequential Thiol-ene/Thiol-yne Reactions

In 2009, Lowe, Hoyle, and co-workers demonstrated the versatility and benefits to employing both the thiol-Michael reaction and the thiol-yne reaction for the synthesis of dendritic molecules.[38] In a preliminary study, the authors reacted propargyl acrylate with mercaptoethylester using DMPP as an initiator. Compound **83** was obtained in quantitative yield and reacted with two equivalents of TG *via* the thiol-yne reaction, employing DMPA as an initiator, to produce compound **84** (Scheme 6.40). Both reactions were performed under ambient and bulk conditions.

After successfully completing the preliminary study, Lowe and co-workers extended the sequential thiol-ene and thiol-yne reactions to the following thiols: 3-mercaptopropionic acid (**85**), 6-mercaptohexanol (**86**), mercaptosuccinic acid (**87**), 3-(trimethoxysilyl)propanethiol (**88**), captopril (**89**), adamantane-1-thiol (**90**), and 3-mercaptopropyl polyhedral oligomeric silsesquioxane (**91**) (Figure 6.5).

Initially, tetrathiol pentaerithrytol tetra(3-mercaptopropionate) was reacted with propargyl acrylate *via* the thiol-Michael reaction to produce compound **92**. Subsequently, compound **92** was reacted with thiols **85–91** using the thiol-yne reaction (Scheme 6.41). The reactions were performed under ambient and bulk conditions with the exception of the reactions performed with thiols **87**, **89**, **90**, and **91**, which required the use of a solvent. Purification of the products, if needed, could be performed by using simple washing/extraction techniques. No yields were reported.

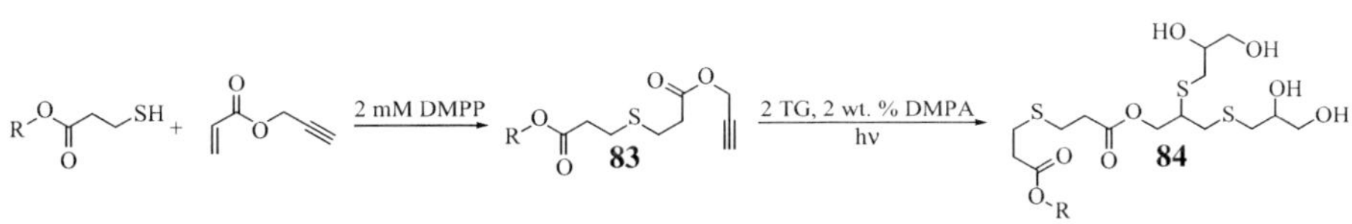


**Scheme 6.40** Preliminary study of the sequential thiol-Michael/thiol-yne reactions.

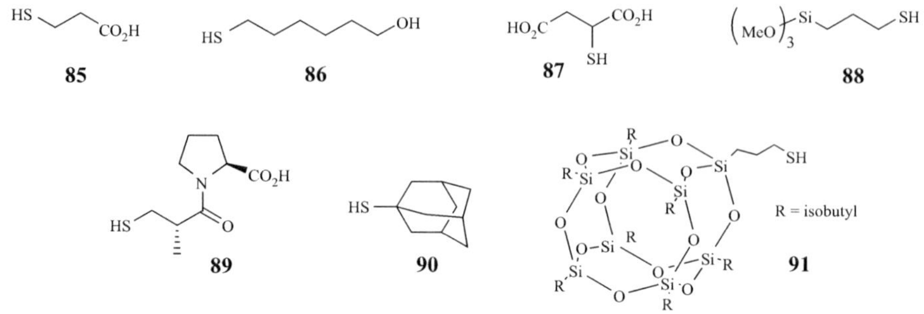


**Figure 6.5** Thiols employed in the sequential thiol-Michael/thiol-yne reactions.

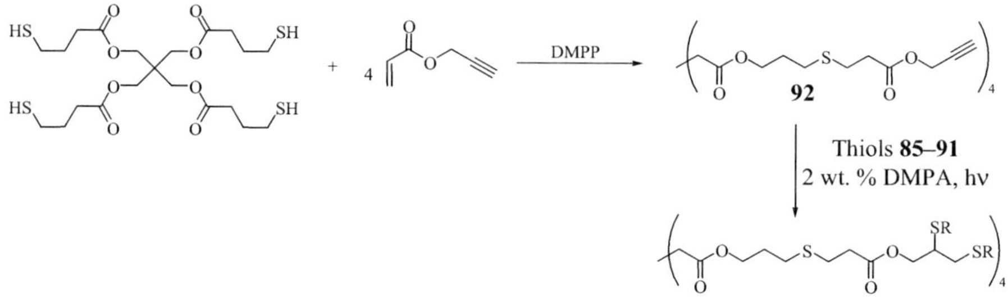


**Scheme 6.41** Synthetic strategy used for the sequential thiol-Michael/thiol-yne reactions with thiols **85–91**.

The compounds synthesized in Scheme 6.41 were characterized by NMR spectroscopy and MALDI-TOF MS. The data obtained from the MALDI-TOF MS spectra showed that the expected molecular weights agreed with those obtained experimentally. Products obtained using thiols **90** and **91** could not be analyzed by MALDI-TOF MS. Thus, FTIR spectroscopy was used to monitor the reactions and showed full consumption of the starting materials.

Another example of sequentially using the thiol-Michael reaction and the thiol-yne reaction was presented in 2011 by Gao *et al.*[39] In this instance, the authors employed the thiol-Michael reaction followed by the free-radical thiol-yne reaction to synthesize hyperbranched polymers. In the first step, 1,6-hexanedithiol or 3,6-dioxaoctane-1,8-dithiol was reacted with propargyl acrylate, using TEA as an initiator, to produce compounds **92** and **93**. After isolation, a solution of AIBN in toluene was used to initiate the free-radical thiol-yne reactions of compounds **92** and **93**, producing hyperbranched polymers (Scheme 6.42).

Using $^1$H NMR spectroscopy, it was determined that after three hours the hyperbranched polymers were 95.4% converted and had a degree of branching greater than 0.5. The PDI's of the hyperbranched polymers were determined by GPC and found to be 14.56 and 3.06 for the hyperbranched polymers synthesized from **92** and **93**, respectively. Compound **92** was also polymerized using DMPA as an initiator using irradiation. After irradiating for 30 min, the resulting hyperbranched polymer had a PDI of 7.62.

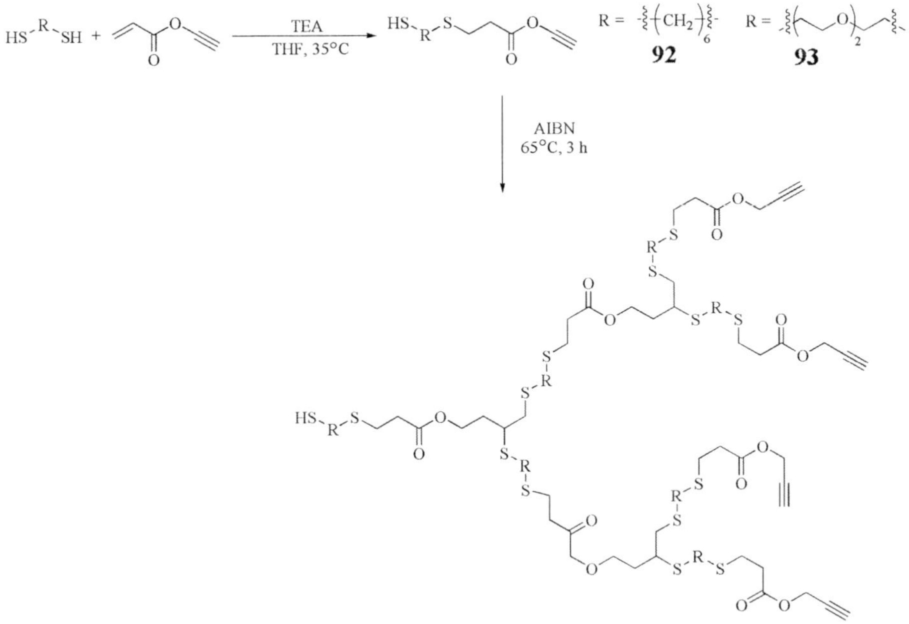

**Scheme 6.42** Synthesis of hyperbranched polymers using the sequential thiol-Michael/thiol-yne reactions.

## 6.6 Conclusions

Many advances have been made in the development, synthesis, and applications of dendritic and hyperbranched materials. The examples presented above demonstrate the versatility of the thiol-Michael, thiol-ene and the thiol-yne "click" reactions for constructing these types of materials. Moreover, some of the benefits observed through the utilization of these thiol-click reactions include mild reaction conditions, high yields of products, simple purification techniques, and functional group tolerance, with respect to more traditional syntheses. Thus, these thiol-click reactions have been shown to be viable choices for the design and synthesis of dendrimers and other hyperbranched materials.

## 6.7 List of Abbreviations

| | |
|---|---|
| Ac-Cl | acryloyl chloride |
| ADMAST | alkyne-dimethylacrylamide styrene thiol |
| AGE | allyl glycidyl ether |
| AMA | alternate multi-addition |
| AOI | 2-isocyanatoethyl acrylate |
| APD | 3-allyloxy-1,2-propanediol |
| ASMs | amphiphilic star-like macromolecules |

| | |
|---|---|
| A*t*BAST | alkyne *tert*-butyl acrylate-styrene thiol |
| ATPST | alkyne terminated polystyrene-thiol |
| BAPA | bis(allyl propionic acid) |
| BEI | 1,1-bis(acryloyloxymethyl)ethyl isocyanate |
| bis-MPA | 2,2-bis(hydroxymethyl)-propionic acid |
| CDDP | *cis*-dichlorodiammineplatinum |
| CS | cysteamine (2-mercaptoethylamine) |
| CS-HCl | cysteamine hydrochloride |
| CuAAC | copper catalyzed azide alkyne cylcoaddition |
| DBTDL | dibutyltin dilaurate |
| DCHA | dicyclohexylamine |
| DMA | dynamic mechanical analysis |
| DMAE | N,N-dimethylaminoethanol |
| DMPA | 2,2-dimethoxy-2-phenylacetophenone |
| DMPP | dimethylphenylphosphine |
| DRI | differential refractive index |
| DSC | differential scanning calorimetry |
| DSL | dynamic light scattering |
| ELISA | enzyme-linked immunosorbent assay |
| ESI | Electrospray ionization |
| Fmoc-SH | 9-fluorenylmethoxycarbonyl |
| FT-IR | Fourier transform infrared spectroscopy |
| GPC | gel-permeation chromatography |
| HexAM | hexylamine |
| IEMA | 2-isocyanatoethyl acrylate |
| LCST | lower critical solution temperature |
| LPEI | linear poly(ethylenimine) |
| MAEA | 2[(methacryloyl)oxy]ethyl acetate |
| MALDI-TOF MS | matrix assisted laser desorption/ionization time-of-flight mass spectrometry |
| ME | 2-mercaptoethanol |
| MPA | 3-mercaptopropionic acid |
| mPEG | methoxy polyethylene glycol |
| NT-SH | 2-naphthalenethiol |
| p(HEMA) | poly(hydroxyethylmethacrylate) |
| PCL | poly(ε-caprolactone) |
| PETA | pentaerythritol tetraacrylate |
| PETKMP | pentaerythritol (tetrakis)mercaptopropionate |
| PGE | propargyl glycidyl ether |
| PLE | porcine liver esterase |
| PYMP | prop-2-ynyl-3-mercaptopropanoate |
| RAFT | reversible addition-fragmentation chain transfer |
| RT | room temperature |
| SEC | size exclusion chromatography |
| TAOT | 2,4,5-troallyloxy-1,3,5-triazine |
| TAT | 1,3,5,-triacryloylhexahydro-1,3,5-triazine |

| | |
|---|---|
| TEA | triethylamine |
| TG | 1-thioglycerol |
| TGA | thermogravimetric analysis |
| TMP | 1,1,1-tris(hydroxymethyl)propane |
| TMPDE | trimethylolpropane diallyl-ether |
| TMPTA | trimethylolpropane triacrylate |
| TMP-triene | trimethylolpropane-triene |
| TPGDA | tripropyleneglycol diacrylate |

## References


1. U. Boas and P. M. H. Heegaard, *Chem. Soc. Rev.*, 2004, **33**, 43.
2. C. R. Yates and W. Hayes, *Eur. Polym. J.*, 2004, **40**, 1257.
3. C.-O. Turrin and A.-M. Caminade, in *Dendrimers*, ed. A.-M. Caminade, C.-O. T., R. Laurent, A. Ouali and B. Delavaux-Nicot, John Wiley & Sons, Ltd, Chichester, 2011, p. 393.
4. B. Forier and W. Dehaen, *Tetrahedron*, 1999, **55**, 9829.
5. H. Ihre, O. L. Padilla De Jesús and J. M. J. Fréchet, *J. Am. Chem. Soc.*, 2001, **123**, 5908.
6. M. C. Parrott, E. B. Marchington, J. F. Valliant and A. Adronov, *J. Am. Chem. Soc.*, 2005, **127**, 12081.
7. A. Dondoni, *Ang. Chem., Int. Ed.*, 2008, **47**, 8995.
8. R. Hoogenboom, *Angew. Chem., Int. Ed. Engl.*, 2010, **49**, 3415.
9. J. Wang, L. Tian, A. Argenti and K. E. Uhrich, *J. Bioact. Compat. Polym.*, 2006, **21**, 297.
10. J. W. Chan, B. Yu, C. E. Hoyle and A. B. Lowe, *Chem. Comm.*, 2008, 4959.
11. J. W. Chan, B. Yu, C. E. Hoyle and A. B. Lowe, *Polymer*, 2009, **50**, 3158.
12. J. J. Fury, E. F. Altenhofer and R. Sedaghat-Herati, *Tetrahedron Lett.*, 2009, **50**, 4205.
13. C. Rim and D. Y. Son, *Tetrahedron Lett.*, 2009, **50**, 4161.
14. X. Ma, J. Tang, Y. Shen, M. Fan, H. Tang and M. Radosz, *J. Am. Chem. Soc.*, 2009, **131**, 14795.
15. Y. Shen, X. Ma, B. Zhang, Z. Zhou, Q. Sun, E. Jin, M. Sui, J. Tang, J. Wang and M. Fan, *Chemistry*, 2011, **17**, 5319.
16. K. i. Aoki and K. Ichimura, *Chem. Lett.*, 2009, **38**, 990.
17. K. i. Aoki and K. Ichimura, *Bull. Chem. Soc. Jpn.*, 2011, **84**, 1215.
18. K. i. Aoki, A. Hiraoka and K. Ichimura, *J. Photopolym. Sci. Technol.*, 2010, **23**, 457.
19. J. Hierold, A. Gray-Weale and D. W. Lupton, *Chem. Commun.*, 2010, **46**, 6789.
20. Y.-M. Bao, X.-H. Liu, X.-L. Tang and Y.-S. Li, *J. Polym. Sci. Part A: Polym. Chem.*, 2010, **48**, 5364.
21. C. D. Heidecke and T. K. Lindhorst, *Chemistry*, 2007, **13**, 9056.
22. K. L. Killops, L. M. Campos and C. J. Hawker, *J. Am. Chem. Soc.*, 2008, **130**, 5062.
23. C. Rissing and D. Y. Son, *Organometallics*, 2009, **28**, 3167.

24. C. Rissing, Synthetic Approaches to New Multidentate Pyridyl Ligands and Sulfur-Containing Organosilicon Dendrimers, Ph.D., Southern Methodist University, August 2009.
25. T. Kang, R. J. Amir, A. Khan, K. Ohshimizu, J. N. Hunt, K. Sivanandan, M. I. Montanez, M. Malkoch, M. Ueda and C. J. Hawker, *Chem. Commun.*, 2010, **46**, 1556.
26. M. I. Montañez, L. M. Campos, P. Antoni, Y. Hed, M. V. Walter, B. T. Krull, A. Khan, A. Hult, C. J. Hawker and M. Malkoch, *Macromolecules*, 2010, **43**, 6004.
27. P. Antoni, M. J. Robb, L. Campos, M. Montanez, A. Hult, E. Malmström, M. Malkoch and C. J. Hawker, *Macromolecules*, 2010, **43**, 6625.
28. M. Lo Conte, M. J. Robb, Y. Hed, A. Marra, M. Malkoch, C. J. Hawker and A. Dondoni, *J. Polym. Sci., A.: Polym. Chem.*, 2011, **49**, 4468.
29. R. A. Ortiz, R. V. G. Flores, A. E. García Valdéz and M. L. B. Duarte, *Prog. Org. Coat.*, 2010, **69**, 463.
30. D. Foix, X. Ramis, A. Serra and M. Sangermano, *Polymer*, 2011, **52**, 3269.
31. M. V. Walter, P. Lundberg, A. Hult and M. Malkoch, *J. Polym. Sci., Part A: Polym. Chem.*, 2011, **49**, 2990.
32. G. Chen, J. Kumar, A. Gregory and M. H. Stenzel, *Chem. Commun.*, 2009, 6291.
33. D. Konkolewicz, A. Gray-Weale and S. Perrier, *J. Am. Chem. Soc.*, 2009, **131**, 18075.
34. D. Konkolewicz, C. K. Poon, A. Gray-Weale and S. Perrier, *Chem. Commun.*, 2011, **47**, 239.
35. D. Konkolewicz, S. Gaillard, A. G. West, Y. Y. Cheng, A. Gray-Weale, T. W. Schmidt, S. P. Nolan and S. Perrier, *Organometallics*, 2011, **30**, 1315.
36. W. Liu and C.-M. Dong, *Macromolecules*, 2010, **43**, 8447.
37. R. J. Amir, L. Albertazzi, J. Willis, A. Khan, T. Kang and C. J. Hawker, *Angew. Chem., Int. Ed. Engl.*, 2011, **50**, 3425.
38. J. W. Chan, C. E. Hoyle and A. B. Lowe, *J. Am. Chem. Soc.*, 2009, **131**, 5751.
39. J. Han, B. Zhao, Y. Gao, A. Tang and C. Gao, *Polym. Chem.*, 2011, **2**, 2175.

CHAPTER 7

# *Dendritic Polymers from Thiol–Yne Reactions*

RAPHAEL BARBEY AND SÉBASTIEN PERRIER*

Key Centre for Polymers & Colloids, School of Chemistry, Building F11, University of Sydney, NSW 2006, Australia
*Email: sebastien.perrier@sydney.edu.au

## 7.1 Introduction

Although developed in the 1930s, the radical-mediated addition of thiols to alkynes has attracted much attention recently following the seminal work on thiol–yne photopolymerisations by Bowman and coworkers in 2009.[1] Similar to the thiol–ene reaction, the thiol–yne mechanism proceeds *via* the addition of thiyl radicals to unsaturated carbon–carbon bonds.[1] As illustrated in Scheme 7.1, the thiol–yne process starts with the addition of a thiyl radical to an alkyne to form a carbon-centred radical vinyl sulphide species. After abstraction of a hydrogen atom from another thiol (which as a consequence regenerates another thiyl radical), the vinyl sulphide moiety can undergo a second, much more rapid thiyl radical addition to its unsaturated bond. Therefore, upon completion and in the presence of a sufficient amount of, each alkyne functional group reacts with two thiols to produce an asymmetric dithioether compound.

The reaction has been reported to proceed to high conversion without the need of any special reaction conditions such as high temperature, anhydrous solvents, or inert atmosphere.[1] The intrinsically branched structure of the final dithioether compound and the 'click' features of the reaction make the

RSC Polymer Chemistry Series No. 6
Thiol-X Chemistries in Polymer and Materials Science
Edited by Andrew B. Lowe and Christopher N. Bowman

Published by the Royal Society of Chemistry, www.rsc.org

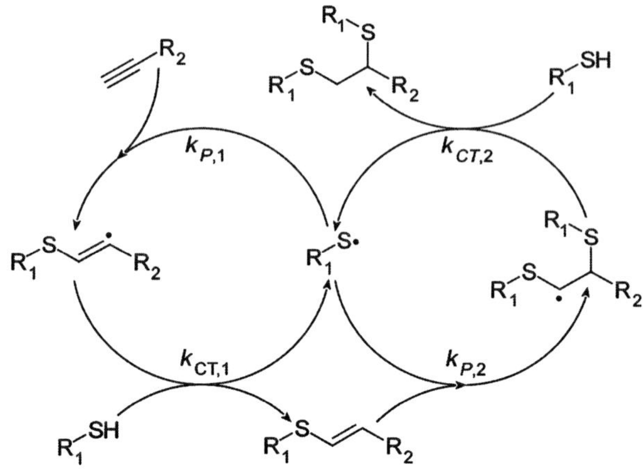


**Scheme 7.1** Thiol–yne reaction mechanism.
Reprinted with permission from ref. 1. Copyright 2009 American Chemical Society.

thiol–yne mechanism especially suitable for the preparation of branched materials, such as (hyper)branched polymers or dendrimers.

This chapter concentrates exclusively on branched materials produced by means of thiol–yne reactions and is an attempt to summarize the state-of-the-art in this field. The following two sections will discuss the synthesis of dendrimers and dendritic polymers *via* thiol–yne chemistry, as well as their functionalisation and potential applications.

## 7.2 Dendrimer Synthesis

Since their discovery in the early 1980s, dendrimers have been thoroughly studied because of their unique branching pattern, shape, structural symmetry, and high chain-end functionalities.[2–4] These structural features translate into specific properties, which render them particularly attractive candidates for a wide range of applications, including in drug delivery, catalysis and biosensing.[5] However, the preparation of such strictly regular generational structures is challenging from a synthetic point-of-view and typically requires tedious, sequential syntheses and intermediate purification,[6] which make their industrial production economically unattractive. Therefore, one of the main challenges for the preparation of dendrimers resides in the development of more practical and more cost-effective synthetic pathways. Amongst others, there is a need for reactions that proceed efficiently and quantitatively, in order to limit the number of purification procedures, without their efficiency being altered by the increasing number of functionalities at each generational step. The recent advent of the highly-efficient and quantitative 'click' reactions, as defined by Sharpless,[7] has opened a new avenue to the fast and controlled production of dendrimers. For example, click-type reactions, including Diels–Alder, Cu$^{I}$-catalysed azide–alkyne cycloaddition (CuAAC), and thiol–ene coupling have recently attracted a lot of interest for the preparation of such macromolecules.[8] Among the techniques employing 'click' chemistry reported to date, and of particular significance to this chapter, is the work by Hawker and coworkers.[9] In their report, the authors described the synthesis of fourth generation polythioether dendrimers *via* a divergent approach by successive sequences of photoinitiated thiol–ene radical reactions and base-catalysed esterification processes (Scheme 7.2).

As with the well-known thiol–ene reaction, thiol–yne chemistry has many of the attributes of 'click' chemistry and can therefore also be utilised for the accelerated synthesis of dendrimers. Advantages of the thiol–ene/yne reactions compared to CuAAC include the absence of a metal catalyst that may contaminate the final product, the ability to undertake reactions at room temperature and the tuning of the reaction by controlling the amount of UV irradiation. A further important advantage offered by the thiol–yne addition in comparison to the other typical 'click' approaches is the ability to introduce twice as many functionalities at each generational step, since each alkyne moiety can react with two thiols.

Stenzel and coworkers took full advantage of these features and adapted the strategy originally developed by Hawker for thiol–ene reactions[9] to prepare a third generation dendrimer with 192 hydroxyl functionalities. The authors used successive photoinitiated radical thiol–yne coupling reactions with 1-thioglycerol, coupled with base-catalysed esterification reactions between the pendant hydroxyl groups and acetylene anhydride (Scheme 7.3).[10] They were able to produce a dendrimer with 48 hydroxyl functional groups in only three steps ([G2]–$OH_{48}$), whereas seven synthetic steps are needed to obtain the same number of functionalities ([G4]–$OH_{48}$) with the thiol–ene-based method (Scheme 7.4). Li and coworkers also reported on the preparation of dendrimers by thiol-yne chemistry. Their approach consisted of the successive aza-Michael addition of but-3-ynyl acrylate to primary amine groups and thiol-yne reaction with cysteamine hydrochloride.[11]

The ability to introduce a high functional group density at the corona of the dendritic structure is clearly advantageous for the development of commercial products such as drug-delivery carriers, catalyst supports, and biosensors. In addition, decreasing the number of synthetic and purification procedures is more cost-effective and less time-consuming, which render dendrimers synthesised by this efficient means more affordable from an industrial point-of-view. Furthermore, the versatility of thiol–yne chemistry permits the introduction of a wide range of functional groups on the corona of the dendrimers by simply choosing the corresponding functional thiol, thus expanding dramatically the number of possible applications. For instance, Stenzel and coworkers introduced carboxylic acid groups on the corona of a dendrimer for application as drug delivery vehicles.[10] Dendrimers decorated with 24 carboxylic acid functionalities ([G2]-$COOH_{24}$) were easily prepared *via* photoinitiated radical thiol–yne reaction between thioglycolic acid and [G1]-$yne_{12}$, a first generation dendrimer containing twelve alkyne end-groups. An anti-cancer drug, *cis*-diamminedichloroplatinum (CDDP, also referred to by its trade name cisplatin), was conjugated to the carboxylic acid-terminated dendrimer. A promising loading capacity of 86% (compared to the theoretical maximal loading capacity) was achieved, making these dendrimers attractive candidates for the development of drug-delivery systems. In another example, the same group

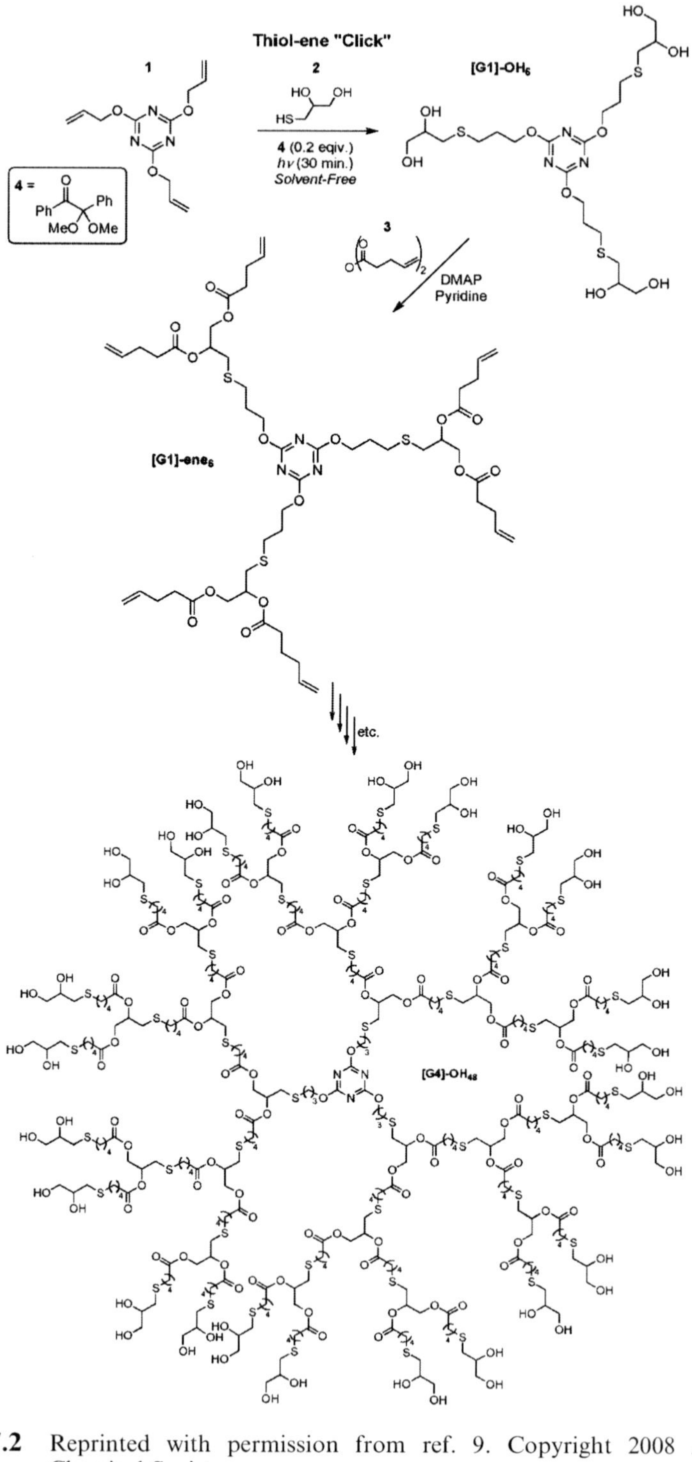


**Scheme 7.2** Reprinted with permission from ref. 9. Copyright 2008 American Chemical Society.

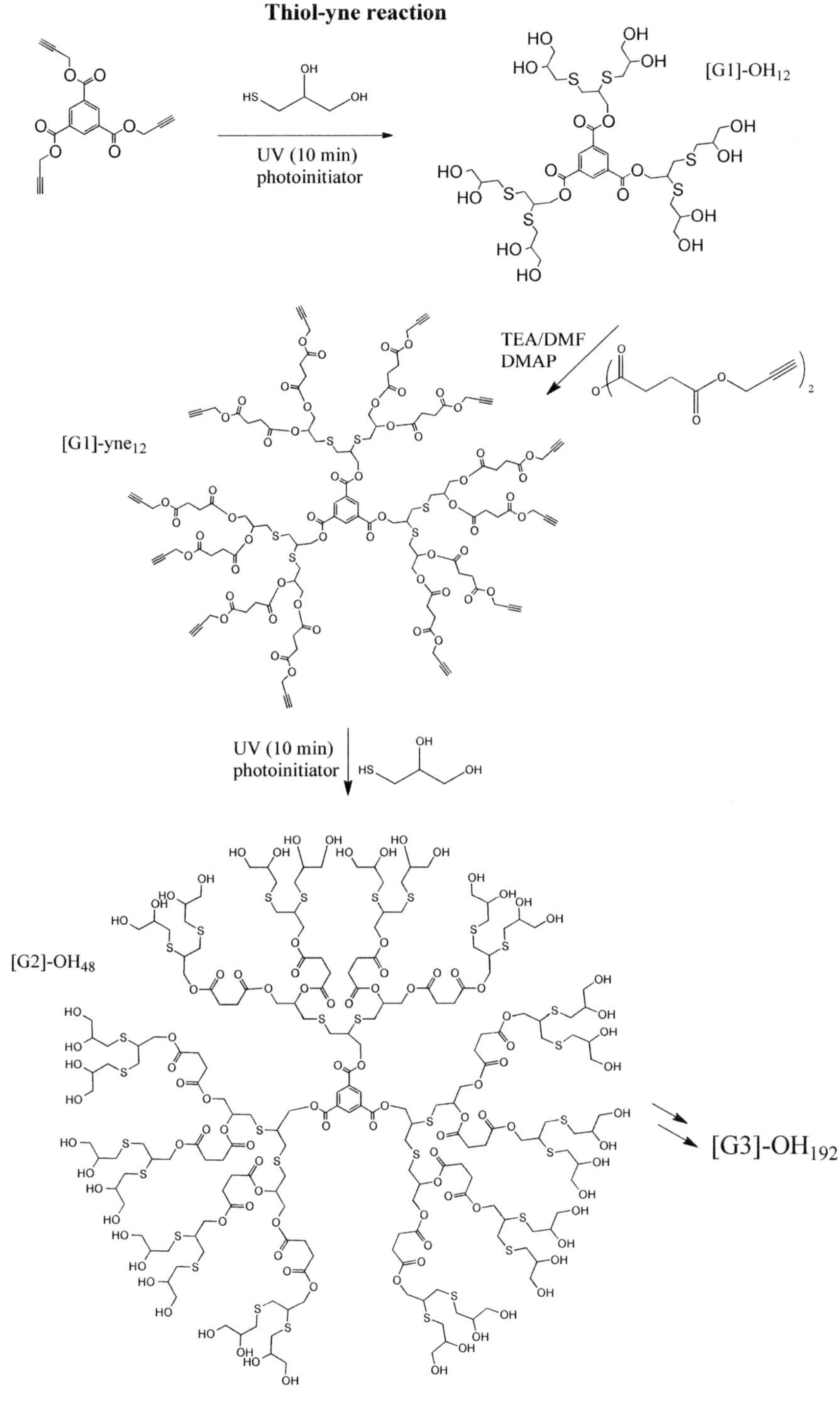


**Scheme 7.3** Reprinted from ref. 10.

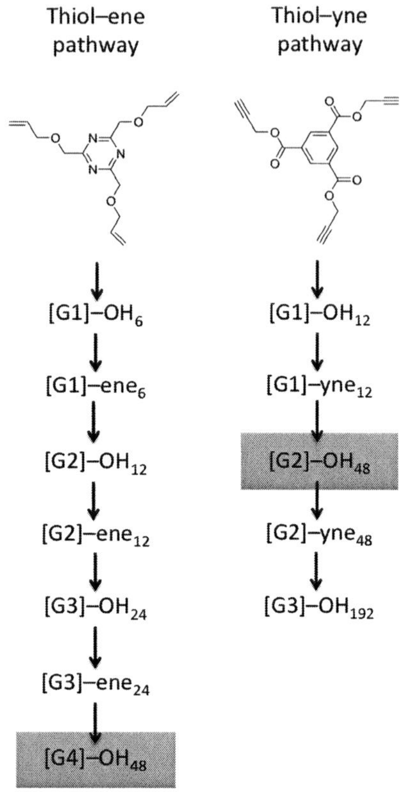


**Scheme 7.4** Comparison between thiol–ene[9] and thiol–yne[10] pathways for the accelerated, divergent synthesis of dendrimers.

developed micelles formed from the self-assembly of a glucose-functionalised dendritic wedge, which were used for the binding of lectins (Concanavalin A).[12] As illustrated in Scheme 7.5, the glycodendritic structure was prepared from an alkyne end-functionalised poly(*tert*-butyl acrylate) *via* a synthetic sequence involving a photoinitiated thiol–yne reaction with 1-thioglycerol, followed first by an esterification with acetylene anhydride and then a thiol–yne reaction with a glucose derivative (β-thioglucose). After micellisation, the binding capacities of the dendritic structure to lectins were compared to that of a linear analogue containing a similar number of glucose residues. As expected, due to their high number of functionalities and possible multivalent interactions, the micelles obtained from the dendritic structures exhibited the highest binding capacity to lectins.

## 7.3 Hyperbranched (Dendritic) Polymer Synthesis

Unlike the regular branching in dendrimers, the branch points in a hyperbranched polymer are randomly distributed throughout the macromolecule. Although hyperbranched polymers are not as well defined as dendrimers, their synthesis is

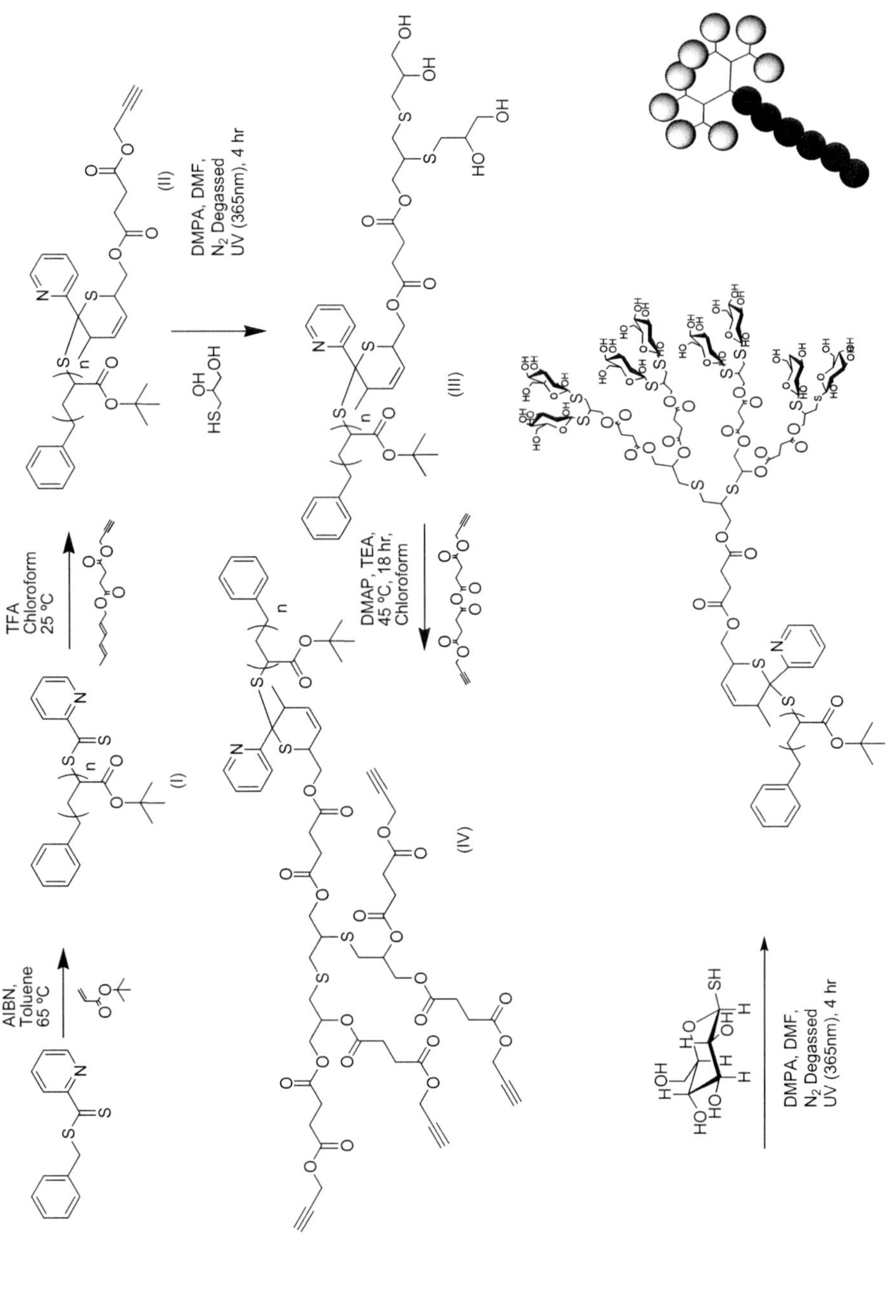


**Scheme 7.5** Reprinted with permission from 12. Copyright 2011 Wiley-VCH Verlag GmbH & Co. KGaA.

typically much simpler, and can often be achieved *via* one-pot procedures.[13,14] The possibility to efficiently add two thiol molecules to one alkyne group *via* thiol–yne chemistry is very attractive for the modular single-step synthesis[15] of hyperbranched polymers. Indeed, a molecule bearing one thiol and one alkyne functionality mimics the typical $AB_2$ type monomer used in hyperbranched polymer synthesis, where A reacts with either of the B components. In this case, the thiol is the A unit, and each π bond in the alkyne is a B unit.

## 7.3.1 Dendritic Polymer from Small Molecule Monomers

Our group was the first to exploit this potential feature of thiol–yne chemistry to synthesise hyperbranched polymers from a thiol/yne-functional small molecule, prop-2-ynyl 3-mercaptopropanoate (PYMP) (Scheme 7.6).[16]

Since a molecule with an alkyne and a thiol has a 2-fold excess of π bonds relative to thiols, and as the thiol–ene reaction is much more rapid than the first thiol–yne addition, each molecule with a thiol and an alkyne at the start of the reaction has a 50% probability of bearing an unreacted alkyne in the final branched polymer. This implies that at any time, the alkyne group of a molecule is either unreacted or fully saturated. Early in the reaction, most of the alkynes are unreacted, while late in the reaction, the probability of an alkyne being fully saturated is 50%. This probability of reaction therefore leads to a distribution of molecular weights within each sample, yielding samples with high molar mass dispersities. One of the most remarkable features of

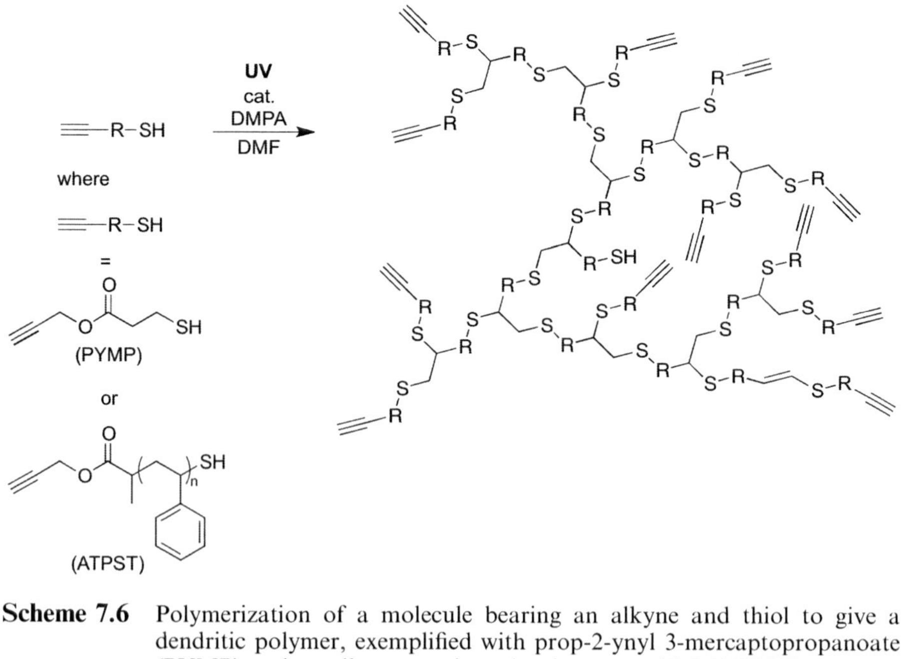


**Scheme 7.6** Polymerization of a molecule bearing an alkyne and thiol to give a dendritic polymer, exemplified with prop-2-ynyl 3-mercaptopropanoate (PYMP) and an alkyne-terminated polystyrene thiol (ATPST). Adapted with permission from ref. 16. Copyright 2009 American Chemical Society.

hyperbranched polymers formed by thiol–yne reactions is that they have degrees of branching (DB) almost as high as those of dendrimers (a perfectly formed dendrimer has no semi-reacted site, and thus has a DB of 1). Since the thiol–yne reaction is near quantitative (no quantifiable amount of alkene can be detected at any point in time), there are very few 'linear' or semi-reacted groups in the hyperbranched polymer. This implies that the DB of a hyperbranched polymer made by the thiol–yne reaction should be very high, with a value close to 1. This is much higher than for conventional $AB_2$ polymerization, which gives a DB of approximately 0.5.[14] To acknowledge this unique structure, hyperbranched polymers produced from a thiol–yne reaction can be referred to as 'dendritic polymers'. In addition to their near perfect degree of branching, dendritic polymers obtained from thiol–yne chemistry offer other unique features. The reaction occurs at room temperature and is very rapid compared with other reaction methods. In addition, since the thiol–yne reaction uses a radical mechanism to transform molecules into hyperbranched polymers, the chemistry is tolerant to almost any functionality.

An elegant procedure to synthesise in a one-pot reaction first the monomer and then the dendritic polymer was proposed by Han *et al.*[17] The authors produce the thiol/alkyne-functional monomer *in situ* by the dropwise addition of one equivalent of propargyl acrylate to a solution of a dithiol in the presence of triethylamine. The Michael addition was then followed by a thiol–yne reaction, initiated either thermally or by UV irradiation, and yielded the dendritic structures. Another approach to prepare thiol/alkyne monomers, which involves the reaction of a dithiol compound with propargyl halides, was described by the same group.[18]

### 7.3.2 Dendritic Polymer from Macromonomers

Beyond small molecule monomers, the versatility of the thiol–yne reaction also allows the production of dendritic structures from macromonomers, thus allowing for tunable functionality. By applying the thiol–yne reaction to linear polymers bearing an alkyne moiety at one chain-end and a thiol group at the other, it is possible to control the overall structure and physical properties of the resulting dendritic polymer, and to incorporate additional functionalities. Our group used reversible addition–fragmentation chain transfer (RAFT) polymerisation to synthesise linear chains with alkyne and thiol functionalities as the α- and ω-end-groups, respectively.[16,19] A RAFT agent (prop-2-ynyl propanoate)yl butyl trithiocarbonate, PYPBTC) was used to mediate the polymerisation of styrene (S) to produce linear chains of well controlled molecular weight and narrow molecular weight distribution exhibiting an alkyne moiety at the α-chain-end. Postpolymerisation aminolysis of the trithiocarbonate moiety with isopropylamine led to ω-thiol-functional polymeric chains. UV irradiation of these polymers in the presence of trace amounts of a photoinitiator (2,2-dimethoxy-2-phenylacetophenone, DMPA) allowed the production of dendritic structures in which each branching unit is linked to the other by the polymeric chain. RAFT polymerisation was also exploited to generate block copolymers thus leading to alternating multiblock dendritic polymers (Scheme 7.7).[19]

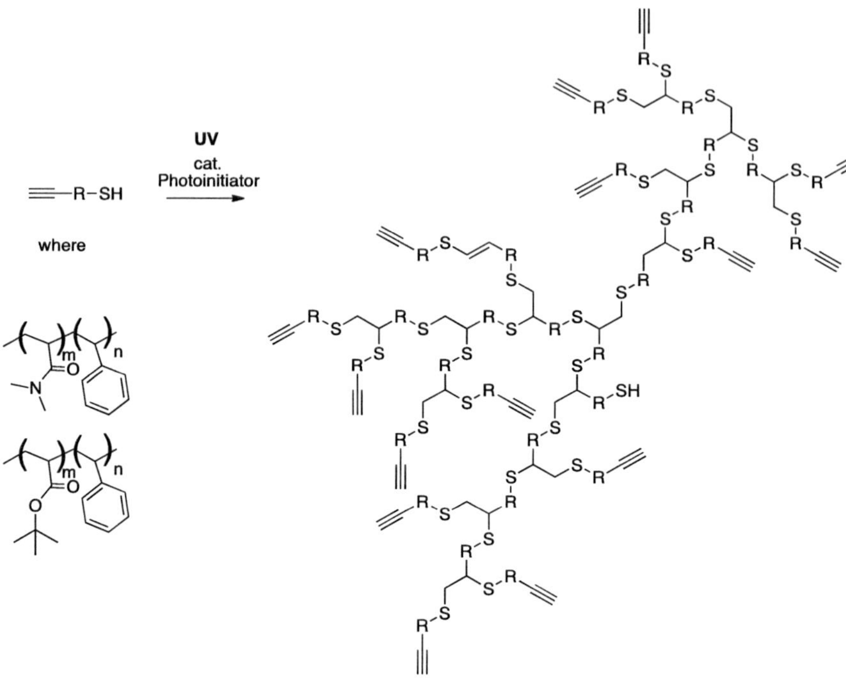


**Scheme 7.7** Dendritic polymer synthesis *via* thiol–yne reaction of block copolymers P(DMA-*b*-S) and P(*t*BA-*b*-S) obtained from subsequent reversible addition–fragmentation chain transfer (RAFT) polymerisation and aminolysis.
Reprinted from ref. 19.

The use of RAFT polymerisation allows the preparation of functional linear 'polymeric' thiol/yne monomers, which as a consequence offers the possibility to introduce functionalities in the core of the final dendritic polymer upon the thiol–yne reaction. For instance, the dendritic structures obtained from the amphiphilic block copolymers described above were observed to self-assemble into micelles in aqueous solution. Dendritic block copolymers based on DMA and styrene gave rise to structures that were significantly larger than those formed by their linear analogues, while the acrylic acid–styrene system showed distinct pH responsiveness. At low pH the system into large, ill-defined structures, while at relatively high pH the system self-assembled into small, possibly uniform particles.[19] The synthesis of dendritic structures from functional thiol/yne polymers has the potential to extend beyond the use of vinyl-based polymers. For instance, Liu and Dong synthesised poly(ε-caprolactone) (PCL) macromonomers with α-thiol and ω-alkyne terminal groups, followed by UV mediated thiol–yne reaction.[20] In their study, the authors observed that crosslinking would occur at high conversions and they attributed this observation to the addition of the focal point thiol (SH) to the intermediate vinyl sulphide (–HC=CH–S–) during the reaction. They reasoned that using a more rigid polymeric chain would prevent the cross-linking, and designed a

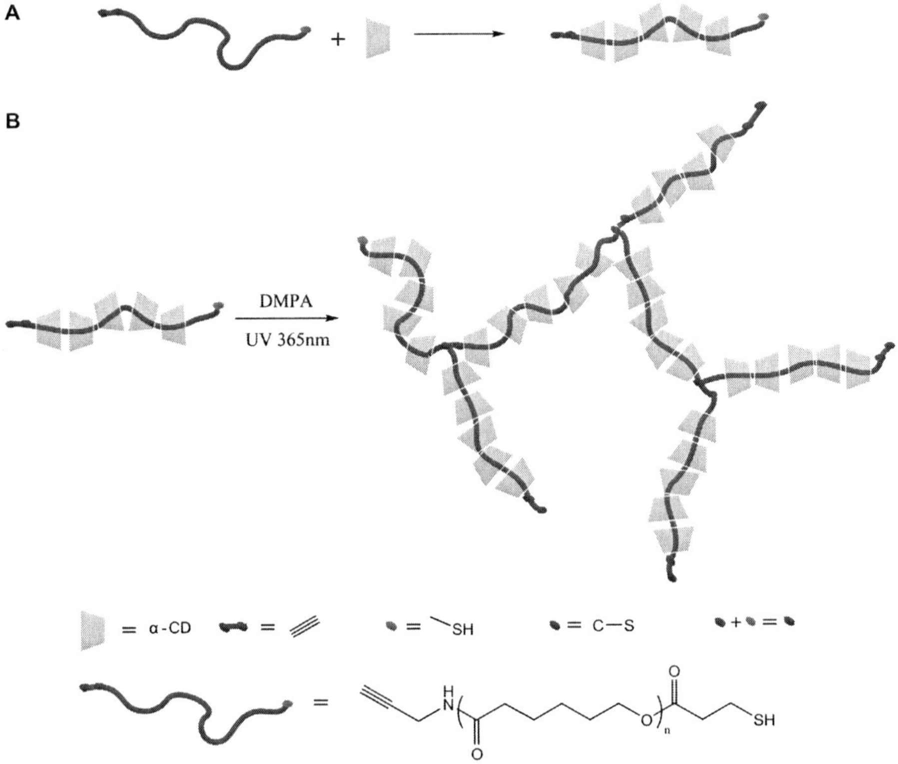


**Scheme 7.8** Synthesis of dendritic polypseudorotaxanes of poly(ε-caprolactone)s threaded with α-cyclodextrin (α-CD).
Reprinted with permission from ref. 20. Copyright 2010 American Chemical Society.

polypseudorotaxane (PPR) of hyperbranched poly(ε-caprolactone)s (HPCLs) threaded with α-cyclodextrin to be used as macromonomer (Scheme 7.8). The thiol–yne reaction of PPR led to dendritic structures with no observable cross-linking. Remarkably, the dendritic polymers of PCL still exhibited melting and crystallisation points, and the degree of crystallisation could be tuned by changing molecular weights through varying UV irradiation times.

### 7.3.3 Functionalisation of Dendritic Polymers *Post* Thiol–Yne Reaction

Dendritic polymers prepared by the thiol–yne reaction have many residual π bonds in their structure, due to the 2 : 1 ratio of π bonds to thiols, with the excess π bonds almost always being alkynes. Hence, each peripheral group in the dendritic polymer except one is a reactive alkyne, which can serve as a potential handle for further functionalisation, including subsequent additional thiol–yne reactions, CuAAC or other alkyne-based derivatisations. For example, the numerous alkyne end-groups of a thiol–yne-generated dendritic

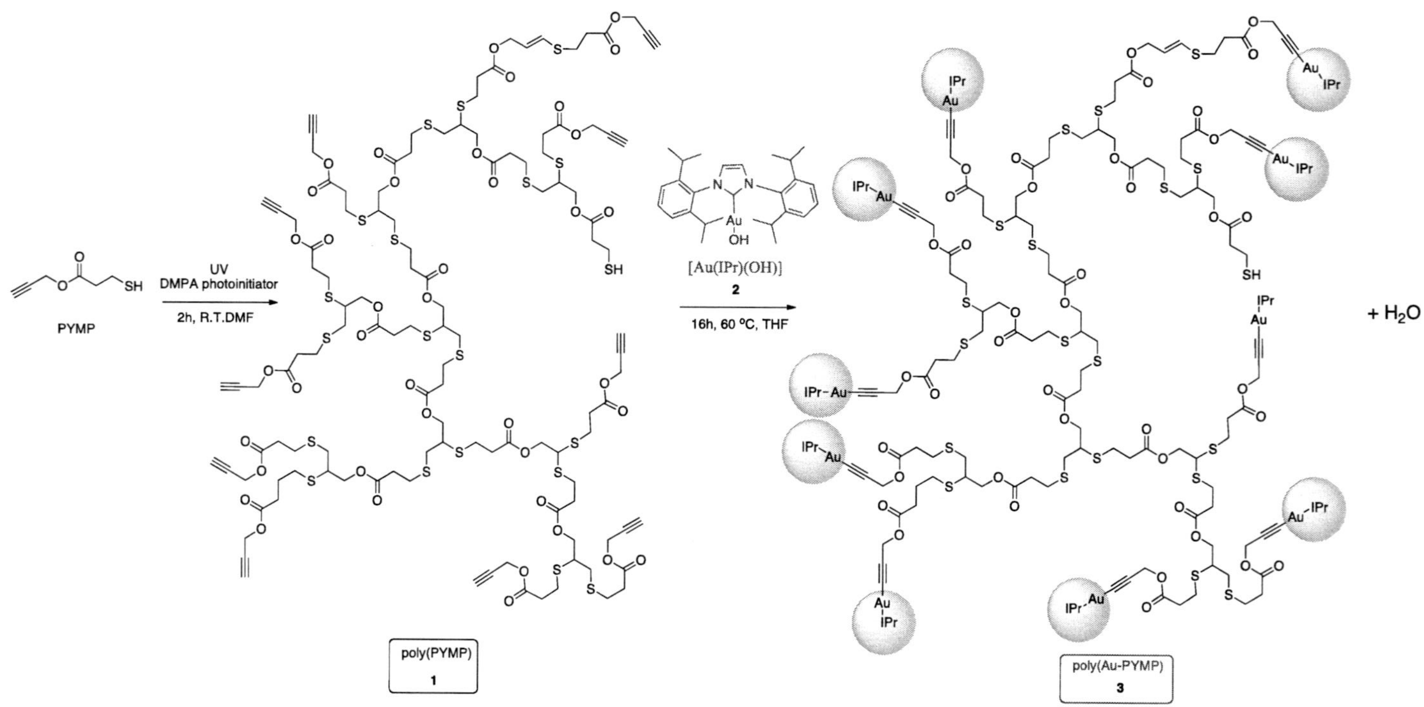


**Scheme 7.9** Synthesis of poly(PYMP) and poly(Au-PYMP) by activation of the terminal alkyne protons by [Au(IPr)] groups. Reprinted with permission from ref. 21. Copyright 2011 American Chemical Society.

structure were reacted with Au(I) to obtain Au(I)-acetylide end-groups *via* a C–H bond activation reaction, to yield materials exhibiting highly luminescent properties (Scheme 7.9).[21] Using dendritic molecules with many gold acetylide groups in close proximity on their corona results in quantum yields much higher than those for typical gold alkynyl complexes. Efficient postpolymerization modification strategies that allow the versatile functionalisation of the focal thiol and peripheral alkyne moieties of an hyperbranched polystyrene were also described by our group.[22]

## 7.4 Conclusions

Thiol–yne addition is inherently an ideal reaction for the design of branched structures, as it readily yields a branch point upon the addition of two thiol moieties to an alkyne functional group. The reaction has been exploited in traditional dendrimer divergent synthesis, to double the number of branching units at each synthetic step, and in the one-pot modular synthesis of hyperbranched polymers. The high efficiency of the reaction yields almost perfect branching patterns, making it a tool of choice for the design of dendritic structures. In addition, this robust radical reaction allows the use of a wide range of functionalities, which can be introduced within the structure or as peripheral end-groups. Although to date the unique properties of the resulting materials have been investigated by only a handful of research groups, there is no doubt they will attract much more interest in the near future and find use in numerous applications as this efficient and versatile method is popularised.

## Acknowledgments

The authors acknowledge the Swiss National Science Foundation (PBELP2-135833) for the provision of a fellowship (RB), and the Australian Research Council (Discovery and Linkage programs) for funding.

## References

1. B. D. Fairbanks, T. F. Scott, C. J. Kloxin, K. S. Anseth and C. N. Bowman, *Macromolecules*, 2009, **42**, 211–217.
2. J. M. J. Fréchet and D. A. Tomalia, (eds.), *Dendrimers and Other Dendritic Polymers*, John Wiley & Sons, Ltd, 2002.
3. G. R. Newkome and C. Shreiner, *Chem. Rev.*, 2010, **110**, 6338–6442.
4. S. M. Grayson and J. M. J. Fréchet, *Chem. Rev.*, 2001, **101**, 3819–3867.
5. D. Astruc, E. Boisselier and C. Ornelas, *Chem. Rev.*, 2010, **110**, 1857–1959.
6. J. M. J. Fréchet, C. J. Hawker, I. Gitsov and J. W. Leon, *J. Macromol. Sci., Part A: Pure Appl. Chem.*, 1996, **A33**, 1399–1425.
7. H. C. Kolb, M. G. Finn and K. B. Sharpless, *Angew. Chem., Int. Ed.*, 2001, **40**, 2004–2021.
8. G. Franc and A. K. Kakkar, *Chem. Soc. Rev.*, 2010, **39**, 1536–1544.

9. K. L. Killops, L. M. Campos and C. J. Hawker, *J. Am. Chem. Soc.*, 2008, **130**, 5062–5064.
10. G. J. Chen, J. Kumar, A. Gregory and M. H. Stenzel, *Chem. Commun.*, 2009, 6291–6293.
11. Y. Shen, Y. N. Ma and Z. B. Li, *J. Polym. Sci., Part A: Polym. Chem.*, 2013, **51**, 708–715.
12. J. Kumar, A. Bousquet and M. H. Stenzel, *Macromol. Rapid Commun.*, 2011, **32**, 1620–1626.
13. M. Jikei and M.-a. Kakimoto, *Prog. Polym. Sci.*, 2001, **26**, 1233–1285.
14. B. I. Voit and A. Lederer, *Chem. Rev.*, 2009, **109**, 5924–5973.
15. D. Konkolewicz, M. J. Monteiro and S. Perrier, *Macromolecules*, 2011, **44**, 7067–7087.
16. D. Konkolewicz, A. Gray-Weale and S. Perrier, *J. Am. Chem. Soc.*, 2009, **131**, 18075–18077.
17. J. Han, B. Zhao, Y. Q. Gao, A. J. Tang and C. Gao, *Polym. Chem.*, 2011, **2**, 2175–2178.
18. J. Han, B. Zhao, A. J. Tang, Y. Q. Gao, C. Gao, Polym. Chem., 2012, **3**, 1918–1925.
19. D. Konkolewicz, C. K. Poon, A. Gray-Weale and S. Perrier, *Chem. Commun.*, 2011, **47**, 239–241.
20. W. Liu and C. M. Dong, *Macromolecules*, 2010, **43**, 8447–8455.
21. D. Konkolewicz, S. Gaillard, A. G. West, Y. Y. Cheng, A. Gray-Weale, T. W. Schmidt, S. P. Nolan and S. Perrier, *Organometallics*, 2011, **30**, 1315–1318.
22. R. Barbey and S. Perrier, ACS Macro Lett., 2013, **2**, 366–370.

CHAPTER 8

# Thiol-X Reactions in Tissue Engineering

DANIEL L. ALGE AND KRISTI S. ANSETH*

Department of Chemical and Biological Engineering, University of Colorado, Boulder, CO 80303-1904, USA
*Email: kristi.anseth@colorado.edu

## 8.1 Introduction

Tissue engineering is a highly interdisciplinary field that spans knowledge from basic cell biology to biochemistry to materials science to clinical science. The ultimate goal of the field is to develop methods or procedures to regenerate or replace tissues that suffer from injury or disease.[1] A myriad of strategies have been described in the literature, but a classic tissue engineering paradigm is to process materials into 3D scaffolds to serve as a template for tissue formation. Cells can be seeded upon the scaffold, encapsulated, or simply allowed to infiltrate from the surrounding tissue. The challenge, however, is to identify appropriate features to incorporate into the scaffold to direct and/or accelerate tissue formation. Both biochemical and biophysical cues often play important roles in directing the regenerative process, and methods to incorporate biological signals at the appropriate dose, time, and spatial location are often critical. This chapter reviews significant advances in the development and application of thiol-ene polymerizations to fabricate hydrogel scaffolds for three-dimensional cell culture and tissue regeneration, as this information provides a foundation to illustrate many principles of scaffold design and also represents a biological area where thiol-X reactions have been extensively applied.

RSC Polymer Chemistry Series No. 6
Thiol-X Chemistries in Polymer and Materials Science
Edited by Andrew B. Lowe and Christopher N. Bowman

Published by the Royal Society of Chemistry, www.rsc.org

In a basic sense, hydrogels are simply hydrophilic macromolecules with strong interactions between the polymer chains (*e.g.*, covalent or physical) that render them insoluble, but swellable, when placed in an aqueous environment.[2] This basic property imparts hydrogels with excellent mass transport properties and soft tissue-like mechanical properties, which make them very attractive materials for cell encapsulation.[3–5] Furthermore, the injectability of hydrogel precursors renders opportunities for the development of *in situ* polymerizable cell delivery systems. The potential for cell delivery is an important advantage considering that cells transplanted without an appropriate material carrier typically have limited efficacy, in part due to extensive cell death that can occur upon implantation, poor engraftment (typically <5% of implanted cells survive and take up residence in the target tissue), and a loss of control over cell fate and function after implantation.[6] As a result, an emerging paradigm in regenerative medicine has been the design of hydrogel cell carriers to overcome some of these problems by providing cells with a 3D matrix to promote their survival and function. Specficially, an extracellular environment that more closely resembles natural tissue architecture, and a material approach to retain cells at the implantation site to facilitate engraftment are critical parameters. An excellent example is Chondron™, which is a natural fibrin gel based technology for autologous chondrocyte implantation that was recently evaluated in clinical trials and found to be effective for decreasing pain and improving knee function in patients with debilitating cartilage damage.[7] Hydrogel based cell delivery strategies are also showing promise in pre-clinical studies for the treatment of myocardial infarctions[8–10] and also central nervous system disorders.[11–13]

In addition to serving as cell delivery vehicles to promote tissue regeneration, hydrogels have also played an important role in fundamental biological research through their use as cell culture substrates. Importantly, cells are conventionally grown on 2D rigid plastic substrates. However, mounting evidence suggests that cells behave very differently when cultured in three dimensions,[14–16] and many researchers are now using hydrogels as a platform to study cell biology in a more physiologically relevant 3D environment *in vitro*.[17–19] Examples of the study and culture of cell-laden hydrogels include: adult and embryonic stem cells,[20–23] fibroblasts,[24,25] chondrocytes,[26–29] neural precursor cells,[30,31] vascular cells,[32–34] and cancer-derived cells.[35–37] Since cells are known to sense their mechanical environment, the effect of matrix stiffness on cell phenotype in both 2D and 3D culture has been a major topic of interest in the field.[38] Another important application of hydrogel cell culture has been studying cell migration in 3D, which more accurately reflects *in vivo* behavior than 2D migration studies because cells in 3D must degrade and remodel their local environment.[39] Through advances in hydrogel design and chemistries, new types of culture platforms are evolving and improving our collective understanding of cell functions that are critical to both developmental and pathological processes.

While several different crosslinking chemistries have been developed to create cell-laden hydrogels for tissue engineering, thiol-based click reactions are playing an increasingly important role in advancing the field by enabling the facile incorporation of biological functionalities into synthetic hydrogels.[40] In this chapter, we briefly review several primary thiol-X reactions that have been reported in the literature for cell delivery, studying cell function in 3D, and promoting tissue regeneration. We also present some of the generalized methods for preparing solutions, encapsulating cells, and characterizing the cell-laden constructs. Specific highlights from recent literature are introduced to illustrate the various ways in which the benefits of thiol-X reactions have been leveraged to create biofunctional hydrogel matrices for tissue engineering. Finally, we conclude with a prospectus and reflect on challenges to the field and how the orthogonality of click reactions, especially photoinitiated click chemistries, might open new opportunities and guide future research efforts.

## 8.2 Important Considerations for Cell Encapsulation

From a purely chemical perspective, a wide range of options is available for synthesizing precursor molecules and crosslinking them to form hydrogels. However, it must be noted that the goal of encapsulating living cells imposes many unique requirements and restrictions that deserve special consideration.

First, and perhaps most obvious, is that the crosslinking reaction must be performed in an aqueous and oxygen rich environment. This requirement, along with the desired method of gelation, informs the selection of catalysts, initiators, and reactive functionalities. However, the cytocompatibility of each individual component must be rigorously assessed. As an example, consider the quintessential copper catalyzed azide-alkyne click reaction, which has been used to form hydrogels,[41,42] but cannot be used for cell encapsulation because of the toxicity of the copper catalyst. Similarly, the reaction conditions themselves must also be mild and cell compatible. Of particular concern are the pH, temperature, and osmolarity, which should be kept close to physiological conditions (*i.e.*, 37 °C, pH = 7.4, ~330 mOsm) to minimize potentially harmful stresses to the cells and prevent cell death; however, small deviations can be tolerated as cells can recover from short exposures to non-physiologic conditions (*e.g.*, minutes). Yet another concern is the molecular weight of the hydrogel precursors, as lower molecular weight macromolecules can be taken up by cells and negatively affect viability. For example, low molecular weight poly(ethylene glycol) (PEG) acrylates are notoriously cytotoxic while their high molecular weight counterparts are not.[43,44] In general, molecular weights between 2–40 kDa are preferred, as lower molecular weights are cytotoxic and larger molecular weights can be more difficult for the body to absorb and excrete, which is important for *in vivo* applications.[45] Finally, the restriction on cell compatible gelation conditions often necessitates the preparation of semi-dilute solutions (*e.g.*, ≤mM), which means that the crosslinking reaction should proceed rapidly (preferably ≤10 minutes), even when the functional group concentration is low. Some of these major requirements are summarized in Table 8.1.

Despite the aforementioned restrictions, several conditions and gelation techniques have been developed for cell encapsulation. Photopolymerization is

**Table 8.1** Important considerations for cell encapsulation in hydrogels.

| *Factor to be considered* | *Comments and guidelines* |
|---|---|
| Reaction environment | • aqueous<br>• oxygen rich |
| Reaction conditions | • must be mild to minimize cell stress<br>• restricted to physiological conditions, which are: pH = 7.4, 37 °C, and ~330 mOsm<br>• ≥365 nm light for photopolymerization |
| Catalysts, initiators, and reactive groups | • can be cytotoxic independent of the reaction conditions and must be screened<br>• reactive group toxicity can vary with molecular weight (*e.g.*, macromolecular monomer design) |
| Precursor molecular weight | • small molecular weight molecules can be taken up by cells (potentially cytotoxic)<br>• higher molecular weights (kDa) are favored, but not beyond 40 kDa for *in vivo* applications to avoid accumulation in the liver, *etc.* |
| Reaction kinetics | • functional group concentrations are often mM or less<br>• gelation and complete polymerization should occur in minutes |

a particularly robust and versatile technique,[46,47] and in Section 8.3.2 we highlight the amenability to photopolymerization as a key advantage of radical mediated thiol-ene reactions. For that reason, the additional factors that must be considered related to the irradiation conditions warrant special consideration here. One of the most critical factors to consider is the wavelength of light used to initiate the photopolymerization. Due to well documented harmful effects of ultraviolet (UV) light on cells,[48] which can cause irreparable DNA damage, the shortest wavelengths of light that are appropriate for photopolymerization in the presence of cells is 350–365 nm. 365 nm light is particularly convenient in the laboratory setting because of the commercial availability of relatively inexpensive light sources that irradiate a large area and enable the preparation of several dozen samples within minutes. However, advances in light emitting diode technology are making photopolymerization with visible wavelengths (*e.g.*, 405 nm, 530 nm) more attractive. In addition to wavelength, dosage is also an important factor to consider, as prolonged exposure to high intensity light can be damaging to cells. For the case of 365 nm light, exposures less than or equal to 10 min at 10 mW $cm^{-2}$ are generally regarded as cytocompatible for this wavelength.

The wavelength and dose requirements for cell encapsulation significantly restrict the selection of photoinitiators. The most widely used UV photoinitiator reported in the biomaterials literature for cell encapsulation has been 2-hydroxy-1-[4-(hydroxyethoxy)phenyl]-2-methyl-1-propanone, which is sold commercially by Ciba under the trade name I2959. Importantly, although I2959 is only sparingly water soluble (up to about 0.5 wt%, ~25 mM), it is cytocompatible at modest concentrations (up to 0.05 wt%, ~2.5 mM) and has been widely reported for photoencapsulation of both primary cells and cell lines in hydrogels.[49,50] Although it is not commercially available, lithium phenyl-2,4,6-trimethylbenzoylphosphinate, also known as lithium acylphosphinate or LAP, has also been reported for its excellent cytocompatibility.[51] Notably, LAP has higher water solubility and cleaves more efficiently at 365 nm than I2959, and can be used to photoinitiate polymerizations at longer wavelengths of light, such as 405 nm visible light. Other visible light photoinitiators that have been used for cell encapsulation include eosin Y (peak absorbance at ~510 nm) and riboflavin (peak absorbance at ~450 nm), but both of these are type II photoinitiators and require a co-initiator such as triethanolamine (TEA).[52,53]

## 8.3 Cytocompatible Thiol-X Chemistries

Due to their fast reaction kinetics, specificity, and high yields, thiol-X reactions are attractive for forming hydrogels *via* step growth polymerization mechanisms. Here, the principal advantage of thiol-X chemistries for tissue engineering applications is that these reactions allow one to readily incorporate thiol-containing biological functionalities (*e.g.*, cysteine-containing peptides or thiolated proteins) into synthetic hydrogels. When combined with synthetic PEG hydrogels, which interact minimally with cells due to the hydrophilicity and non-protein binding properties of PEG,[18,54] thiol-X chemistries can be leveraged to engineer cell-instructive matrices by integrating biological signals that mimic key aspects of the extracellular matrix (ECM) in which cells naturally reside. Notably, peptide incorporation is facile with this approach, since cysteine residues are readily incorporated during solid phase peptide synthesis (SPPS) and post-synthetic modification of the oligo sequences is not needed. The thiolation of proteins is also routine,[55] but care must be taken when assessing the effects of modification and reaction conditions on protein activity.

To leverage the advantages of thiol-X chemistries, several thiol based crosslinking reactions have been described in the literature for hydrogel formation and cell encapsulation. These can generally be classified as either Michael addition or radical mediated thiol-ene addition. A generalized schematic for cell encapsulation using these chemistries is presented in Figure 8.1.

### 8.3.1 Michael Addition

The application of the Michael addition reaction between a thiol and an electron deficient ene to hydrogel crosslinking and the formation of cell-laden tissue engineering matrices was pioneered by Hubbell, Lutolf, and colleagues. Building upon their previous work with hydrogels formed by chain polymerization of PEG diacrylates, but seeking to form more ideal networks that were susceptible to proteolytic degradation by cell-secreted enzymes, they used di-cysteine peptides and acrylated or vinyl sulfone functionalized PEG to form peptide crosslinked PEG hydrogels *via* Michael addition. Notably, the addition

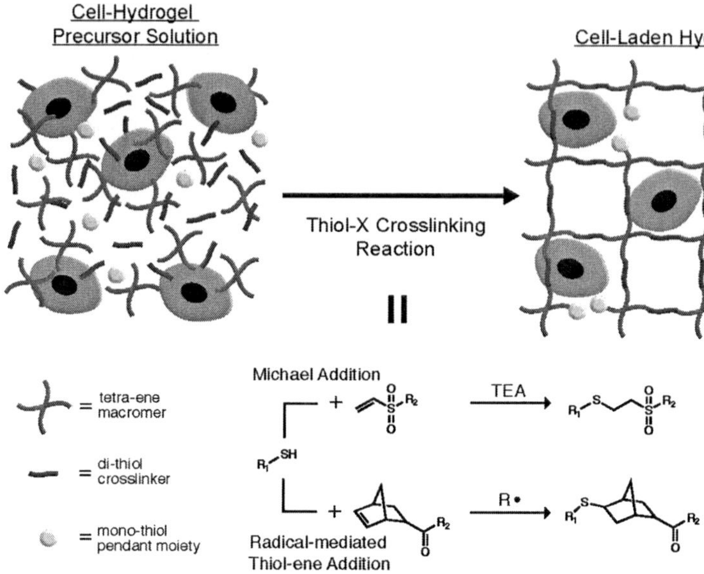


**Figure 8.1** Schematic for forming cell-laden hydrogels using a thiol Michael addition or radical mediated thiol-ene crosslinking reaction. Note: cells are not drawn to scale.

of thiols across acrylate functional groups renders the ester group hydrolytically unstable,[56,57] so depending on the desired time scale of degradation, non-specific hydrolysis can become a competing reaction with the more targeted enzymatic degradation. To overcome this limitation, subsequent studies employed vinyl sulfone functional groups, which form a hydrolytically stable network, to create PEG hydrogels with proteolytically degradable crosslinks and containing integrin binding peptide functional groups to promote cell attachment and survival.[58–60] These landmark studies were the first to demonstrate the incorporation of biofunctional ECM mimetic peptides into synthetic PEG hydrogels to create cell-instructive/responsive matrices for tissue engineering. More recently, the thiol-maleimide Michael addition, which has faster reaction kinetics and requires less catalyst compared to the thiol-acrylate/vinyl sulfone reactions, has also been demonstrated as a cytocompatible crosslinking chemistry.[61] Some of the major functional groups suitable for cell encapsulation *via* Michael addition reactions that have been reported in the literature are summarized in Table 8.2.

One important advantage of encapsulating cells in hydrogels using Michael addition reactions is that the mechanism allows for the formation of thick, homogenous samples, which is important for many tissue regeneration applications. In addition, it taught the field about the robustness and versatility of including peptides to recapitulate essential features of the native ECM, like integrin binding, enzymatic degradability, and affinity binding of proteins. Despite the successful encapsulation of multiple cell types with high levels of

**Table 8.2** Functional groups for reaction with cysteine containing peptides or thiolated PEGs and cell encapsulation *via* Michael addition.

| *Functional group* | | *Comments* |
|---|---|---|
| Acrylate | 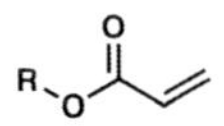 | • highly reactive<br>• resulting network can be hydrolytically unstable |
| Vinyl sulfone | 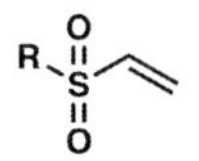 | • moderately reactive<br>• forms a hydrolytically stable network |
| Maleimide | 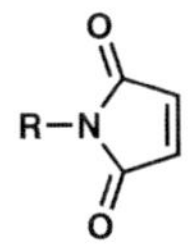 | • highly reactive<br>• requires less TEA catalyst than acrylate and vinyl sulfone |

post encapsulation viability (*e.g.*, fibroblasts,[58] mesenchymal stem cells (MSCs),[62] smooth muscle cells,[63] hepatocytes,[64] dorsal root ganglia,[65] and C2C12 myofibroblasts[61]), the requirement of a catalyst such as TEA, which has been used almost exclusively in the aforementioned cell encapsulations, is often viewed as a disadvantage due to its potential negative effects on the viability of highly sensitive cell types. The basicity of TEA can also encourage the formation of disulfides, which can be disadvantageous, because disulfides introduce non-idealities into the network structure. HEPES (4-(2-hydroxyethyl)-1-piperazineethanesulfonic acid) buffer may be a more mild alternative to TEA,[66] but any improvements in cell survival come at the expense of the reaction rate. Finally, while the Michael addition reaction is versatile, it lacks spatio-temporal control over the reaction, which can be desirable for many tissue engineering applications. These limitations have motivated research into the application of thiol-ene photopolymerizations as a complementary platform to create cell scaffolds with similar structures and chemistries to the Michael addition networks, but through a distinctly different reaction mechanism. The thiol-ene photopolymerization is discussed in greater detail in the next section.

## 8.3.2 Radical Mediated Thiol-Ene Addition

The second type of thiol-X reaction used for cell encapsulation in hydrogels is the radical mediated thiol-ene addition. Importantly, because the reaction is a radical-mediated polymerization, when paired with an appropriate photoinitiator molecule and polymerized under cytocompatible conditions (summarized in Section 8.2), spatio-temporal control over gelation and network formation is possible. Although other free radical initiators can also be used (*e.g.*, ammonium persulfate/*N*,*N*,*N*′,*N*′-tetramethylethylenediamine[67]), this amenability to photopolymerization is a key advantage of thiol-ene reactions over the Michael addition.

Since the 1990s, the photoinitiated chain polymerization of macromolecular monomer solutions has been used for cell encapsulation methods and directed at tissue engineering applications. Typically, (meth)acrylated precursors of water soluble polymers (*e.g.*, PEG,[68–70] poly(vinyl alcohol),[71,72] hyaluronic acid[73–75]) were photopolymerized under strictly controlled photoinitiation conditions to create cell-laden hydrogels, but users were critically aware of limitations with oxygen inhibition, light attenuation, the potential for radical damage to the encapsulated cells or biological molecules. The introduction of the radical mediated thiol-ene reaction for cellular encapsulation represented an alternative approach to create hydrogels that circumvented many of these issues, while simultaneously offering some distinct advantages.

The first reported application of thiol-ene photpolymerization for cell encapsulation used (meth)acrylates as the ene source.[76–79] Similar to Michael addition, thiol-(meth)acrylate photoinitiated polymerizations allow for the facile incorporation of cysteine containing peptides. An important limitation of thiol-(meth)acrylate polymerizations, however, is that these chemistries result in a mixed mode polymerization due to simultaneous homopolymerization of the (meth)acrylate moieties. Further, the presence of poly(meth)acrylate kinetic chains produces a heterogenous and non-ideal polymer network. A natural evolution to research in material development with thiol-acrylate polymerizations was the subsequent development of thiol-ene polymerized hydrogels based on norbornene functionalized macromolecular precursors.[80] Importantly, the high reactivity of the ring strained norbornene, which does not homopolymerize, coupled with a $k_P/k_{CT}$ reactivity ratio for the thiyl radical propagation and chain transfer steps that is near unity, facilitates the rapid formation of a uniform step growth network[81] that is analogous to those reported using Michael additions. The functional groups reported in the literature that are suitable for cell encapsulation *via* radical-mediated thiol-ene reactions are summarized in Table 8.3.

**Table 8.3** Functional groups for reaction with cysteine containing peptides or thiolated PEGs and cell encapsulation *via* radical mediated thiol-ene additions.

| *Functional group* | | *Comments* |
|---|---|---|
| Acrylate | 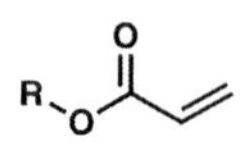 | • highly reactive<br>• mixed mode polymerization produces non-ideal networks |
| Methacrylate | 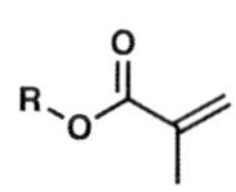 | • less reactive than acrylates, but more cytocompatible<br>• mixed-mode polymerization |
| Norbornene | 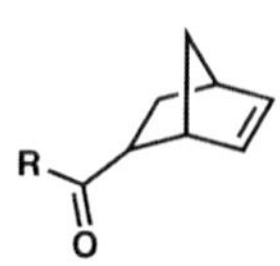 | • highly reactive<br>• forms an ideal step growth network, analogous to the Michael addition reactions |

Radical mediated thiol-ene additions have several important advantages related to forming hydrogels in the presence of tissue, cells and proteins. First, is that the reaction kinetics favor polymerization and gel formation rather than cell/protein damage. This advantage can be seen by considering a simplified perspective of chain transfer events, in which the reactivity ratio ($R$) is proportional to the ratio of the propagation kinetic constant relative to the chain transfer constant ($k_P/k_{CT}$) and the concentration of monomer functional groups relative to the concentration of the encapsulant ($[M]/[encap]$) (eqn (8.1)).

$$R \propto \frac{k_P[M]}{k_{CT}[encap]} \tag{8.1}$$

Importantly, most proteins lack free thiols for the chain transfer reaction, and other potentially reactive amino acids are two orders of magnitude less efficient as chain transfer agents.[82] Nevertheless, one must be especially careful in monitoring the polymerization reaction. At later stages of polymerization, it is critical to cease photoinitiation upon complete network formation, as the encapsulated biologic is highly susceptible to damage at this stage of the reaction. An additional advantage of thiol-ene reactions in general is that they are not inhibited by oxygen.[83] Consequently, lower initiator concentrations can be used and the total number of radicals generated to complete polymerization is often much lower compared to traditional free radical polymerizations. These benefits, combined with the extremely short lifetime of thiyl radicals during thiol-ene polymerizations,[84] have been indirectly observed to reduce protein and cell damage during encapsulation. Most notably, Lin and coworkers, showed a striking improvement in the survival of MIN6 beta cells when exposed to non-gelling thiol-norbornene reactions or photoencapsulated in PEG gels formed using thiol-norbornene chemistry compared to conventional vinyl polymerization chemistry (Figure 8.2).[85]

Overall, radical mediated thiol-ene reactions, particularly those using norbornene as the ene, offer exceptional versatility for tailoring the biological functionality of hydrogel scaffolds and directing specific cell functions. However, despite mounting evidence showing the many advantages of radical mediated thiol-ene reactions to encapsulate and culture cells, especially compared to hydrogels formed *via* traditional chain radical polymerizations, the radical-mediated mechanism is still viewed as a drawback and careful analyses must be completed to ensure cytocompatibililty. Nevertheless, similar to the Michael addition crosslinking reactions, a wide range of primary cells have been encapsulated using photoinitiated thiol-ene reactions, including fibroblasts,[80] MSCs,[20] valvular interstitial cells,[86,87] fibrosarcoma cells,[36] and islets.[85] A second and critical consideration when using photoinitiated thiol-ene reactions is that light attenuation and scattering through the gel can prevent the formation of uniform three-dimensional materials. While one must be acutely aware of conditions that can lead to heterogeneous initiation and polymerization, materials several centimetres thick can be formed in the absence of cells and, in the presence of cells, thicknesses of several millimeters are readily achieved.

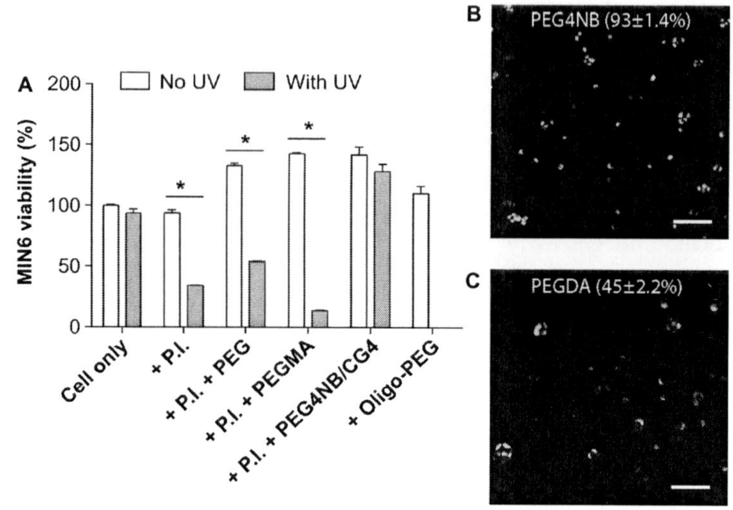


**Figure 8.2** Cytocompatibility of thiol-ene photopolymerizations enhances cell viability compared to conventional acrylate chemistry. (A) To compare cytocompatibility of the reaction conditions, a MIN6 β-cell line was exposed to non-gelling reactions and cell survival was assessed *via* a viability assay. Following light exposure, significant decreases in cell viability were observed when cells were exposed to photoinitiator alone (denoted P.I. in the figure), a mixture of photoinitiator and a non-reactive PEG, and a solution of photoinitiator and PEG monoacrylate (PEGMA). However, viability of MIN6 exposed to photoinitiator and tetra-functional PEG-norbornene (PEG4NB) and mono-cysteine peptide (CG4) and irradiation was not significantly reduced. Note: * indicates statistical significance ($\alpha < 0.05$). (B) and (C) Percent viability quantified from Live/Dead staining showed that viability of MIN6 cells photo-encapsulated with thiol-ene chemistry was $93 \pm 1.4\%$, which was significantly higher than $45 \pm 2.2\%$ viability calculated for cells encapsulated in diacrylate crosslinked PEG hydrogels ($\alpha < 0.05$; scale bars = 100 μm).
Adapted from ref. 85 with permission.

## 8.4 Protocol for Cellular Encapsulation

Having discussed the basic parameters to consider when designing functional macromolecules and the cytocompatible chemistries that have been published, we now present a generalized protocol for cell encapsulation in hydrogels using thiol-X crosslinking reactions. This protocol is derived from methods that have been published by numerous research groups, but has been adapted to include practical advice based on decades of experience in our laboratory. For the sake of brevity, we have assumed that hydrogel precursors with the required functional groups are in hand (for the synthesis of vinyl sulfone and norbornene functionalized polymers we recommend ref. 58 and 80, respectively). We have also assumed that mono- and di-cysteine containing peptides are being used as the thiol source, although thiol functionalized polymers can also be used.

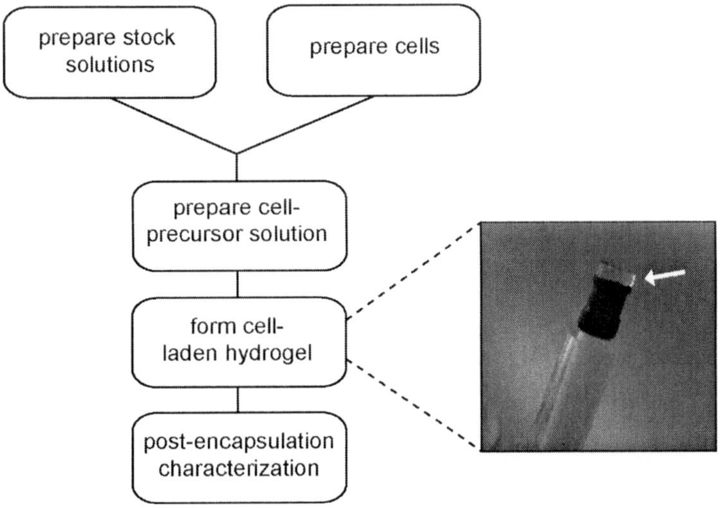


**Figure 8.3** Basic schematic summarizing the steps involved in preparing cell-laden hydrogels. The image shows a hydrogel that was formed in a syringe tip mold (indicated by arrow).

For additional information on cell encapsulation, we recommend ref. 88 and 89 as useful references. A basic schematic summarizing the steps involved in the preparation of cell-laden hydrogels is presented in Figure 8.3.

## 8.4.1 Preparation of Stock Solutions

Due to the nature of cell encapsulation experiments, which often involve the preparation of dozens of hydrogel samples for multiple experimental groups, it is most convenient to work from aliquots of pre-made stock solutions. All of the stock solutions should be prepared in cell culture grade phosphate buffered saline (PBS) at a pH of 7.4), sterilized by filtration through a sterile 0.2 μm syringe filter, and then stored in autoclave sterilized microtubes at $\leq$–20 °C for future use.

1. Prepare a concentrated stock solution of a multi-arm polymer functionalized with the desired ene moiety (*e.g.*, acrylate, methacrylate, vinyl sulfone, maleimide, or norbornene). 20–30 wt% is a typical concentration.
2. Prepare concentrated stock solutions of all peptides being incorporated into the hydrogel network. Concentrations of 100 mM or greater are recommended for crosslinking peptides. For pendant peptides, which are typically incorporated at much lower concentrations, 50 mM is recommended.
   a. Thiol concentrations in the peptide solutions can be quantified using Ellman's reagent 5,5′-dithio-bis-(2-nitrobenzoic acid), which reacts with free thiols to produce the chromophore 2-nitro-5-thiobenzoate.[90]
   b. Concentration can be determined by measuring the absorbance at 405 nm and comparing to a freshly prepared cysteine standard curve.

3. Prepare a concentrated stock solution of catalyst or initiator according to the desired mechanism of gelation.
   a. Michael addition: TEA is the most commonly used catalyst. Working concentrations ranging from 4 to 400 mM have been reported. A 10× stock solution is recommended.
   b. Photoinitiated thiol-ene addition: working concentrations for photoinitiators typically range from 1–5 mM. A 10× stock solution is recommended. (Note: to prepare concentrated solutions of I2959, gentle heating and vigorous mixing are often required. However, once solubilized the solutions can be stored at room temperature. This is not required for LAP due to its high solubility in water.)

## 8.4.2 Preparation of the Cell-precursor Solution

To encapsulate cells in hydrogels, targeted cells must first be mixed into and suspended in the liquid hydrogel precursor solution. During this process it is critical to adhere to standard tissue culture practices in order to prevent contamination by microbial and fungal organisms. For a thorough review of tissue culture techniques, we recommend ref. 91. Furthermore, to minimize stress to the cells being encapsulated during preparation of the cell-precursor solution, the cells should be kept at room temperature for the minimum amount of time possible, but no more than 1 hour.

1. Based on the desired gel volume and composition (*i.e.*, polymer weight percent, stoichiometry, concentration of pendent peptides, and initiator/catalyst concentration), calculate the required volumes of each stock solution. Note that cells suspended in PBS will be used to make up the remainder of the gel volume that is not accounted for by the stock solutions.
2. Evaluate the volumes calculated in step one for each experimental condition and prepare a master-mix solution using the lowest required volume of each stock solution.
   a. Preparing hydrogels from a master-mix solution is *highly* recommended as it minimizes gel-to-gel variability within an experiment.
   b. Example: if 10 gels are needed and each gel requires 15 μl of PEG tetra-norbornene, 5 μl of peptide crosslinker, and 2 μl of photoinitiator, then the master-mix solution should contain 150 μl of PEG tetra-norbornene, 50 μl of peptide crosslinker, and 20 μl of photoinitiator.
   c. If photoinitiated thiol-ene polymerization is being used to form the gel, the photoinitiator can be included at this step, but one should be careful to protect the solution from light to prevent premature gelation. However, if the gel is being crosslinked by a Michael addition reaction, we recommend leaving the di-cysteine crosslinker out of the master-mix.

3. Re-suspend the cells that are being encapsulated in the master-mix solution at the desired cell density. Depending on the cell type and the application, cell densities can range from $5 \times 10^5$–$5 \times 10^7$ cells $mL^{-1}$.
   a. For adherent cell cultures, detach the cells to be encapsulated from their two dimensional culture dish or flask by treating them with 0.05% trypsin-ethylenediaminetetraacetic acid (EDTA) for 2–5 min at 37 °C.
   b. Transfer the cells to an appropriately sized sterile tube, centrifuge at 1000 × g for 5 min, and carefully aspirate the medium from the resulting cell pellet.
   c. Re-suspend the cells in PBS so that the desired cell density is obtained upon mixing the volume calculated in step 1 with the master-mix solution. Alternatively, the cells can be re-suspended in cell culture medium rather than PBS. However, because cell culture medium generally contains an undefined cocktail of serum proteins that will be co-encapsulated with the cells, this approach is not preferred.
   d. Combine the cell-PBS solution with the master-mix solution and gently mix by trituration with a micropipette.

### 8.4.3 Formation of the Cell-laden Hydrogel

After preparing the cell-precursor solution, crosslinking can be initiated to form cell-laden hydrogels with virtually any desirable geometry. We have found that 1 ml syringes with the tips removed serve as excellent molds for the preparation of disk shaped samples for *in vitro* experiments. Simple molds can also be fabricated out of rubber gasket material sandwiched between glass slides. To prevent the hydrogel from sticking to the glass slides, they should be pre-treated with a non-adherent coating such as Sigmacote. Alternatively, the hydrogel can be attached to one of the glass slides simply by functionalizing the glass surface with a mercaptosilane, which can be convenient for certain experiments. All molds should be sterilized prior to each use by treatment with 70% ethanol followed by at least 30 min exposure to the UV germicidal light in the tissue culture hood.

1. Add the liquid cell-precursor solution to the desired mold and initiate gelation.
   a. For photoinitiated thiol-ene polymerizations, an appropriate light source should be set up in the tissue culture hood. The photo-polymerization conditions should be optimized based on the photoinitiator selected and also the gel formulation. For UV photo-polymerizations, we recommend 3–5 minutes at 365 nm and 5–10 $mW\,cm^{-2}$ as a starting point.
   b. For Michael addition polymerizations, the di-cysteine crosslinker should be added to the cell-precursor solution immediately prior to addition of the solution to the mold. The batch size will likely require optimization based on gelation kinetics.

2. After gelation is complete, remove the cell-laden hydrogel from the mold, quickly immerse it in cell culture medium, and then transfer it to a cell culture incubator for equilibration at 37 °C, 5% $CO_2$.

### 8.4.4 Post-encapsulation Characterization

Following encapsulation, the cell distribution within the hydrogel can be readily assessed by phase contrast microscopy due to the transparency of the gel. However, it is often desirable to also evaluate the post-encapsulation cell viability. The commercially available Live/Dead staining kit provides a convenient means for post-encapsulation characterization of cell viability.

1. Because serum in cell culture medium contains esterases, which can interfere with staining, for the best results, the cell-laden hydrogels should first be transferred to several milliliters of PBS and incubated for approximately 10 min at room temperature with gentle agitation.
2. Remove the gels being stained from the PBS and immerse them in Live/Dead staining solution, containing ethidium homodimer and calcien AM. Incubate the gels in the staining solution at room temperature with gentle agitation and protected from light. Typically, 30–45 min is sufficient. However, the time required for staining may vary depending on the dimensions of the gel, as diffusion time scales depend on length scales squared.
   a. Ethidium homodimer is a non-membrane permeable fluorescent dye that labels nucleic acids (528 nm excitation peak, 617 nm emission peak). Because it cannot penetrate the cell membrane, it should only stain the nuclei of dead cells.
   b. Calcein AM is a non-fluorescent substrate that is membrane permeable and is cleaved by intracellular esterases to produce the fluorophore calcein (494 nm excitation peak, 517 nm emission peak) and stain the cytoplasm of living cells.
   c. The manufacturer's recommendation of 4 μM ethidium homodimer and 2 μM calcein AM in PBS works well for most cell types. However, optimization of the concentrations and staining time may be necessary.
3. Remove the gels from the staining solution and transfer to fresh PBS. Incubate for 5–10 min at room temperature with gentle agitation to remove excess dyes from the gel and reduce background.
4. Cell viability in the hydrogels can be assessed by epifluorescence or confocal microscopy, as desired. Percent viability can be quantified using basic image analysis software such as Image J.

## 8.5 Imparting Biofunctionality to Synthetic Hydrogels

In Section 8.3 we emphasized that the principal advantage of thiol-X chemistries for tissue engineering applications is that they enable the facile incorporation of biological functionalities that recapitulate key aspects of the

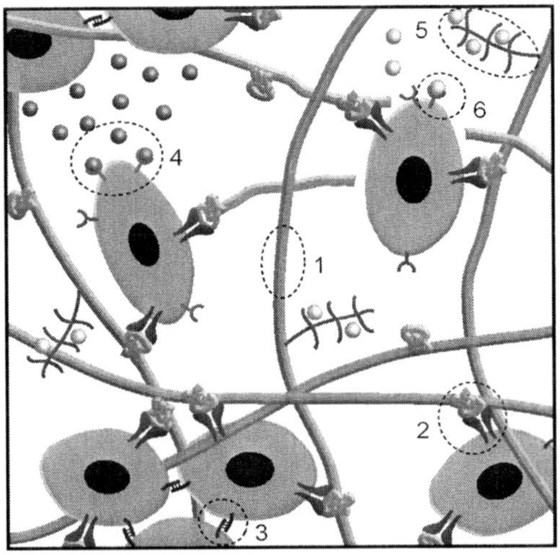


**Figure 8.4** Illustration of the native ECM microenvironment. The ECM presents cells with critical cues, some of which can be recapitulated in biomaterials designed to serve as ECM mimics: (1) Proteolytically degradable biopolymer, (2) cell–matrix interactions *via* integrin binding, (3) cell–cell interactions *via* cell adhesion molecules, (4) cell–cell interactions *via* paracrine signaling, (5) growth factor sequestering through affinity binding, (6) growth factor signaling through tailored diffusion.

ECM, which are depicted in Figure 8.4, into synthetic hydrogels. We have emphasized this point because a key advantage of synthetic hydrogels is their ability to provide cells with a defined microenvironment. This concept is powerful because it enables tissue engineering researchers to deconstruct the roles of biochemical and biophysical cues in directing cell function. PEG hydrogels have been particularly important for this purpose because PEG is a classic non-fouling polymer that interacts minimally with cells. Thus, a PEG hydrogel is effectively a "blank slate" which can be engineered with biomimetic components from the bottom up to provide cells with a microenvironment that is fully defined in terms of its biochemical and biophysical properties.[18,17,54] However, biofunctionalization was not always trivial in the field. It was not until the implementation of thiol click chemistries that facile incorporation of biofunctional components became more standard. Today thiol-X reactions are routinely used to create cell-instructive hydrogels that promote specific cell functions like integrin mediated cell attachment, enzymatic degradation of the hydrogel, cell migration, and even directed stem cell differentiation. In this section we highlight specific examples from the literature to illustrate the impact that thiol-X chemistries have had on the field of tissue engineering.

## 8.5.1 Creating Enzymatically Degradable Polymer Networks

Biodegradability is a key requirement for scaffold materials in tissue engineering. This design goal is often approached using hydrolytically degradable polymers. However, the complexities of biological environments can lead to significant departures from predicted degradation profiles and hydrolytic degradation may not appropriately correspond with ECM production by encapsulated cells, leading to sub-optimal outcomes. Recognizing the limitations of hydrolytically degradable networks, researchers began incorporating

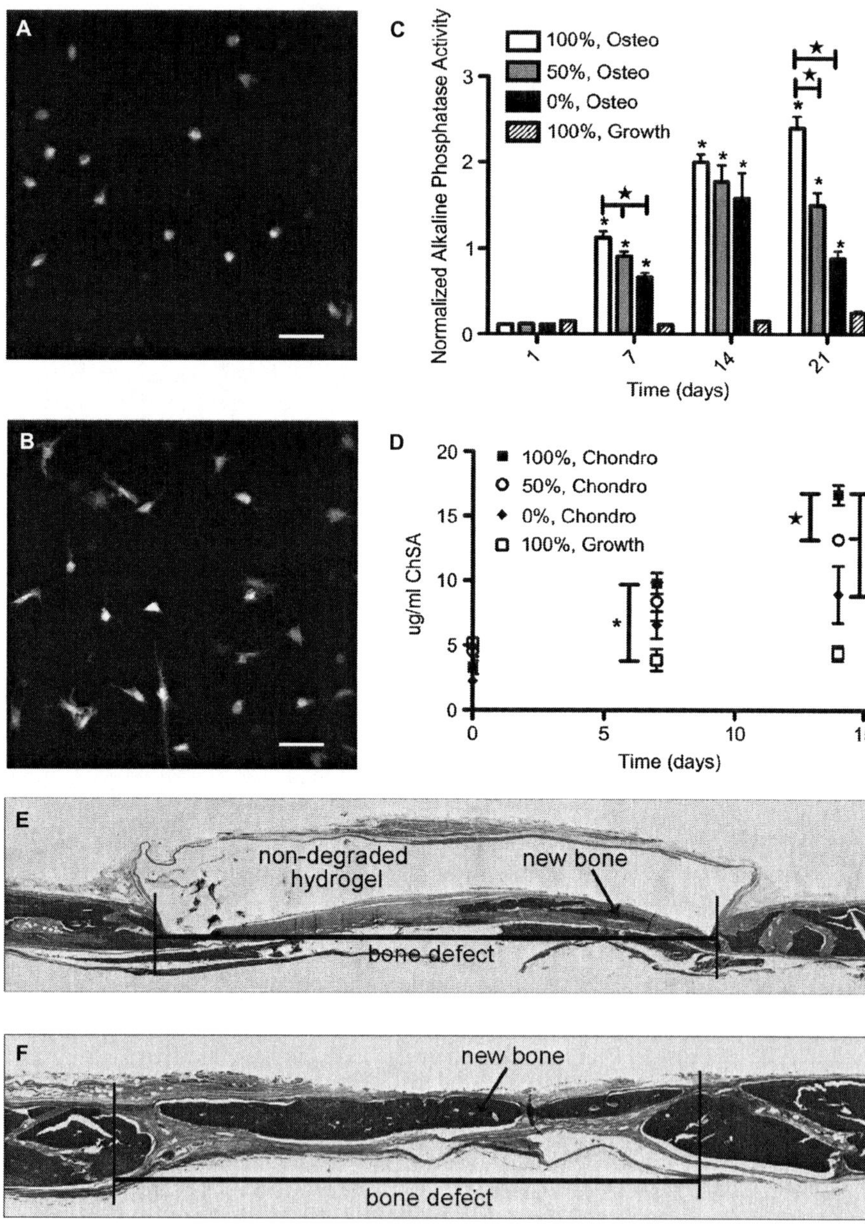

enzymatically degradable peptides to allow for cell-mediated degradation of the hydrogel network. Due to the widespread use of chain polymerization methods at the time, enzymatically degradable peptides were originally incorporated as acrylate-PEG-peptide-PEG-acrylate crosslinkers.[92] Although effective, these macromolecules are not trivial to synthesize and are also costly because they require a hetero-bifunctional PEG acrylate (*e.g.*, acrylate-PEG-succinimidyl ester). Thus, the concept of using thiol-ene click chemistry to crosslink a hydrogel with a di-cysteine containing enzymatically degradable peptide sequence and create a cell-degradable polymer network was a key advancement in the field.

A primary interest in tissue engineering has been the creation of matrix metalloproteinase (MMP) sensitive peptide crosslinked hydrogels. This research direction is motivated by the fact that the MMP family of enzymes is naturally utilized by cells to degrade and remodel their ECM. Some of the useful MMP degradable peptide sequences which have been reported in the literature are presented in Table 8.4.[93] One important application of MMP degradable hydrogels is studying cell migration *in vitro*. Raeber *et al.* were the first to demonstrate thiol-X crosslinked MMP degradable PEG hydrogels as a model platform for studying cell migration in 3D using dermal fibroblasts.[94] Schwartz *et al.* built upon this precedent by using a similar MMP degradable hydrogel system to elucidate the mechanisms of fibrosarcoma cell migration.[36] However, applications beyond studying cell migration have also been demonstrated. For example, Shepard *et al.* recently used MMP degradable hydrogels as a platform to study and promote neurite extension from dorsal root ganglia.[95] In an application to stem cell biology, Anderson *et al.* demonstrated that MMP degradability enabled human MSCs to not only proliferate in 3D culture, but also enhanced the expression of lineage specific differentiation markers compared to non-degradable gels (Figure 8.5A–D).[20] For *in vivo* applications, susceptibility of the hydrogel to proteolytic degradation allows for cells to invade the artificial matrix and accelerates tissue healing. Terella *et al.* illustrated this advantage in a rat calvarial bone defect model by showing

**Figure 8.5** Enhancing tissue engineering efficacy by mimicking the enzymatic degradability of the ECM. Human MSCs encapsulated in non-degradable PEG hydrogels exhibit a rounded morphology after 14 days (A), whereas cells in MMP degradable gels are able to degrade the hydrogel and adopt a spread morphology (B). Degradability also improves the ability of MSCs to differentiate down the osteogenic (C) and chondrogenic (D) lineages, as measured by alkaline phosphatase activity and chondroitin sulfate (*i.e.*, ChSA) production, respectively (note: the legends indicate percent degradability of the hydrogel; growth media is a negative control for differentiation). Finally, degrability also enhances *in vivo* tissue formation, as non-degradable PEG hydrogels remain in a rat calvarial bone defect after 6 weeks (E), whereas MMP degradable hydrogel promoted healing and were replaced by large amounts of new bone (F). Scale bars in (A) and (B) are 50 μm.
Adapted from ref. 20 and 96 with permission.

that MMP degradable PEG hydrogels formed by thiol-norbornene photopolymerization were replaced by new bone, while non-degradable PEG-diacrylate gels remained in the defect and prevented healing (Figure 8.5E,F).[96] Similar results were also reported by Rizzi *et al.*[97]

While MMP degradability has been a major focus in the field, thiol-X chemistries can also be used to engineer hydrogels that are susceptible to proteolytic degradation by other enzymes. For example, Aimetti *et al.* used thiol-ene chemistry to crosslink hydrogels with a human neutrophil elastase degradable peptide sequence for enzymatically triggered protein release, which could be useful because of the rapid recruitment of neutrophils to an injury site during tissue repair.[98] Plasmin sensitive hydrogels have also been described by Patterson and Hubbell.[99] Finally, in addition to bulk hydrogel degradation, cell-mediated release of a bioactive drug molecule can also be achieved using thiol-ene chemistry and an enzymatically cleavable tether. Yang *et al.* recently demonstrated this possibility using MMP releasable dexamethasone to induce osteogenic differentiation of encapsulated mesenchymal stem cells.[100] A similar approach using other proteolytically degradable tethers can easily be envisioned and is a topic of broad interest to the biomaterials and tissue engineering communities.

### 8.5.2 Mimicking Cell–ECM and Cell–Cell Interactions

Synthetic materials lack biological functionalities recognized by cells. Thus, in order to promote cell-material interactions, pendant bioactive moieties that mimic cell-ECM interactions are often incorporated into synthetic hydrogels. These are most often synthetic peptide eptitopes derived from ECM proteins, which are readily incorporated using thiol-X chemistry to promote integrin mediated cell attachment (Figure 8.6). The most widely used ECM mimetic peptide is RGD, which is a ubiquitous cell adhesion motif found in multiple ECM proteins including fibronectin, vitronectin, collagen, and laminin.[101] With the exception of ref. 98, which only investigated protein release, all of the studies cited in the previous section used CRGDS or a similar peptide sequence to promote cell attachment along with matrix degradation. However, several other integrin binding peptides that mimic specific ECM proteins are known and can be used to tailor the microenvironment. Peptide sequences that have been used to mimic cell-ECM interactions and promote integrin mediated cell attachement to synthetic hydrogels are summarized in Table 8.4.[102–104]

For some cell types, cell–cell interactions can be equally important or more important than cell–matrix interactions. Two examples are β-cell clusters found in islets and embryonic stem cells, which naturally grow as embryoid bodies rather than as single cells. Importantly, because chemical gradients and hypoxic conditions can exist within a cluster and ultimately lead to heterogeneities in the cell population, biofunctional hydrogels are beginning to prove very useful for better understanding how these cells can be used for regenerative medicine because they could enable single cell culture.

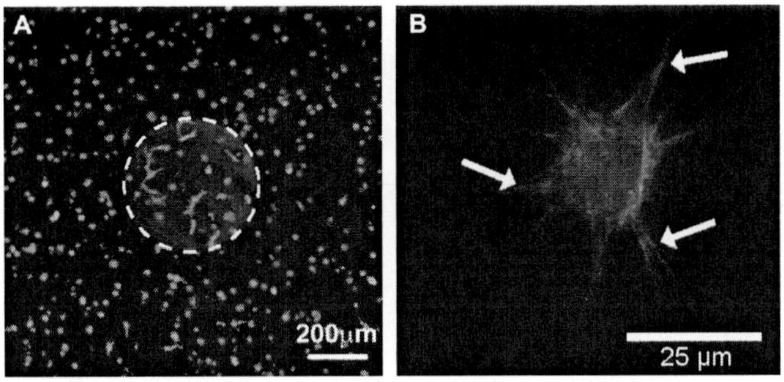


**Figure 8.6** Mimicking cell-ECM interactions with integrin binding peptides. Mono-cysteine containing integrin binding peptides can be incorporated in synthetic PEG hydrogels to promote cell attachment and spreading. (A) 3T3 fibroblasts in PEG hydrogels are only able to adopt a spread morphology in the circular region highlighted by the dashed line, which contains fluorescently labeled RGD peptide. (B) Immunostaining of human MSCs in PEG hydrogels with RGD shows the formation of focal adhesions (indicated by arrows) and the presence of distinct actin stress fibers in the cytoskeleton.
(A) is adapted from ref. 80 with permission.

**Table 8.4** Synthetic peptides for creating enzymatically degradable hydrogels and promoting cell adhesion. ↓ indicates the cleavage site of the enzymatically degradable peptide sequences.

| | *Sequence* | *Comments* |
|---|---|---|
| Enzymatically degradable peptides[a] | GPQGI↓AGQ | wild type MMP cleavage sequence derived from type I collagen |
| | GPQGI↓WGQ | enhanced degradation rate due to Trp substitution |
| | VPMS↓MRGG | optimized for MMP-1 |
| | SGESPAY↓YTA | optimized for MMP-2 |
| | GPLG↓LWAR | optimized for MMP-13 |
| | YAAPV↓RGG | cleaved by human neutrophil elastase |
| Cell adhesion peptides | RGDS | derived from fibronectin; RGD is a ubiquitous adhesion motif[102] |
| | PHSRN | derived from fibronectin; acts as a synergy sequence with RGD[102] |
| | DGEA | derived from collagen[103] |
| | YIGSR | derived from laminin[102] |
| | IKVAV | derived from laminin[102] |
| | CREETAWAC | identified from phage display to bind integrin $\alpha5\beta1$[104] |

[a]All enzymatically degradable sequences are from ref. 93.

Mimicking cell–cell interactions is theoretically feasible using a similar approach to cell–matrix interactions, although the availability of synthetic peptide mimics for cell–cell signaling is currently a limiting factor.

Alternatively, it is also possible to functionalize the polymer network with thiolated proteins. Protein thioloation can be simply achieved by converting primary amines to thiols using 2-iminothiolane,[55] which is commercially available as Traut's reagent. Lin and Anseth used this approach along with a mixed-mode thiol-acrylate polymerization to functionalize PEG hydrogels with EphA5-Fc receptor and ephrinA5-Fc ligand, which are naturally involved in cell–cell signaling between β-cells found in islets of the pancreas.[105] Importantly, this biomimetic approach promoted the survival and function of the pancreatic β-islet cell line MIN6, as well as dissociated primary islet cells at lower cell densities than was otherwise possible (Figure 8.7). While this result demonstrated the utility of mimicking cell–cell interactions, extending the concept to other cell populations and also mimicking other types of cell–cell interactions could be very impactful. The cadherin family of cell adhesion molecules is a particularly interesting target. For example, e-cadherin has been shown to have a crucial role in maintaining embryonic stem cell pluripotency in culture.[106]

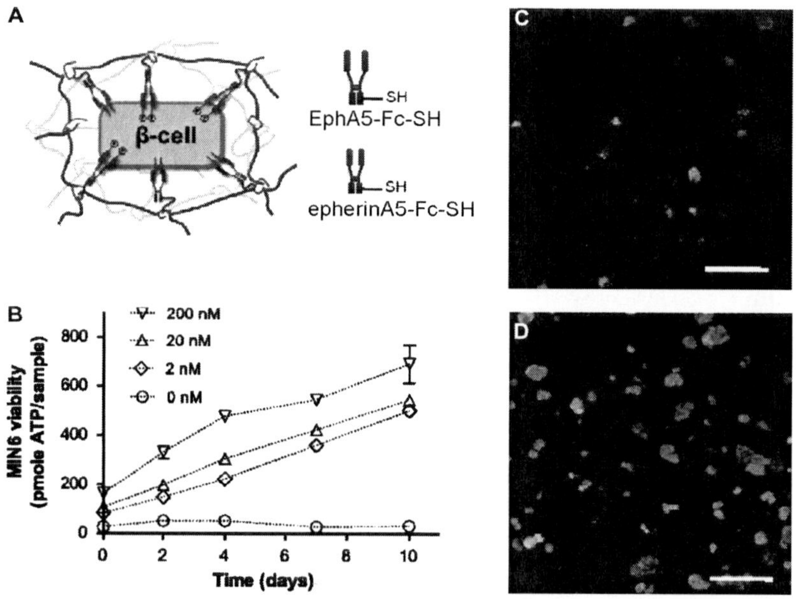


**Figure 8.7** Enhancing tissue engineering by mimicking cell–cell interactions. (A) To mimic cell–cell signaling between beta cells within islet clusters, PEG hydrogels were functionalized with the EphA5-Fc receptor and epherin A5-Fc ligand. (B) Increasing the protein concentration from 0 to 200 nM significantly improved MIN6 β-cell viability over 10 days *in vitro*. This result was confirmed by Live/Dead staining at day 10, which showed poor viability for cells encapsulated without protein (C) but excellent cell viability and proliferation when the cells were encapsulated with 200 nM protein (D). Scale bars in (C) and (D) are 200 μm.
Adapted with permission from ref. 105.

### 8.5.3 Growth Factor and Cytokine Functionalization

Growth factors and cytokines are key elements in many tissue engineering strategies because of their natural roles in tissue morphogenesis, homeostasis, and healing. Conventionally, growth factor and cytokine delivery have been achieved by simply entrapping proteins in cell-laden biomaterial matrices and relying on diffusion to control their local and temporal presentation. Although this approach can be effective, it is often inefficient due to an initial burst release during which a large amount of the therapeutic is released within hours. For many applications, it is desirable to control the protein release locally and for prolonged periods of time, all while protecting activity. With this goal in mind, a very interesting approach is to use non-covalent affinity binding interactions, which effectively slow the rate of diffusion, increasing protein retention time within a hydrogel and sustaining release.[107,108] Alternatively, as discussed in Section 8.3, the bioorthogonality of thiol-ene click reactions also makes it possible to covalently tether proteins into polymer networks with minimal loss of bioactivity during the polymerization reaction. In contrast to controlled release strategies, covalently tethered proteins have the potential to promote localized and sustained signaling.

As mentioned in Section 8.5.2, proteins thiolated with 2-iminothiolane can be used to functionalize synthetic hydrogels. Hume *et al.* leveraged this approach to facile protein incorporation to engineer immunosuppressive PEG hydrogels functionalized with transforming growth factor-β1 (TGF-β) and interleukin-10, which effectively reduced dendritic cell maturation and T-cell activation *in vitro*.[109] Similarly, McCall *et al.* used thiolated TGF-β growth factor to create chondrogenic synthetic hydrogel niches for delivering MSCs and promoting cartilage tissue engineering.[110] Importantly, their results showed that tethered TGF-β not only remained bioactive, but it was as effective as soluble growth factor at inducing encapsulated human MSCs to undergo chondrogenic differentiation and synthesize a cartilage-like proteoglycan matrix (Figure 8.8). In a more spatially defined approach based on protein engineering, Wylie and Shoichet used a thiol-maleimide Michael addition to immobilize human serum albumin to discrete regions of an agarose hydrogel and subsequently tethered a fusion protein of fibroblast growth factor-2 and albumin binding domain, which binds with high affinity to human serum albumin.[111] In subsequent work, they used a similar approach to simultaneously immobilize fusion proteins of sonic hedgehog and ciliary neurotrophic factor by utilizing the orthogonal physical binding pairs barnase-barstar and streptaviding-biotin.[112] Importantly, because their approach is based on covalent tethering to photo-deprotected thiols, high fidelity spatio-temporal patterning was achieved.

## 8.6 Future Directions and Opportunities in Tissue Engineering

Thiol-X click chemistry has provided tissue engineering researchers with versatile hydrogel platforms for fundamental *in vitro* studies and also *in vivo*

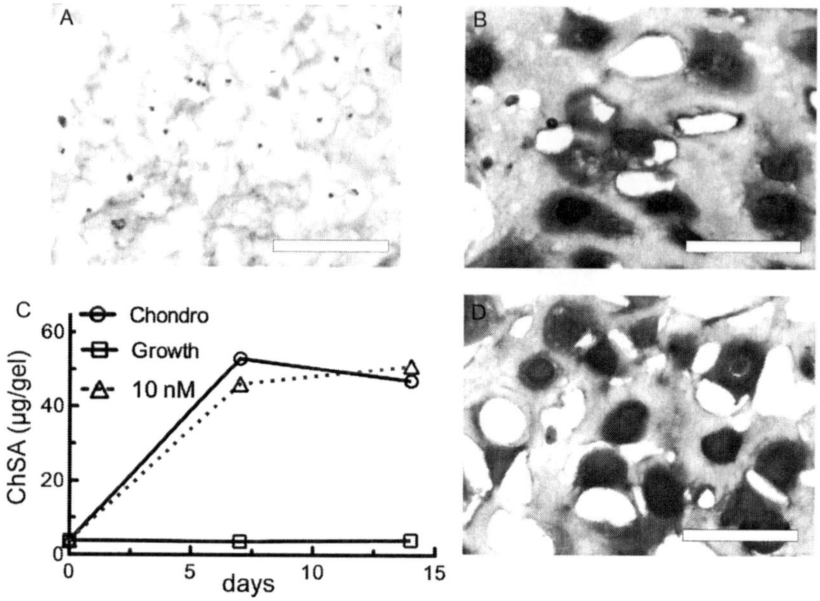


**Figure 8.8** Enhancing tissue engineering through growth factor immobilization in hydrogel matrices. TGF-β is a potent growth factor that drives the chondrogenic differentiation of MSCs. Histological analysis of human MSCs encapsulated in PEG hydrogels and cultured in growth medium (*i.e.*, a negative control) for 3 weeks showed minimal proteoglycan matrix deposition, as indicated by safranin O staining (A). In contrast, MSCs cultured in chondrogenic medium containing soluble TGF-β (*i.e.*, a positive control) differentiated into chondrocytes and secreted a high density of proteoglycan matrix in the pericellular region, which appears dark (B). Importantly, when TGF-β was covalently tethered to the hydrogel network at 10 nM, rather than delivered solubly, MSCs were able to secrete chondroitin sulfate (ChSA), which is a major component of cartilage tissue, at a level comparable to the positive control (C). Histological staining with safaranin O revealed that the persistent tethered TGF-β signal led to extensive proteoglycan deposition similar to what was observed for the positive control (D; image is for gels with 10 nM tethered TGF-β). Scale bars in (A), (B) and (D) are 100 μm.
Adapted with permission from ref. 110.

application. While this impact on the field has been both broad and deep, there are still many important challenges that are critical to the advancement of the field. In this section, we provide a prospectus for future research by outlining some of the key challenges facing the field of tissue engineering and speculate on how thiol-X chemistries might be exploited to address some of these issues.

### 8.6.1 Creating Complex Peptide Architectures with Enhanced Bioactivity

As highlighted in Section 8.5, synthetic peptides have been widely used in conjunction with thiol-X chemistry to create biofunctional hydrogels for tissue engineering. The widespread use of synthetic peptides stems primarily from ease of cysteine incorporation during SPPS to create mono- and di-thiolated peptides. Peptides can also be viewed as minimal functional units of proteins, which are attractive for deconstructing the roles of biochemical cues in directing cell function and understanding which cues are important for tissue engineering. Nevertheless, synthetic peptides often have significantly reduced bioactivity compared to the proteins from which they are derived. For example, the GRGDSP hexapeptide derived from the cell-binding domain of fibronectin is approximately 1000 times less effective at mediating cell attachment than fibronectin itself.[101] Similarly, peptide mimics of growth factors of interest for tissue engineering have been published in the literature (*e.g.*, bone morphogenetic protein-2), but are far less bioactive than the full length protein.[113,114]

Because the protein–protein interactions that mediate receptor-ligand binding are often structurally complex and involve protein folding, the reduced efficacy of synthetic peptides is generally attributed to a lack of conformational stability in short linear sequences, which reduces the likelihood of receptor binding.[115] To address this limitation, a major focus in peptide research has been the introduction of macrocycles to constrain conformation and enhance bioactivity. Thiol-ene chemistry has been demonstrated as a powerful tool for this purpose,[116] which is important considering the general interest in techniques for synthesizing cyclized peptide therapeutics.[117] Perhaps even more powerful, however, is the application of thiol-X chemistry to the synthesis of complex, multivalent peptide architectures. Motivated by the natural occurrence of integrin clustering during cell binding to ECM proteins,[118] Aimetti *et al.* used a photoinitiated thiol-yne reaction, which is a di-addition across the -yne, to create multimers of cyclic RGD that had enhanced bioactivity compared to monomeric peptides.[119] This study demonstrated proof of concept, but it is easy to envision these and other methods, such as peptide stapling to stabilize α-helical confirmations,[120] being applied more broadly across the field. One particularly intriguing possibility is in the design of growth factor mimetic peptides. In addition to enhancing bioactivity through a multivalent design, thiol-X reactions could be used to design peptides conducive to receptor dimerization, which is often a key step in growth factor signaling.[121]

### 8.6.2 User-directed Manipulation of the Cellular Microenvironment

ECM mimetic hydrogels have emerged as powerful tools for studying cells in a more physiologically relevant 3D environment *in vitro*. However, to answer critical questions related to complex biological processes like stem cell differentiation, tissue morphogenesis, and cancer metastasis, the ability to introduce and remove biological cues with spatio-temporal control in a dynamic, user-directed manner is highly desirable. Furthermore, as we synthesize information gained from fundamental studies, such 4D control over the cellular microenvironment could also be leveraged to direct the fate and function of encapsulated cells and orchestrate tissue formation *in vitro*, which is an intriguing prospect.

Photochemical reactions offer a diverse and powerful toolkit for creating dynamic hydrogels that can direct the function of encapsulated cells. For example, photoinitiated thiol-ene polymerizations can be used to pattern biological signals into the hydrogel with spatio-temporal control,[42,122] whereas photolabile tethers can be used to remove a biochemical cue or release a soluble factor at a precise time.[123–125] In addition, hydrogels can be crosslinked with a peptide or macromer containing a photolabile moiety to enable predictable tuning of the matrix stiffness or even the introduction of physical features like channels, all in the presence of cells.[89,123,126,127] Finally, through the strategic placement of a photolabile moiety on a peptide or drug molecule, spatio-temporally controlled photo-activation of a "caged" biological cue is possible.[128–130]

While thiol-ene photopatterning, photodegradation/release, and caging have been demonstrated independently to be very powerful tools for tissue engineering, combining these tools into one system is an important challenge that will expand our ability to interrogate the cellular microenvironment and enable more sophisticated experiments. DeForest and Anseth have made significant progress towards this lofty goal by capitalizing on the orthogonality of eosin Y photoinitiaton at 490–650 nm (860 nm for two photon) and nitrobenzyl ether photodegradation at 365 nm (740 nm for two photon) to combine thiol-ene photopatterning and photodegradation into one biomaterial system. In two separate publications they showed that it was possible to photopattern an RGD peptide and then either completely degrade a region of the hydrogel (*e.g.*, to create a path for cell migration)[126] or release the peptide to remove the signal from discrete regions of the gel.[125] Importantly, while this work demonstrated the power of combining orthogonal chemistries, only two photochemical reactions were used. An important challenge for future research will be implementing additional orthogonal chemistries, multiple wavelengths, and reversibility to provide researchers with diverse means to tune the cellular microenvironment. Furthermore, integrating these elegant chemistries to direct a specific cellular response is also an important direction for future research.

### 8.6.3 Beyond Scaffolds

The primary focus of this chapter has been the application of thiol-X chemistries for the creation of cell-laden hydrogel scaffolds. However, opportunities beyond scaffolds exist for the application of thiol-X chemistries in tissue engineering and regenerative medicine. For example, drug delivery is inherently linked to tissue engineering, and just as thiol-X crosslinked hydrogel scaffolds can be designed to deliver drugs and proteins to encapsulated cells, these bio-orthogonal chemistries can be leveraged to manufacture advanced drug delivery platforms. For example, Tibbitt *et al.* recently developed a photodegradable hydrogel microparticle platform for on demand release of bioactive proteins.[131] A key element of their material design, however, was thiol-X chemistry, as the protein payload was entrapped during microparticle crosslinking *via* a thiol Michael addition. An additional example of how thiol-X chemistries can be applied is the preparation of hydrogel microarrays for high throughput diagnostics and screening of biomolecules, as demonstrated by Hawker, Gupta and coworkers.[132] These two examples are powerful illustrations of how the utility of thiol-X chemistries goes beyond scaffolds. Ultimately, creative applications such as these will come full circle and advance the field of tissue engineering by enabling the design and implementation of better biomaterials based strategies.

## References

1. W. M. Saltzman, *Tissue engineering: engineering principles for the design of replacement organs and tissues*, Oxford University Press, Oxford, 2004.
2. N. A. Peppas, J. Z. Hilt, A. Khademhosseini and R. Langer, *Adv. Mater.*, 2006, **18**, 1345–1360.
3. C. N. Salinas and K. S. Anseth, *J. Dent. Res.*, 2009, **88**, 681–692.
4. J. A. Burdick, *Biomed. Mater.*, 2012, **7**, 020201–020201.
5. G. D. Nicodemus and S. J. Bryant, *Tissue Eng., Part B*, 2008, **14**, 149–165.
6. D. J. Mooney and H. Vandenburgh, *Cell Stem Cell*, 2008, **2**, 205–213.
7. N.-Y. Choi, B.-W. Kim, W.-J. Yeo, H.-B. Kim, D.-S. Suh, J.-S. Kim, Y.-S. Kim, Y.-H. Seo, J.-Y. Cho, C.-W. Chun, H.-S. Park, A. A. Shetty and S.-J. Kim, *BMC Musculoskelet. Disord.*, 2010, **11**, 103.
8. Z. Liu, H. Wang, Y. Wang, Q. Lin, A. Yao, F. Cao, D. Li, J. Zhou, C. Duan, Z. Du, Y. Wang and C. Wang, *Biomaterials*, 2012, **33**, 3093–3106.
9. M. Habib, K. Shapira-Schweitzer, O. Caspi, A. Gepstein, G. Arbel, D. Aronson, D. Seliktar and L. Gepstein, *Biomaterials*, 2011, **32**, 7514–7523.
10. G. Zhang, Q. Hu, E. A. Braunlin, L. J. Suggs and J. Zhang, *Tissue Eng. Part A*, 2008, **14**, 1025–1036.
11. D. R. Nisbet, K. E. Crompton, M. K. Horne, D. I. Finkelstein and J. S. Forsythe, *J. Biomed. Mater. Res. Part B Appl. Biomater.*, 2008, **87**, 251–263.
12. L. Pan, Y. Ren, F. Cui and Q. Xu, *J. Neurosci. Res.*, 2009, **87**, 3207–3220.
13. H. Kim, M. J. Cooke and M. S. Shoichet, *Trends Biotechnol.*, 2012, **30**, 55–63.
14. V. A. Spencer, R. Xu and M. J. Bissell, *J. Mammary Gland Biol. Neoplasia*, 2010, **15**, 65–71.
15. K. S. M. Smalley, M. Lioni and M. Herlyn, *In Vitro Cell. Dev. Biol. Anim.*, 2006, **42**, 242–247.
16. J. S. Harunaga and K. M. Yamada, *Matrix Biol.*, 2011, **30**, 363–368.
17. M. C. Cushing and K. S. Anseth, *Science*, 2007, **316**, 1133–1134.
18. M. W. Tibbitt and K. S. Anseth, *Biotechnol. Bioeng.*, 2009, **103**, 655–663.
19. D. Huh, G. A. Hamilton and D. E. Ingber, *Trends Cell Biol.*, 2011, **21**, 745–754.

20. S. B. Anderson, C.-C. Lin, D. V. Kuntzler and K. S. Anseth, *Biomaterials*, 2011, **32**, 3564–3574.
21. S. Gerecht, J. A. Burdick, L. S. Ferreira, S. A. Townsend, R. Langer and G. Vunjak-Novakovic, *Proc. Natl. Acad. Sci. U. S. A.*, 2007, **104**, 11298–11303.
22. E. Demange, Y. Kassim, C. Petit, C. Buquet, V. Dulong, D. L. Cerf, G. Buchonnet and J.-P. Vannier, *J. Tissue Eng. Regener. Med.*, 2012.
23. B. M. Rüger, J. Breuss, D. Hollemann, G. Yanagida, M. B. Fischer, I. Mosberger, A. Chott, I. Lang, P. F. Davis, P. Höcker and M. Dettke, *Differentiation*, 2008, **76**, 772–783.
24. S. Rhee, *Exp. Mol. Med.*, 2009, **41**, 858–865.
25. F. Grinnell, *Trends Cell Biol.*, 2003, **13**, 264–269.
26. S. C. Skaalure, I. L. Milligan and S. J. Bryant, *Biomed Mater*, 2012, **7**, 024111.
27. G. D. Nicodemus, S. C. Skaalure and S. J. Bryant, *Acta Biomater*, 2011, **7**, 492–504.
28. S. Degala, R. Williams, W. Zipfel and L. J. Bonassar, *J. Orthop. Res.*, 2012, **30**, 793–799.
29. E. Schuh, S. Hofmann, K. S. Stok, H. Notbohm, R. Müller and N. Rotter, *J. Tissue Eng. Regen. Med.*, 2011.
30. J. P. Frampton, M. R. Hynd, M. L. Shuler and W. Shain, *Biomed. Mater.*, 2011, **6**, 015002.
31. S. K. Seidlits, Z. Z. Khaing, R. R. Petersen, J. D. Nickels, J. E. Vanscoy, J. B. Shear and C. E. Schmidt, *Biomaterials*, 2010, **31**, 3930–3940.
32. S. K. Seidlits, C. T. Drinnan, R. R. Petersen, J. B. Shear, L. J. Suggs and C. E. Schmidt, *Acta Biomater.*, 2011, **7**, 2401–2409.
33. S. J. Grainger and A. J. Putnam, *PLoS ONE*, 2011, **6**, e22086.
34. J. E. Leslie-Barbick, J. E. Saik, D. J. Gould, M. E. Dickinson and J. L. West, *Biomaterials*, 2011, **32**, 5782–5789.
35. M. A. Cichon, V. G. Gainullin, Y. Zhang and D. C. Radisky, *Integr. Biol.*, 2012, **4**, 440.
36. M. P. Schwartz, B. D. Fairbanks, R. E. Rogers, R. Rangarajan, M. H. Zaman and K. S. Anseth, *Integr. Biol.*, 2010, **2**, 32–40.
37. J.-M. Nam, Y. Onodera, M. J. Bissell and C. C. Park, *Cancer Res.*, 2010, **70**, 5238–5248.
38. J. A. Pedersen and M. A. Swartz, *Ann. Biomed. Eng.*, 2005, **33**, 1469–1490.
39. F. Biname, G. Pawlak, P. Roux and U. Hibner, *Mol. BioSyst.*, 2010, **6**, 648.
40. C. M. Nimmo and M. S. Shoichet, *Bioconjug. Chem.*, 2011, **22**, 2199–2209.
41. M. Malkoch, R. Vestberg, N. Gupta, L. Mespouille, P. Dubois, A. F. Mason, J. L. Hedrick, Q. Liao, C. W. Frank, K. Kingsbury and C. J. Hawker, *Chem. Commun.*, 2006, 2774–2776.
42. B. D. Polizzotti, B. D. Fairbanks and K. S. Anseth, *Biomacromolecules*, 2008, **9**, 1084–1087.
43. E. Yoshii, *J. Biomed. Mater. Res.*, 1997, **37**, 517–524.

44. H. Shin, J. S. Temenoff and A. G. Mikos, *Biomacromolecules*, 2003, **4**, 552–560.
45. C. R. Nuttelman, M. A. Rice, A. E. Rydholm, C. N. Salinas, D. N. Shah and K. S. Anseth, *Prog. Polym. Sci.*, 2008, **33**, 167–179.
46. J. L. Ifkovits and J. A. Burdick, *Tissue Eng.*, 2007, **13**, 2369–2385.
47. K. T. Nguyen and J. L. West, *Biomaterials*, 2002, **23**, 4307–4314.
48. G. P. Pfeifer, Y.-H. You and A. Besaratinia, *Mutat. Res.*, 2005, **571**, 19–31.
49. S. J. Bryant, C. R. Nuttelman and K. S. Anseth, *J. Biomater. Sci. Polym. Ed.*, 2000, **11**, 439–457.
50. C. G. Williams, A. N. Malik, T. K. Kim, P. N. Manson and J. H. Elisseeff, *Biomaterials*, 2005, **26**, 1211–1218.
51. B. D. Fairbanks, M. P. Schwartz, C. N. Bowman and K. S. Anseth, *Biomaterials*, 2009, **30**, 6702–6707.
52. C. S. Bahney, T. J. Lujan, C. W. Hsu, M. Bottlang, J. L. West and B. Johnstone, *Eur. Cell Mater.*, 2011, **22**, 43–55; discussion 55.
53. J. Hu, Y. Hou, H. Park, B. Choi, S. Hou, A. Chung and M. Lee, *Acta Biomater.*, 2012, **8**, 1730–1738.
54. J. Zhu, *Biomaterials*, 2010, **31**, 4639–4656.
55. M. Brinkley, *Bioconjugate Chem.*, 1992, **3**, 2–13.
56. D. L. Elbert, A. B. Pratt, M. P. Lutolf, S. Halstenberg and J. A. Hubbell, *J. Control. Release*, 2001, **76**, 11–25.
57. A. Metters and J. Hubbell, *Biomacromolecules*, 2005, **6**, 290–301.
58. M. P. Lutolf, J. L. Lauer-Fields, H. G. Schmoekel, A. T. Metters, F. E. Weber, G. B. Fields and J. A. Hubbell, *Proc. Natl. Acad. Sci. U. S. A.*, 2003, **100**, 5413–5418.
59. M. P. Lutolf, G. P. Raeber, A. H. Zisch, N. Tirelli and J. A. Hubbell, *Adv. Mater.*, 2003, **15**, 888–892.
60. M. P. Lutolf and J. A. Hubbell, *Biomacromolecules*, 2003, **4**, 713–722.
61. E. A. Phelps, N. O. Enemchukwu, V. F. Fiore, J. C. Sy, N. Murthy, T. A. Sulchek, T. H. Barker and A. J. García, *Adv. Mater.*, 2012, **24**, 64–70.
62. S. Q. Liu, Q. Tian, J. L. Hedrick, J. H. Po Hui, P. L. R. Ee and Y. Y. Yang, *Biomaterials*, 2010, **31**, 7298–7307.
63. C. Adelöw, T. Segura, J. A. Hubbell and P. Frey, *Biomaterials*, 2008, **29**, 314–326.
64. M. Kim, J. Y. Lee, C. N. Jones, A. Revzin and G. Tae, *Biomaterials*, 2010, **31**, 3596–3603.
65. A. Conovaloff and A. Panitch, *J. Neural Eng.*, 2011, **8**, 056003.
66. A. B. Pratt, F. E. Weber, H. G. Schmoekel, R. Müller and J. A. Hubbell, *Biotechnol. Bioeng.*, 2004, **86**, 27–36.
67. W. Zhu and J. Ding, *J. Appl. Polym. Sci.*, 2006, **99**, 2375–2383.
68. C. R. Nuttelman, M. C. Tripodi and K. S. Anseth, *Matrix Biol.*, 2005, **24**, 208–218.
69. D. Wang, C. G. Williams, Q. Li, B. Sharma and J. H. Elisseeff, *Biomaterials*, 2003, **24**, 3969–3980.

70. A. S. Gobin and J. L. West, *Biotechnol. Prog.*, 2003, **19**, 1781–1785.
71. R. H. Schmedlen, K. S. Masters and J. L. West, *Biomaterials*, 2002, **23**, 4325–4332.
72. C. R. Nuttelman, S. M. Henry and K. S. Anseth, *Biomaterials*, 2002, **23**, 3617–3626.
73. J. Baier Leach, K. A. Bivens, C. W. Patrick Jr and C. E. Schmidt, *Biotechnol. Bioeng.*, 2003, **82**, 578–589.
74. K. S. Masters, D. N. Shah, L. A. Leinwand and K. S. Anseth, *Biomaterials*, 2005, **26**, 2517–2525.
75. D. L. Nettles, T. P. Vail, M. T. Morgan, M. W. Grinstaff and L. A. Setton, *Ann. Biomed. Eng.*, 2004, **32**, 391–397.
76. C. N. Salinas, B. B. Cole, A. M. Kasko and K. S. Anseth, *Tissue Eng.*, 2007, **13**, 1025–1034.
77. C. N. Salinas and K. S. Anseth, *Macromolecules*, 2008, **41**, 6019–6026.
78. A. E. Rydholm, C. N. Bowman and K. S. Anseth, *Biomaterials*, 2005, **26**, 4495–4506.
79. A. E. Rydholm, N. L. Held, D. S. W. Benoit, C. N. Bowman and K. S. Anseth, *J. Biomed. Mater. Res. A*, 2008, **86**, 23–30.
80. B. D. Fairbanks, M. P. Schwartz, A. E. Halevi, C. R. Nuttelman, C. N. Bowman and K. S. Anseth, *Adv. Mater.*, 2009, **21**, 5005–5010.
81. N. B. Cramer, S. K. Reddy, A. K. O'Brien and C. N. Bowman, *Macromolecules*, 2003, **36**, 7964–7969.
82. A. Valdebenito, P. Espinoza, E. A. Lissi and M. V. Encinas, *Polymer*, 2010, **51**, 2503–2507.
83. A. K. O'Brien, N. B. Cramer and C. N. Bowman, *J. Polym. Sci., Part A: Polym. Chem.*, 2006, **44**, 2007–2014.
84. S. K. Reddy, N. B. Cramer and C. N. Bowman, *Macromolecules*, 2006, **39**, 3673–3680.
85. C.-C. Lin, A. Raza and H. Shih, *Biomaterials*, 2011, **32**, 9685–9695.
86. S. T. Gould, N. J. Darling and K. S. Anseth, *Acta Biomater.*, 2012.
87. J. A. Benton, B. D. Fairbanks and K. S. Anseth, *Biomaterials*, 2009, **30**, 6593–6603.
88. S. Khetan and J. Burdick, *J. Vis. Exp.*, 2009.
89. A. M. Kloxin, M. W. Tibbitt and K. S. Anseth, *Nat. Protoc.*, 2010, **5**, 1867–1887.
90. G. L. Ellman, *Arch. Biochem. Biophys.*, 1959, **82**, 70–77.
91. M. C. Phelan, *Curr. Protoc. Cell Biol.*, 2007, **Chapter 1**, Unit 1.1.
92. A. S. Gobin and J. L. West, *FASEB J.*, 2002, **16**, 751–753.
93. J. Patterson and J. A. Hubbell, *Biomaterials*, 2010, **31**, 7836–7845.
94. G. P. Raeber, M. P. Lutolf and J. A. Hubbell, *Biophys. J.*, 2005, **89**, 1374–1388.
95. J. A. Shepard, A. C. Stevans, S. Holland, C. E. Wang, A. Shikanov and L. D. Shea, *Biotechnol. Bioeng.*, 2012, **109**, 830–839.
96. A. Terella, P. Mariner, N. Brown, K. Anseth and S.-O. Streubel, *Arch. Facial Plast. Surg.*, 2010, **12**, 166–171.

97. S. C. Rizzi, M. Ehrbar, S. Halstenberg, G. P. Raeber, H. G. Schmoekel, H. Hagenmüller, R. Müller, F. E. Weber and J. A. Hubbell, *Biomacromolecules*, 2006, **7**, 3019–3029.
98. A. A. Aimetti, A. J. Machen and K. S. Anseth, *Biomaterials*, 2009, **30**, 6048–6054.
99. J. Patterson and J. A. Hubbell, *Biomaterials*, 2011, **32**, 1301–1310.
100. C. Yang, P. D. Mariner, J. N. Nahreini and K. S. Anseth, *J. Control. Release*, 2012, **162**, 612–618.
101. E. Ruoslahti, *Annu. Rev. Cell Dev. Biol.*, 1996, **12**, 697–715.
102. G. Delaittre, A. M. Greiner, T. Pauloehrl, M. Bastmeyer and C. Barner-Kowollik, *Soft Matter*, 2012, **8**, 7323–7347.
103. W. D. Staatz, K. F. Fok, M. M. Zutter, S. P. Adams, B. A. Rodriguez and S. A. Santoro, *J. Biol. Chem.*, 1991, **266**, 7363–7367.
104. Z. Hamidouche, O. Fromigué, J. Ringe, T. Häupl, P. Vaudin, J.-C. Pagès, S. Srouji, E. Livne and P. J. Marie, *Proc. Natl. Acad. Sci. U. S. A.*, 2009, **106**, 18587–18591.
105. C.-C. Lin and K. S. Anseth, *Proc. Natl. Acad. Sci. U. S. A.*, 2011, **108**, 6380–6385.
106. M. Nagaoka, U. Koshimizu, S. Yuasa, F. Hattori, H. Chen, T. Tanaka, M. Okabe, K. Fukuda and T. Akaike, *PLoS ONE*, 2006, **1**, e15.
107. C.-C. Lin and K. S. Anseth, *Adv. Funct. Mater.*, 2009, **19**, 2325.
108. C.-C. Lin, P. D. Boyer, A. A. Aimetti and K. S. Anseth, *J. Control. Release*, 2010, **142**, 384–391.
109. P. S. Hume, J. He, K. Haskins and K. S. Anseth, *Biomaterials*, 2012, **33**, 3615–3625.
110. J. D. McCall, J. E. Luoma and K. S. Anseth, *Drug Delivery Transl. Res.*, 2012, **2**, 305–312.
111. R. G. Wylie and M. S. Shoichet, *Biomacromolecules*, 2011, **12**, 3789–3796.
112. R. G. Wylie, S. Ahsan, Y. Aizawa, K. L. Maxwell, C. M. Morshead and M. S. Shoichet, *Nat. Mater.*, 2011, **10**, 799–806.
113. Y.-J. Seol, Y.-J. Park, S.-C. Lee, K.-H. Kim, J.-Y. Lee, T.-I. Kim, Y.-M. Lee, Y. Ku, I.-C. Rhyu, S.-B. Han and C.-P. Chung, *J. Biomed. Mater. Res. A*, 2006, **77**, 599–607.
114. A. Saito, Y. Suzuki, S. Ogata, C. Ohtsuki and M. Tanihara, *Biochim. Biophys. Acta*, 2003, **1651**, 60–67.
115. J. Eichler, *Curr. Opin. Chem. Biol.*, 2008, **12**, 707–713.
116. A. A. Aimetti, R. K. Shoemaker, C.-C. Lin and K. S. Anseth, *Chem. Commun.*, 2010, **46**, 4061–4063.
117. C. J. White and A. K. Yudin, *Nat. Chem.*, 2011, **3**, 509–524.
118. D. Lepzelter, O. Bates and M. Zaman, *Langmuir*, 2012, **28**, 5379–5386.
119. A. A. Aimetti, K. R. Feaver and K. S. Anseth, *Chem. Commun.*, 2010, **46**, 5781–5783.
120. L. K. Henchey, A. L. Jochim and P. S. Arora, *Curr. Opin. Chem. Biol.*, 2008, **12**, 692–697.
121. M. A. Lemmon and J. Schlessinger, *Cell*, 2010, **141**, 1117–1134.

122. C. A. DeForest, B. D. Polizzotti and K. S. Anseth, *Nat. Mater.*, 2009, **8**, 659–664.
123. A. M. Kloxin, A. M. Kasko, C. N. Salinas and K. S. Anseth, *Science*, 2009, **324**, 59–63.
124. D. R. Griffin, J. T. Patterson and A. M. Kasko, *Biotechnol. Bioeng.*, 2010, **107**, 1012–1019.
125. C. A. DeForest and K. S. Anseth, *Angew. Chem., Int. Ed. Engl.*, 2012, **51**, 1816–1819.
126. C. A. DeForest and K. S. Anseth, *Nat. Chem.*, 2011, **3**, 925–931.
127. M. W. Tibbitt, A. M. Kloxin, K. U. Dyamenahalli and K. S. Anseth, *Soft Matter*, 2010, **6**, 5100–5108.
128. H.-M. Lee, D. R. Larson and D. S. Lawrence, *ACS Chem. Biol.*, 2009, **4**, 409–427.
129. A. R. Morckel, H. Lusic, L. Farzana, J. A. Yoder, A. Deiters and N. M. Nascone-Yoder, *Development*, 2012, **139**, 437–442.
130. S. Petersen, J. M. Alonso, A. Specht, P. Duodu, M. Goeldner and A. del Campo, *Angew. Chem., Int. Ed. Engl.*, 2008, **47**, 3192–3195.
131. M. W. Tibbitt, B. W. Han, A. M. Kloxin and K. S. Anseth, *J. Biomed. Mater. Res. A*, 2012, **100**, 1647–1654.
132. N. Gupta, B. F. Lin, L. M. Campos, M. D. Dimitriou, S. T. Hikita, N. D. Treat, M. V. Tirrell, D. O. Clegg, E. J. Kramer and C. J. Hawker, *Nature Chem.*, 2010, **2**, 138–145.

CHAPTER 9

# Thiolactones as Functional Handles for Polymer Synthesis and Modification

PIETER ESPEEL, FABIENNE GOETHALS AND FILIP E. DU PREZ*

Department of Organic Chemistry, Polymer Chemistry Research Group, Ghent University, Krijgslaan 281 S4-bis, B-9000 Ghent, Belgium
*Email: Filip.duprez@Ugent.be

## 9.1 Introduction

The continuous demand for custom materials, devised for specific and high tech applications in a wide area of academic domains, undeniably challenges polymer chemists. Original synthetic strategies should encompass straightforward, mild and widely applicable methods. Robust and efficient organic reactions, generally labeled as 'click' chemistry[1–6] and originally developed and applied in the synthesis of low-molecular-weight compounds, infiltrated the toolbox of synthetic macro- and supramolecular research groups during the last decade.[2,7,8] The 'click' approach has led to the extensive development and evaluation of new types of chemistry and polymer materials with a broad range of applications (medicine,[9–12] electronics,[13–15] bio-conjugation,[16–20] labelling,[21–25] *etc.*). The decoration of polymeric materials with reactive functional handles and their subsequent post-polymerization modification,[26] ideally *via* 'click' chemistry, ultimately lead to a plethora of potential materials, possessing unique physicochemical properties. The synergy between

RSC Polymer Chemistry Series No. 6
Thiol-X Chemistries in Polymer and Materials Science
Edited by Andrew B. Lowe and Christopher N. Bowman

Published by the Royal Society of Chemistry, www.rsc.org

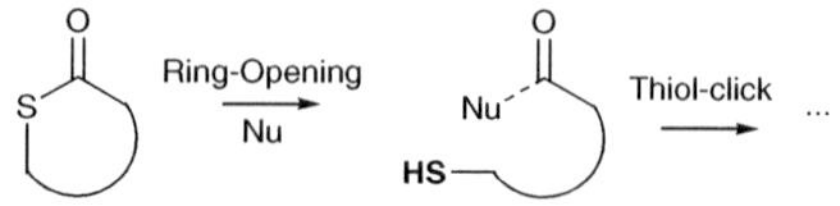


**Scheme 9.1** A thiolactone as a latent thiol functionality: the thiol is released by nucleophilic ring-opening and subsequently a thiol-click can occur.

functionality and modularity is the key aspect of this popular approach.[27] Established methods are generally employed for the incorporation of these reactive entities *via* customized monomers and initiators.[28] Research was recently more focused on the nature of 'click' reactions, more specifically the synthetic availability of the 'clickable' functionalities and the reaction conditions (stoichiometry, catalyst, and temperature).[6,7] This resulted in the exploration of a myriad of possible click reactions and, while some orthogonal reactions are amenable to multi-step one-pot sequences, new paths to multifunctional materials *via* accelerated synthetic protocols became available.[29,30] Moreover, the use of renewable resources as starting materials is a strong driving force for extensive worldwide research and development.[31] Polymer synthesis based on these alternative building blocks generally requires adapted or new synthetic methods. Ideally, these do not rely on the use of toxic or hazardous chemicals and guarantee a minimal amount of generated waste and energy consumption.

Conjugation and modification reactions involving thiols have lately gained momentum as metal-free alternatives to the heavily exploited copper assisted azide–alkyne cyclo-addition (CuAAC).[32–34] In addition to the nucleophilic nature of thiols in the presence of alkyl halides, epoxides, isocyanates and Michael acceptors, thiyl radicals readily react with double and triple bonds.[35–39] In order to solve some of the thiol-related issues (smell, shelf life and synthetic availability), the reactivity of a thiolactone (a cyclic thioester) as a latent thiol functionality has been explored: the thiol is released by nucleophilic ring-opening and subsequently reacts with a thiol 'scavenger' (Scheme 9.1). The chemistry of thiolactones and its potential in polymer science will be reviewed in this chapter.

## 9.2 Chemistry of Thiolactones and Derivatives

### 9.2.1 Chemical Structure and Properties of Thiolactones

Thiolactones are cyclic esters of mercapto-acids.[40,41] The most representative members of this class of sulfur-containing compounds are β-, γ- and δ-thiolactones, respectively four-, five- and six-membered rings. α-Thiolactones, being strained three-membered ring compounds, are considered unstable (except for α,α-disubstituted analogues[42]) and are described as intermediates.[43] Additional structural classification distinguishes unsaturated and saturated thiolactones; only the latter will be further discussed. The most widely used synthetic approach for the preparation of saturated γ- and δ-thiolactones is the

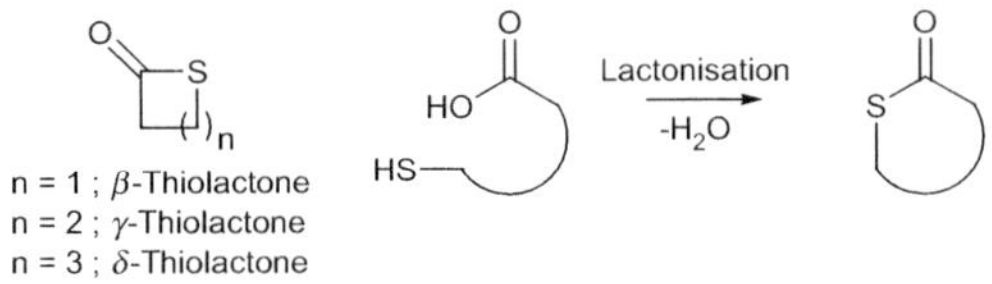


**Scheme 9.2** General structure of saturated thiolactones and schematic depiction of the preparation of thiolactones by dehydration of mercapto-acids.

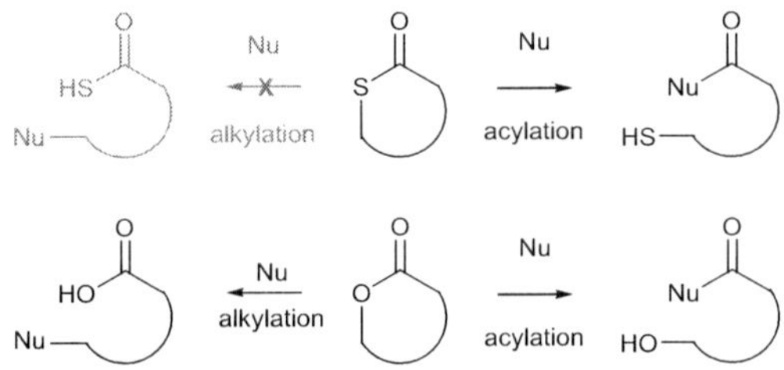


**Scheme 9.3** General alkylation and acylation pathways for the ring opening of thiolactones (top) and lactones (bottom).

direct lactonisation of the corresponding mercapto-acid (Scheme 9.2).[40,41,44–50] Dehydration in diluted media is accelerated by heating and the use of an acid catalyst. Dehydrating agents like carbodiimides and phosphorus pentoxides enable the synthesis of β-thiolactones. In selected cases, cyclization of thiol-containing esters[51,52] and amides[53,54] similarly allows for thiolactone formation.

The most important reactivity-related property of thiolactones is lysis of the ring by the action of a nucleophile. Thiolactones are more sensitive for ring opening than their corresponding lactones, except for β-thiolactones.[40] In addition to their different ring stability, a clear distinction regarding the reactive nature of lactones and thiolactones can be made. Thiolactones are split by nucleophilic agents exclusively at the *S*-acyl bond (nucleophilic addition–elimination), whereas their oxygen analogues are opened by the same agents at either the *O*-acyl or *O*-alkyl (nucleophilic substitution) bonds. Thus, thiolactones only behave as acylating agents, in contrast to lactones that have both acylating and alkylating activity (Scheme 9.3).[40]

Due to the inherent ring strain, the susceptibility of thiolactones towards lysis decreases in following order: β- > γ- > δ-thiolactones. Generally, the stability of the ring increases with the number of substituents by steric congestion of the nucleophilic attack.[40]

Several nucleophiles for the ring opening can be considered; water, alcohols and amines are most prevalent. The hydrolysis and alcoholysis are only significant in basic medium, while aminolysis requires no additives. Aminolysis can thus be performed in aqueous medium, although hydrolysis is an important side reaction. The concomitant high pH-value of aqueous aminolysis should be lowered by buffering the solution. Additionally, the hydrolytic opening can be suppressed in favor of aminolysis by using low temperatures, since the temperature coefficient for the aminolysis rate is exceptionally small.[41,55]

The use of other nucleophiles like thiols and carbanions for the ring opening is far less documented. β-Thiolactones can be opened by treatment with $H_2S$ and $Et_3N$,[48,56–59] while no reports on thiolysis of γ-thiolactones have been found so far. Interestingly, ε-thiolactones (seven-membered thiol equivalent of caprolactone) and β-thiolactones can be polymerized in the absence of water and air to linear polymers by a base-catalyzed ring-opening reaction.[41,60] However, this requires high temperature and prolonged reaction times. Under these conditions, δ-thiolactones react slowly while γ-thiolactones are inert.

Stabilized carbanions such as sodium dimethylmalonate[61] and ethyl lithiodiazoacetate[62–64] can open both γ- and δ-thiolactones. Hydride reduction of γ-thiolactones is a convenient way to prepare α,ω-hydroxyalkanethiols.[65,66]

### 9.2.2 Structural Features, Synthesis and Properties of Homocysteine-γ-thiolactone

α-Amino-γ-butyrothiolactone or homocysteine-γ-thiolactone **1**, a cyclic thioester of homocysteine, is γ-thiolactone with a primary amine in α-position (Scheme 9.4). Homocysteine **4**, a by-product of cellular methylation reactions on dietary methionine **2**, is perhaps the most reactive amino acid in biological systems. Therefore, homocysteine **4** and many of its *in vivo* formed metabolites, including homocysteine-γ-thiolactone **1**, interfere with vital processes, rendering them a subject for intense biomedical and biochemical studies.[67]

Chemical preparation of **1** requires an acid-catalyzed intramolecular condensation of methionine **2**,[68] or homocysteine **4**.[69] The thiolactone **1** absorbs UV-light ($\lambda_{max} = 240$ nm and $\varepsilon \sim 5000\ M^{-1}\ cm^{-1}$ in water). The low $pK_a$ value of 6.67 for the amino group in **1** can be explained by the electron-withdrawing effects of the sulfur atom.[70] The hydrochloric acid salt of the racemate **1** is a white solid (mp 199–203 °C), indefinitely stable at room temperature and readily available (low cost). Under physiological conditions (pH 7.4, 37 °C), the thiolactone **1** has a half-life of one day.[67]

### 9.2.3 Assessment of the Dual Reactivity of Homocysteine-γ-thiolactone

Because of its dual aminoacyl-thioester character, the reactivity of homocysteine-γ-thiolactone **1** is of exceptional practical and theoretical interest.

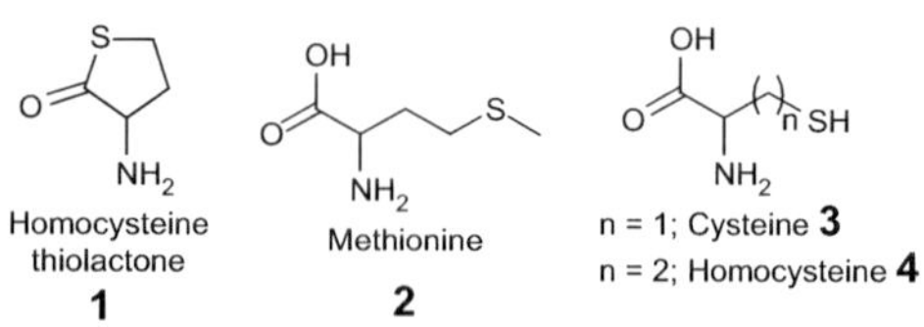


**Scheme 9.4** Chemical structures of homocysteine thiolactone **1**, methionine **2**, cysteine **3** and homocysteine **4**.

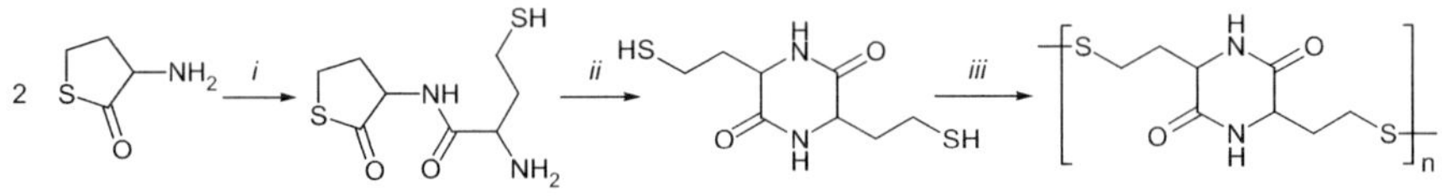


**Scheme 9.5** Mutual aminolysis (an inter- (i), followed by an intramolecular (ii) reaction) of homocysteine thiolactone **1** leads to diketopiperazine (**6**) formation, which polymerizes by disulfide formation (iii).

In fact, it is susceptible for both nucleophilic (*vide supra*) and electrophilic attack. Self-condensation of 2 molecules of **1** occurs in an attempt to obtain the free amino-group, and mutual aminolysis (first intermolecular and second intramolecular aminolysis) takes place with the formation of 3,6-di-(β-mercaptoethyl)-2,5-dioxopiperazine **6**. This dithiol **6** readily polymerizes by disulfide formation (Scheme 9.5).[40]

Homocysteine-γ-thiolactone **1** is a valuable synthetic building block. Due to the intrinsic instability of the neutral homocysteine-γ-thiolactone **1**, an efficient reaction between the amino group in α-position and an electrophile is required to stabilize **1**. Amidation reactions are most frequently used: conjugation with acid halides,[71–74] (*in situ*) activated carboxylic acids[71,75–80] and anhydrides[81] enable the synthesis of homocysteine-γ-thiolactone derivatives. A carbamate linkage is formed by treatment with chloroformates.[82] Another important transformation reaction is imine formation, resulting from the condensation with aldehydes.[70,83,84] The remarkable reactivity of **1** with aldehydes is most likely the result of the low $pK_a$ value of the α-amino group, which is 2–3 units lower than the $pK_a$ values of amino acids. Alkylation reactions of the α-amino group are less common. Reductive amination is, however, reported to be a method to monoalkylate **1**.[85]

The availability of enantiomerically pure homocysteine-γ-thiolactone **1** *via* chiral enrichment comprises resolution of the racemate with a chiral organic acid wherein one isomer precipitates out of the solution as a diastereomeric salt and the remaining isomer is racemized with a catalytic amount of an aromatic aldehyde and resubmitted to optical resolution.[86]

### 9.2.4 Synthetic Applications of Derivatives of Homocysteine-γ-thiolactone

The most commonly used homocysteine-γ-thiolactone derivative is *N*-acetylhomocysteine thiolactone or citiolone **8** (Scheme 9.6), a commercial compound that was introduced as a thiolating agent for proteins. This thiolation consists of the aminolysis of the water-soluble *N*-acetylhomocysteine thiolactone **8** by the ε-$NH_2$ groups of lysine residues and can be accelerated by the addition of silver ions (and imidazole[87]), allowing the reaction to proceed at physiologic pH.[55,88,89] Thiolation of a large variety of macromolecular biochemical systems has been reported.[72,89–99] Citiolone **8** is also a drug that is used as a mucolytic

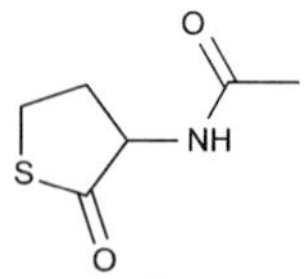

8

**Scheme 9.6** Chemical structure of *N*-acetylhomocysteine thiolactone **8**.

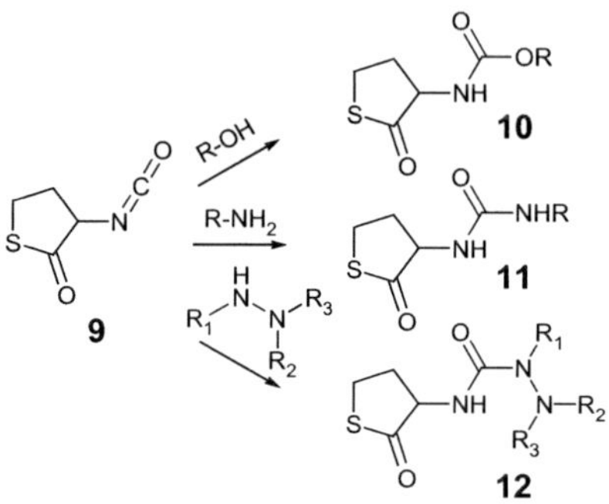

**Scheme 9.7** Derivatization of α-isocyanato-γ-thiolactone **9** by nucleophilic treatment.

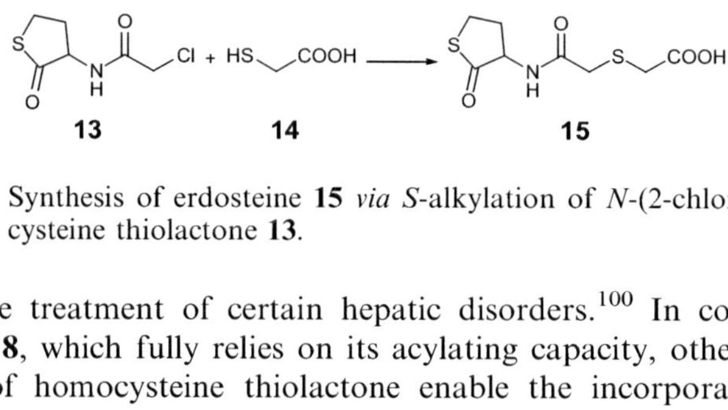

**Scheme 9.8** Synthesis of erdosteine **15** *via* *S*-alkylation of *N*-(2-chloroacetyl)homocysteine thiolactone **13**.

agent for the treatment of certain hepatic disorders.[100] In contrast to the reactivity of **8**, which fully relies on its acylating capacity, other noteworthy derivatives of homocysteine thiolactone enable the incorporation of thiolactones in (macro)molecules. The primordial requirements are the conversion of the α-amino group of **1** to a functional handle and, most importantly, the conjugation to the reactive system of interest. Both steps should proceed while maintaining the integrity of the thiolactone ring. Several of these dual reactive compounds have been synthesized. α-Isocyanato-γ-thiolactone **9** is obtained by phosgene treatment of homocysteine thiolactone **1**. In the presence of alcohols, amines or hydrazines, this isocyanate **9** is converted to the corresponding carbamate **10**, urea **11** or semicarbazide **12** (Scheme 9.7).[101]

Other substrates susceptible for chemoselective nucleophilic attack are thiolactone-containing halides. For example, *N*-(2-chloroacetyl)homocysteine thiolactone **13** reacts with thioglycolic acid **14**, yielding erdosteine **15**, a mucolytic agent for the treatment of bronchitis (Scheme 9.8).[102,103] *N*-(3-Bromopropionyl)homocysteine thiolactone is a useful intermediate for the preparation of several anionic derivatives of homocysteine thiolactone.[72]

A vinylic thiolactone monomer, *N*-methacryloyl homocysteine-γ-thiolactone, has been incorporated in synthetic liposomes by free radical copolymerization in order to induce pH sensitivity.[78]

# 9.3 Thiolactones as Functional Handles for Polymer Synthesis and Modification

## 9.3.1 One-Pot Multistep Reactions Based on Thiolactones: Amine-Thiol-Ene Conjugation

The combination of robust, efficient, and orthogonal conjugation chemistries, resulting in the development of several elegant one-pot, multistep strategies, enables the implementation of accelerated synthetic protocols.[30] An important class of one-pot, multistep reactions is characterized by the fact that all chemicals are initially mixed and that consecutive reactions depend on each other. For example, a potentially explosive azide can be generated *in situ*, starting from the corresponding bromide *via* nucleophilic substitution[104] or amine *via* diazotransfer,[105,106] and can subsequently be converted to a triazole *via* the copper-catalyzed azide-alkyne cycloaddition (CuAAC) in the same pot.

As Malkoch *et al.* expressed the need to increase the range of available 'click' reactions that can be achieved without metal catalysts and to develop libraries of compatible reactions,[30] the development of efficient one-pot processes based on metal-free conjugation chemistries in order to modify or prepare polymeric materials is of particular interest. In addition to the fact that the commercial availability of thiols as starting materials is rather limited, thiols usually have an unpleasant smell and might have a poor shelf life due to oxidation reactions. Therefore, thiolactones are considered to be valuable thiol precursors, potentially offering solutions for some of these thiol-related issues. First of all and most importantly, the thiolactone ring opening, resulting in the release of a free sulfhydryl group, can be achieved by the means of a wide variety of nucleophiles in an orthogonal way (*vide supra*). These latent thiols are not smelly and are recognized to be stable compounds. Moreover, thiolactone substrates can be subjected to one-pot multi-step reaction sequences. This concept has been evaluated mainly by combining hydrolysis (or alcoholysis) and *S*-alkylation. The lysis of thiolactones under basic conditions is often carried out in the presence of an alkylating agent. As a result of the high nucleophilicity of sulfur, *S*-alkylated compounds are the final products of these reactions.[41] Methanolysis of homocysteine-γ-thiolactone hydrochloride **1** and subsequent alkylation by treatment with an alkyl halide in the same pot has been reported as a simple method for the synthesis of *S*-alkylhomocysteines (Scheme 9.9).[107,108]

The basic hydrolysis of homocysteine thiolactone, accompanied by the alkylation of the thiol group, is the key step in the synthesis of several homocysteine derivatives.[109–111]

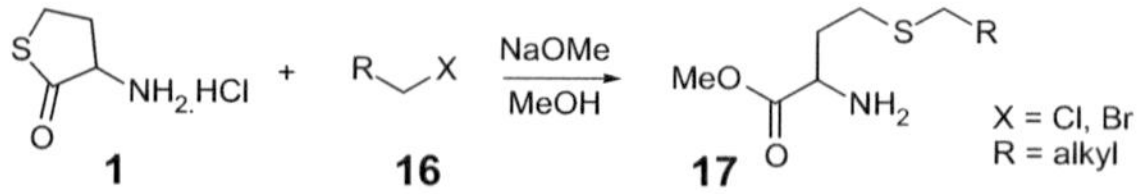


**Scheme 9.9** Two-step one-pot sequence (methanolysis and *S*-alkylation) based on homocysteine-γ-thiolactone **1** enables the synthesis of *S*-alkylhomocysteines **17**.

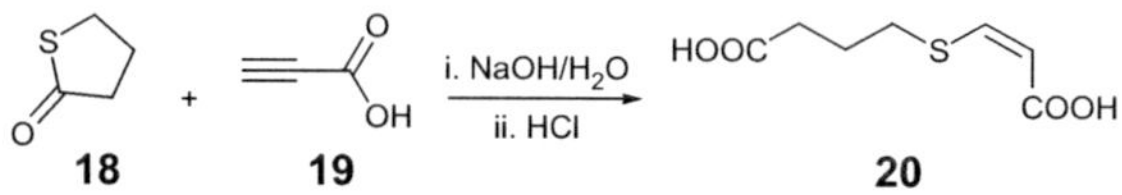


**Scheme 9.10** Treatment of γ-thiobutyrolactone **18** with propiolic acid **19** in basic aqueous medium generates the acrylic acid **20**.

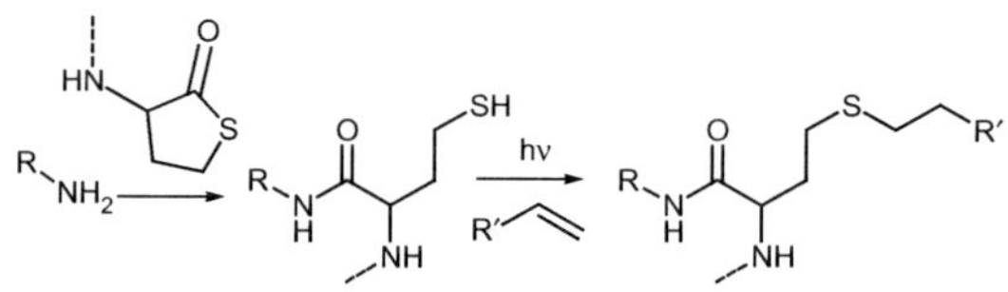


**Scheme 9.11** Schematic depiction of the metal-free one-pot reaction: aminolysis of a thiolactone, followed by a UV-initiated radical thiol-ene conjugation. Scheme adapted from ref. 114.

Other combinations of two-step one-pot reactions have been explored. As a result of sequential hydrolysis and conjugate addition (thio-Michael addition), the reaction of γ-thiobutyrolactone **18** with propiolic acid **19** under basic conditions generates the corresponding acrylic acid **20** (Scheme 9.10).[112]

Other electrophiles, like alkylisothiocyanates,[113] have been employed similarly.

Aminolysis as a start of a one-pot multi-step reaction sequence based on thiolactones is not straightforward, as in many cases orthogonality issues would arise due the reactive nature of amines, but it provides the opportunity to introduce a functional group in the reaction product *via* the amine. It has been demonstrated that the side reactions, as a result of inadequate orthogonality, can be prevented or at least reduced significantly, by combining aminolysis and the radical thiol-ene process, which proceeds *via* a fundamentally different reaction mechanism, in a one-pot fashion. This amine-thiol-ene conjugation as a simple, efficient, and modular linking process has therefore been considered to be a relevant extension of the currently quite popular thiol-ene chemistry, especially in polymer science (Scheme 9.11).[114]

The one-pot two-step sequence initially has been performed on a mixture of benzylamine **21**, *N*-acetylhomocysteine thiolactone **8** and norbornene **22** as low-molecular-weight model compounds in order to master the reaction conditions. The solution was irradiated by an external UV-light source, and 2,2-dimethoxy-2-phenyl acetophenone (DMPA) was selected as an efficient photoinitiator for thiol-ene conjugation.[115] Because of the extremely high reactivity of the norbornene double bond toward thiyl radicals,[35] the first step of the reaction sequence was considered to be rate-determining. An online $^1$H NMR experiment pointed out that **8** was fully consumed after being in the presence of a 2-fold excess of **21** and 10 mol % of DMAP for 6 h. Furthermore, thorough analysis of the reaction mixture obtained after the two-step reaction, revealed the formation of side products originating from the reaction between **21** and radical fragments of DMPA. However, using optimal conditions (no photoinitiator) and after a straightforward chromatographic purification, the model reaction yielded the conjugation compound **23** with an isolated yield of 80% (Scheme 9.12).[114]

Consequently, a monomer containing both a double bond and a thiolactone unit has been prepared. Perrier *et al.*[116,117] reported on thiol-yne chemistry for the synthesis of hyperbranched structures using $AB_2$ type (macro)monomers, molecules bearing a thiol at one end of the molecule and an alkyne at the other. As a thiolactone moiety can be considered as a precursor of a thiol functionality, the combination with a double bond results in an AB′ type monomer. The valorization of the reactivity of this monomer, *N*-(allyloxy)carbonylhomocysteine thiolactone (Alloc-TL) **24**, upon aminolysis and subsequent UV-curing, would enable the development of convenient and accelerated protocols for the synthesis of novel polymeric architectures. A scalable and high-yielding synthesis of the proposed ene-thiolactone monomer **24** consisted of the treatment of homocysteine thiolactone **1** with allyl chloroformate, introducing the corresponding (allyloxy)carbonyl or alloc group, a popular amino-protecting group (Scheme 9.13).

Radical (photo)polymerization reactions between a thiol and an alkene occur in a stepwise fashion. Typically, a gradual increase of the molecular weight with

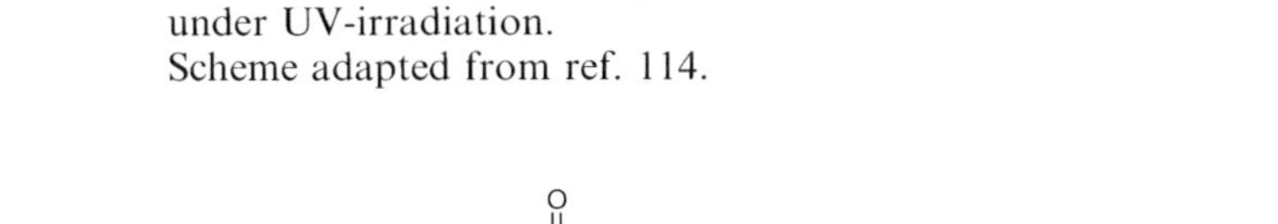


**Scheme 9.12** Model amine-thiol-ene conjugation: one-pot reaction between benzylamine **21**, *N*-acetylhomocysteine thiolactone **8**, and norbornene **22** under UV-irradiation.
Scheme adapted from ref. 114.

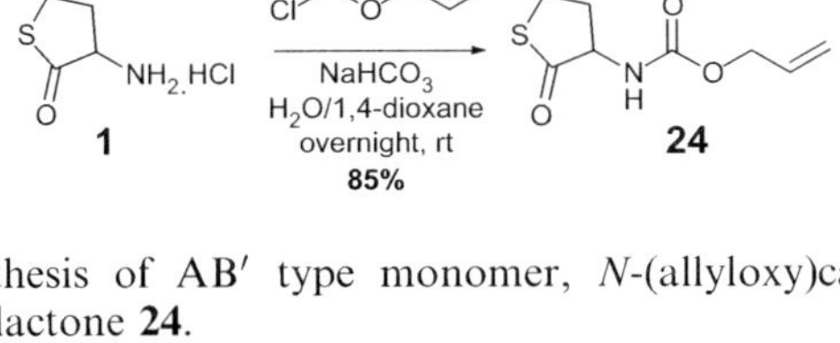


**Scheme 9.13** Synthesis of AB′ type monomer, *N*-(allyloxy)carbonylhomocysteine thiolactone **24**.
Scheme adapted from ref. 114.

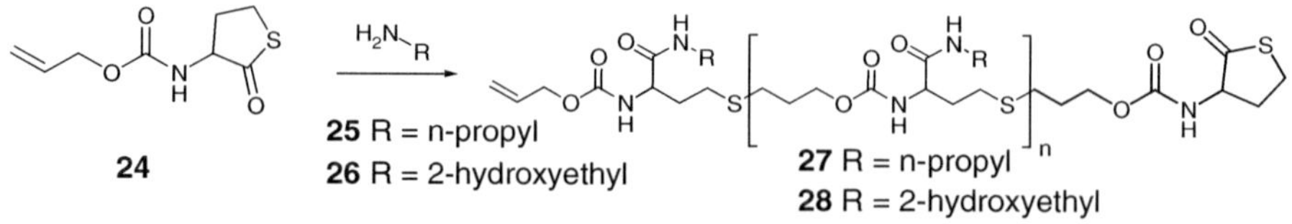


**Scheme 9.14** Stepwise radical polymerization of the Alloc-TL Monomer **24** in a one-pot process yielding a linear polymer with a polythioether/polyurethane backbone.
Scheme adapted from ref. 114.

**Table 9.1** The reaction conditions of the stepwise radical polymerization of the Alloc-TL Monomer **24** and analysis of the obtained polymers.

| *Entry* | *Amine* | *Initiator* | *Temperature/°C* | *Reaction time/h* | $M_{n,GPC}/kDa$ | $M_{n,NMR}/kDa$ | *PDI* |
|---|---|---|---|---|---|---|---|
| 1 | *n*-Propylamine | DMPA | 40 | 0.5 | 4.2 | 3.8 | 1.9 |
| 2 | *n*-Propylamine | AIBN | 70 | 17 | 3.6 | 3.0 | 1.7 |
| 3 | Ethanolamine | DMPA | 40 | 1 | 22 | 7.8 | 1.6 |

increasing conversion is observed. In order to obtain high-molecular-weight polyaddition compounds, the ratio between the involved functional groups should equal one. The aminolysis of the thiolactone in the Alloc-TL monomer **24** should therefore reach full conversion. The radical amine-thiol-ene polymerization of **24** has been performed in the presence of two equivalents of different amines, respectively *n*-propylamine **25** and ethanolamine **26** (Scheme 9.14).

The reaction conditions and analysis of the obtained polymers are summarized in Table 9.1. In all cases, polyaddition occurs and after isolation of the obtained polymer 27/28, the SEC chromatograms display a unimodal distribution, which allows for the determination of $M_n$ and the PDI. The quantification of $M_n$ *via* end group analysis (double bond protons) in the $^1$H NMR spectrum confirms the trends observed by SEC analysis. Interestingly, polymers with comparable chain length are obtained *via* the thermal initiated polymerization using AIBN (entry 2) in comparison with the UV-initiated reaction (entry 1).[114]

This mild and efficient one-pot polyaddition process yielded a polymer **28** with a polythioether/polyurethane backbone and pendant hydroxyl groups. Under neutral conditions, hydroxyl functions are unable to open the thiolactone ring (*vide supra*) and alcohols do not interfere with radical thiol-ene reactions.[38] Standard synthetic methods for the synthesis of hydroxyl functionalized polyurethanes would certainly require a protection/deprotection strategy. Generalization of this reaction concept emphasizes the fact that, as long as the additional functional group of the multifunctional amine does not interfere with either reactions in the one-pot multistep process (aminolysis and radical thiol-ene), linear polymers can be obtained with direct introduction of side chain functional groups, prone to post-polymerization modification.[26]

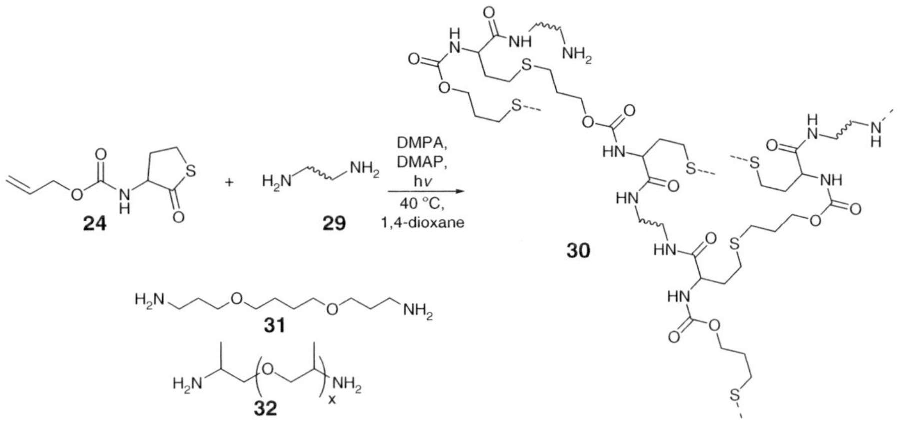


**Scheme 9.15** Schematic depiction of the network formation: one-pot reaction between the Alloc-TL monomer **24** and a diamine cross-linker **29** results in a polythioether polyurethane network **30**.
Scheme adapted from ref. 114.

To further extend the scope of this method in material science, polymer networks **30** based on the AB′ type monomer **24** have been targeted. Polymer film formation occurs under UV-irradiation of a homogeneous reaction mixture of **24** and a diamine **29** (Scheme 9.15). The choice of the diamine cross-linker **29** regarding structure and molecular weight proved to be critical. While 1,6-hexanediamine was insoluble in the reaction mixture, the use of the more polar 4,9-dioxadodecanediamine **31** as a cross-linker yielded a clear, non-tacky network film with good mechanical properties after UV-curing for 3 h. The use of the Jeffamine D series **32** (D-400, D-2000, and D-4000) as macromolecular cross-linkers was also attempted, but poor film formation was observed, probably because the amine group is located on a secondary carbon atom, thus sterically hampering cross-linking.[114]

## 9.3.2 Double Modular Modification of Thiolactone-containing Polymers

Thiols are attractive sites for chemical modification and conjugation, yet their reactive nature renders direct incorporation into polymer systems very challenging. The interference of free thiols in most of the polymerization processes is a major issue and controlled/living polymerization[118–124] reactions especially suffer from thiol-induced side reactions. Thiols can react with (vinylic) monomers through radical or Michael additions and will also induce chain transfer reactions with propagating radicals.[125] Therefore, research activities regarding thiol-based conjugation methods have been mainly focused on linear polymers containing functional groups, capable of trapping low-molecular-weight monothiols. These thiol scavengers can either be situated at the chain end or as side chain functionalities. For example, double bonds on the side chain of polyenes like 1,2-polybutadiene[126,127] and in poly[2-(3-butenyl)-2-oxazoline][128] have been modified *via* a UV-initiated radical thiol-ene reaction, while aliphatic polyurethanes containing thiol-reactive maleimide groups on their side chains have been synthesized by utilization of a diol containing a masked maleimide.[129] A synthetic platform for the preparation of various norbornenyl (Nb) containing reversible addition-fragmentation chain transfer (RAFT) agents,[130] permits rapid and efficient post-polymerization thiol-ene conjugation at the α-chain end of various linear polymer scaffolds.[131] In another example, Sumerlin and co-workers used RAFT polymerization of *N*-isopropylacrylamide (NIPAM) followed by aminolysis and subsequent coupling with a bismaleimide to obtain maleimido-functionalized PNIPAM, permitting near-quantitative end group functionalization with thiols.[132]

The aminolysis of the RAFT end group is a popular route towards (semi-) telechelics bearing a chain end thiol,[133,134] which can subsequently be employed as a nucleophile in a thiol-bromo conjugation[135] or Michael addition[136–138] or as a thiyl radical in thiol-ene[138,139] and thiol-yne[116,117] conjugation. In some cases, the aminolysis and modification preferentially occurs in the same reaction medium (one-pot), avoiding the undesired oxidation to the corresponding disulfide that leads to bimodal polymer populations.[138]

While the synthetic availability of various thiol end-capped polymers has been extensively documented, reports on the synthesis of macromolecules containing multiple thiols (polythiols) on their side chains are scarce. The strong interest in the development of such multi-thiol containing polymers finds its origin in the fact that they are potential substrates for the incorporation of various chemical functionalities *via* a thiol-click reaction of choice. Indeed, depending on the chemical environment and eventual application, a radical or nucleophilic reaction can be selected, producing either a non-cleavable (thio-ether from thiol-ene-, thiol-bromo- or thiol-epoxide-conjugation) or a cleavable (thio-carbamate from thiol-isocyanate-conjugation) linkage.

However, there are two major issues regarding linear polymers with unprotected thiols. A first drawback is the interference of free thiols in the polymerization processes (*vide supra*). Consequently, reported synthetic routes towards polythiols require a protection/deprotection strategy, *i.e.* a detrimental approach in terms of atom efficiency and overall yield. A mercapto group can be incorporated along the backbone of linear polymers masked with different protecting groups. Acid-labile groups like a *p*-methoxybenzyl thioether,[140–142] a trityl thioether[143,144] and a *tert*-butyldimethylsilyl thioether[145] require a strong acid treatment for deprotection. Thioacetates,[146–148] thiobenzoates[149] and *S*-alkyl-*O*-ethyl xanthates[150] can be cleaved by aminolysis. Na/liquid ammonia reduction liberates thiols from their corresponding benzyl thioethers.[151] The photodeprotection of thiol protecting groups based on UV-labile 2-nitrobenzyl moieties allows for the subsequent thiol-ene modification of the released thiols.[152,153] Concomitant disulfide formation during and after deprotection is sometimes problematic, resulting in a polymer network.[143,152] In selected cases however, polyesters with pendant thiols have been prepared either *via* direct enzymatic[154] or chemoselective[155] polycondensation of unprotected thiol monomers, based on thiomalic acid.

Secondly, it should be stressed that the hard-to-avoid oxidative cross-linking of linear polythiols is a major hurdle to take, further hampering their more widespread use. The thiol-disulfide conversion is reversible and various chemical agents are used for the scission of disulfide linkages, either *via* a thiol exchange or a reduction. Some elegant approaches toward synthetic polythiols are based on this reversibility.[156–159] For example, Tsarevsky *et al.* employed a disulfide containing difunctional methacrylate as comonomer (cross-linker) for the preparation of degradable PMMA-based gels by atom-transfer radical polymerization (ATRP). Treatment of the networks with tri-*n*-butylphosphine ($Bu_3P$) yielded a relatively narrow-disperse linear polythiol.[160] Thiol-exchange in polymers with pendant pyridyldisulfide moieties[159,161–164] is a mild and efficient method, circumventing the mentioned issues regarding the preparation and manipulation of linear polythiols.

Endo and co-workers reported the synthesis of functional polymers based on another latent thiol moiety: a 5-membered cyclic dithiocarbonate group is introduced in linear poly(methacrylate)s[165] and poly(norbornene)s[166] as a masked thiol functionality. Similar to thiolactones, dithiocarbonates are sensitive to aminolysis, liberating thiols along the polymer chain. However, the stability of these dithiocarbonate monomers is an issue and a chemoselectivity issue arises due to the reaction between the released thiol and a dithiocarbonate in basic medium.[165] Moreover, in all reported attempts,[165,166] it proved to be impossible to preserve the linear nature of the polymer chains during the aminolysis of pendant cyclic dithiocarbonates, *i.e.* disulfide formation resulted in (partially) cross-linked materials.

Thiolactones as functional handles along the backbone of a variety of linear polymers are sites along the linear backbone where a double modification/functionalization, a prime example of post-polymerization modification, can occur: first, a wide variety of amines can be employed for the aminolysis, followed by a thiol-click reaction of choice (Scheme 9.16).

The synthesis of a thiolactone containing monomer starts from homocysteine-γ-thiolactone hydrochloride **1**. The most straightforward synthesis of a styrenic analogue **34** is the reaction between **1** and 4-vinylbenzenesulfonyl chloride **33**, the corresponding sulfonyl chloride of styrene sulfonic acid (Scheme 9.17).[167] The stable sulfonamide linkage in **34** avoids unwanted detachment of the thiolactone unit during the post-polymerization modification.

Successful copolymerization of this stable monomer **34** with styrene or methyl methacrylate (MMA), *via* RAFT or NMP,[168] yields linear polymers with tunable thiolactone content (4–25%) and controlled molecular weight, although PDIs are relatively high ($\sim$1.5) in the case of PMMA synthesized by NMP.[167]

The obtained series of RAFT copolymers **35** are consequently subjected to double modular modification (Scheme 9.18). In a first step, the reaction with a series of amines (benzylamine **21**, *n*-propylamine **25**, ethanolamine **26** and Jeffamine M-1000) generates thiols as pendant groups on the polymer chain. Disulfide formation could be suppressed using an excess amount of

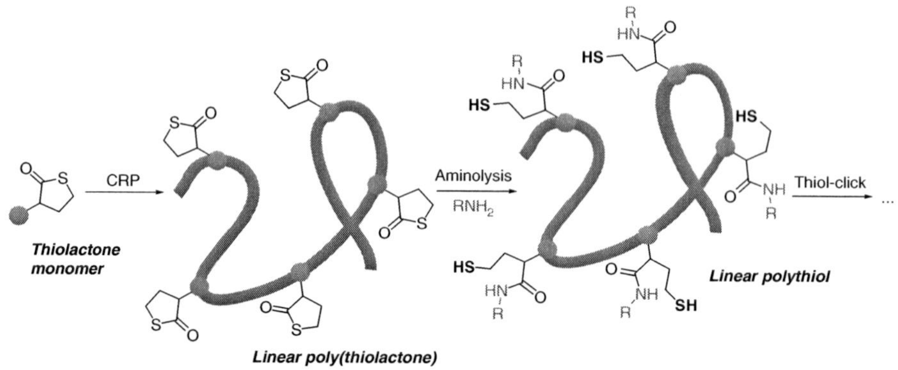


**Scheme 9.16** Schematic depiction of double modification of pendent thiolactone moieties. After controlled radical copolymerization (CRP) of a stable thiolactone monomer, aminolysis of the linear poly(thiolactone) yields a linear polythiol, *i.e.* a reactive polymer scaffold for thiol-click modification. Scheme adapted from ref. 167.

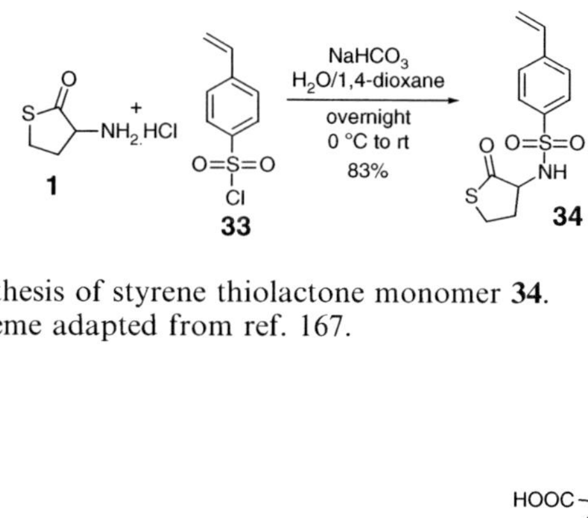


**Scheme 9.17** Synthesis of styrene thiolactone monomer **34**. Scheme adapted from ref. 167.

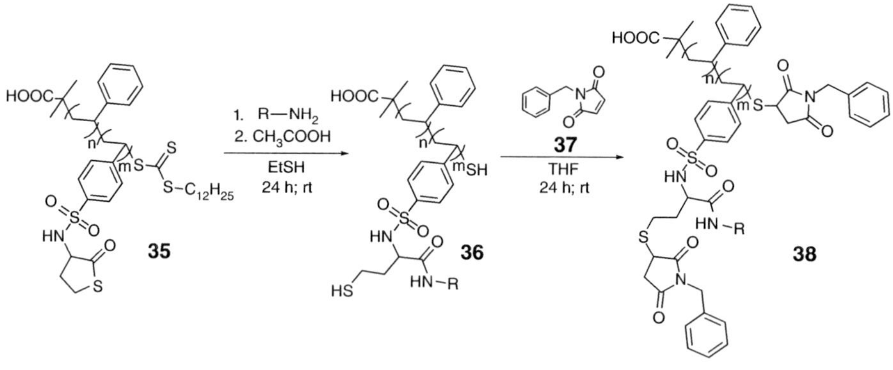


**Scheme 9.18** Double modification reaction of thiolactone-containing linear polymers **35**. Nucleophilic ring-opening of pendent thiolactone moieties with an amine, generating polythiols **36**. Modification of the polythiols **36** with *N*-benzylmaleimide **37** *via* a Michael addition reaction. Scheme adapted from ref. 167.

low-molecular weight thiol as reducing agent. Acidic work-up before precipitation guarantees the formation of a series of polythiols **36**, with varying thiol content. After isolation of these polythiols **36**, conjugate addition (Michael reaction) with *N*-benzylmaleimide **37** yields the modified polymers **38**. This result demonstrates that the thiols introduced on the polymer backbone can serve as functional handles for subsequent thio-click reactions, permitting a double modification of the polymer. The reaction conditions are adapted during all stages of the process in order to completely suppress undesired disulfide formation. The narrow polydispersity of all linear polymers **35**, **36** and **38**, as indicated by SEC analysis, is a proof of the success of this approach. Similar results were obtained when starting the double modification sequence with thiolactone-containing PMMA.[167]

## 9.4 Conclusion and Outlook

In conclusion, thiolactones are valuable functional handles for polymer synthesis and modification. A thiolactone entity can serve as a precursor for thiols in a one-pot amine-thiol-ene reaction: the thiolactone ring opens upon aminolysis and the *in situ* generated thiol reacts with a double bond that is already present in the same pot. An AB′ type monomer, consisting of a double bond and a thiolactone moiety enables the preparation of functional linear polymers and networks *via* a mild and facile radical photopolymerization process. An elegant thiolactone approach as a synthetic tool for double, modular post-polymerization modification (aminolysis and thiol-click) has been elaborated. The copolymerization of a stable thiolactone-containing styrenic monomer, mediated by RAFT or NMP, yields a series of linear polymers, based on styrene and MMA, with a tuneable content of pendant thiolactone units. These poly(thiolactone)s are precursors for narrow-disperse polythiols, given the fact that disulfide formation is completely suppressed. Aminolysis with various (functional) amines enables production of polythiols with narrow PDI. The subsequent thiol-click conjugation has been demonstrated by the treatment of the obtained polythiols with *N*-benzylmaleimide, the ultimate evidence for the efficient double modification of pendant thiolactone moieties. Other thiol-X reactions could be performed in a similar way, starting from the same backbone.

The potential of thiolactone chemistry in polymer science has only been addressed very recently, yet the first reports clearly demonstrate its virtues regarding versatility and synthetic scope. Extensive research efforts, targeting the preparation of advanced macromolecular architectures (dendrimers, (hyper)-branched and cyclic structures, telechelics, *etc.*) and multi-functional materials, should further reveal the possibilities and limitations of this methodology.

## References

1. H. C. Kolb, M. G. Finn and K. B. Sharpless, *Angew. Chem., Int. Ed.*, 2001, **40**, 2004–2021.
2. C. J. Hawker and K. L. Wooley, *Science*, 2005, **309**, 1200–1205.

3. C. J. Hawker, V. V. Fokin, M. G. Finn and K. B. Sharpless, *Aust. J. Chem.*, 2007, **60**, 381–383.
4. W. H. Binder and R. Sachsenhofer, *Macromol. Rapid Commun.*, 2007, **28**, 15–54.
5. W. H. Binder and R. Sachsenhofer, *Macromol. Rapid Commun.*, 2008, **29**, 952–981.
6. C. R. Becer, R. Hoogenboom and U. S. Schubert, *Angew. Chem., Int. Ed.*, 2009, **48**, 4900–4908.
7. B. S. Sumerlin and A. P. Vogt, *Macromolecules*, 2009, **43**, 1–13.
8. P. Golas and K. Matyjaszewski, *Chem. Soc. Rev.*, 2010, **39**, 1338–1354.
9. H. C. Kolb and K. B. Sharpless, *Drug Discovery Today*, 2003, **8**, 1128–1137.
10. C. D. Hein, X.-M. Liu and D. Wang, *Pharm. Res.*, 2008, **25**, 2216–2230.
11. M. van Dijk, D. T. S. Rijkers, R. M. J. Liskamp, C. F. van Nostrum and W. E. Hennink, *Bioconjugate Chem.*, 2009, **20**, 2001–2016.
12. O. Veiseh, J. W. Gunn and M. Zhang, *Adv. Drug Delivery Rev.*, **62**, 284–304.
13. S. Ciampi, G. Le Saux, J. B. Harper and J. J. Gooding, *Electroanalysis*, 2008, **20**, 1513–1519.
14. T. S. Hansen, A. E. Daugaard, S. Hvilsted and N. B. Larsen, *Adv. Mater.*, 2009, **21**, 4483–4486.
15. K. Huang, F. Duclairoir, T. Pro, J. Buckley, G. Marchand, E. Martinez, J.-C. Marchon, B. De Salvo, G. Delapierre and F. Vinet, *ChemPhysChem*, 2009, **10**, 963–971.
16. A. H. El-Sagheer and T. Brown, *Chem. Soc. Rev.*, **39**, 1388–1405.
17. A. S. Hoffman and P. S. Stayton, *Prog. Polym. Sci.*, 2007, **32**, 922–932.
18. J. Nicolas, G. Mantovani and D. M. Haddleton, *Macromol. Rapid Commun.*, 2007, **28**, 1083–1111.
19. J.-F. Lutz and H. G. Boerner, *Prog. Polym. Sci.*, 2008, **33**, 1–39.
20. J.-F. Lutz and Z. Zarafshani, *Adv. Drug Delivery Rev.*, 2008, **60**, 958–970.
21. J. M. Antos and M. B. Francis, *Curr. Opin. Chem. Biol.*, 2006, **10**, 253–262.
22. J. M. Baskin and C. R. Bertozzi, *QSAR Comb. Sci.*, 2007, **26**, 1211–1219.
23. C. P. R. Hackenberger and D. Schwarzer, *Angew. Chem., Int. Ed.*, 2008, **47**, 10030–10074.
24. S. H. Weisbrod and A. Marx, *Chem. Commun.*, 2008, 5675–5685.
25. M. D. Best, *Biochemistry*, 2009, **48**, 6571–6584.
26. M. A. Gauthier, M. I. Gibson and H. A. Klok, *Angew. Chem., Int. Ed.*, 2009, **48**, 48–58.
27. C. Barner-Kowollik and A. J. Inglis, *Macromol. Chem. Phys.*, 2009, **210**, 987–992.
28. U. Mansfeld, C. Pietsch, R. Hoogenboom, C. R. Becer and U. S. Schubert, *Polym. Chem.*, 2010, **1**, 1560–1598.
29. M. Malkoch, R. J. Thibault, E. Drockenmuller, M. Messerschmidt, B. Voit, T. P. Russell and C. J. Hawker, *J. Am. Chem. Soc.*, 2005, **127**, 14942–14949.

30. P. Lundberg, C. J. Hawker, A. Hult and M. Malkoch, *Macromol. Rapid Commun.*, 2008, **29**, 998–1015.
31. P. N. R. Vennestrøm, C. M. Osmundsen, C. H. Christensen and E. Taarning, *Angew. Chem., Int. Ed.*, 2011, **50**, 10502–10509.
32. V. V. Rostovtsev, L. G. Green, V. V. Fokin and K. B. Sharpless, *Angew. Chem., Int. Ed.*, 2002, **41**, 2596–2599.
33. J. F. Lutz, *Angew. Chem., Int. Ed.*, 2007, **46**, 1018–1025.
34. M. Meldal and C. W. Tornoe, *Chem. Rev.*, 2008, **108**, 2952–3015.
35. C. E. Hoyle, T. Y. Lee and T. Roper, *J. Polym. Sci., Part A: Polym. Chem.*, 2004, **42**, 5301–5338.
36. J. W. Chan, C. E. Hoyle, A. B. Lowe and M. Bowman, *Macromolecules*, 2010, **43**, 6381–6388.
37. C. E. Hoyle, *Chem. Soc. Rev.*, 2010, **39**, 1355–1387.
38. C. E. Hoyle and C. N. Bowman, *Angew. Chem., Int. Ed.*, 2010, **49**, 1540–1573.
39. R. Hoogenboom, *Angew. Chem., Int. Ed.*, 2011, **49**, 3415–3417.
40. M. G. Linkova, N. D. Kuleshova and I. L. Knunyants, *Russ. Chem. Rev.*, 1964, **33**, 493–507.
41. Z. Paryzek and W. Skiera, *Org. Prep. Proc. Int.*, 2007, **39**, 203–296.
42. E. Schaumann and U. Behrens, *Angew. Chem., Int. Edit. Engl.*, 1977, **16**, 722–723.
43. M. B. Finkelstein, W. Dekant, A. S. Kende and M. W. Anders, *J. Am. Chem. Soc.*, 1995, **117**, 9590–9591.
44. B. J. Garbiras and S. Marburg, *Synthesis*, 1999, 270–274.
45. J. Hausler, *Monatsh. Chem.*, 1993, **124**, 1071–1075.
46. F. Korte and H. Christoph, *Chem. Ber. Recl.*, 1961, **94**, 1966–1976.
47. J. Martens, J. Kintscher and W. Arnold, *Synthesis*, 1991, 497–498.
48. G. Pattenden and A. J. Shuker, *Tetrahedron Lett.*, 1991, **32**, 6625–6628.
49. I. Polec, L. Lutsen, D. Vanderzande and J. Gelan, *Eur. J. Org. Chem.*, 2002, 1033–1036.
50. D. Vegh, J. Morel, B. Decroix and P. Zalupsky, *Synth. Commun.*, 1992, **22**, 2057–2061.
51. L. S. Jeong, H. O. Kim, H. R. Moon, J. H. Hong, S. J. Yoo, W. J. Choi, M. W. Chun and C. K. Lee, *J. Med. Chem.*, 2001, **44**, 806–813.
52. A. H. Li, S. Moro, N. Forsyth, N. Melman, X. D. Ji and K. A. Jacobson, *J. Med. Chem.*, 1999, **42**, 706–721.
53. G. A. Dafforn and D. E. Koshland, *J. Am. Chem. Soc.*, 1977, **99**, 7246–7257.
54. R. S. McDonald, P. Patterson, J. Rodwell and A. Whalley, *Can. J. Chem.*, 1992, **70**, 62–67.
55. R. Benesch and R. E. Benesch, *J. Am. Chem. Soc.*, 1956, **78**, 1597–1599.
56. A. H. F. Lee, A. S. C. Chan and T. Li, *Tetrahedron*, 2003, **59**, 833–839.
57. A. H. F. Lee, J. Chen, A. S. C. Chan and T. Li, *Phosphorus, Sulfur Silicon Relat. Elem.*, 2003, **178**, 1163–1174.
58. G. Pattenden and A. J. Shuker, *Synlett*, 1991, 717–718.

59. G. Pattenden and A. J. Shuker, *J. Chem. Soc., Perkin Trans. 1*, 1992, 1215–1221.
60. C. G. Overberger and J. K. Weise, *J. Am. Chem. Soc.*, 1968, **90**, 3533–3537.
61. H. Matsumura, T. Bando and M. Sunagawa, *Heterocycles*, 1995, **41**, 147–159.
62. C. J. Moody and R. J. Taylor, *Tetrahedron Lett.*, 1987, **28**, 5351–5352.
63. C. J. Moody and R. J. Taylor, *Tetrahedron Lett.*, 1988, **29**, 6005–6008.
64. C. J. Moody and R. J. Taylor, *Tetrahedron*, 1990, **46**, 6501–6524.
65. L. E. Iglesias, A. Baldessari and E. G. Gros, *Org. Prep. Proced. Int.*, 1996, **28**, 319–324.
66. F. Zanardi, L. Battistini, L. Marzocchi, D. Acquotti, G. Rassu, L. Pinna, L. Auzzas, V. Zambrano and G. Casiraghi, *Eur. J. Org. Chem.*, 2002, 1956–1965.
67. H. Jakubowski, *Cell. Mol. Life Sci.*, 2004, **61**, 470–487.
68. H. D. Baernstein, *J. Biol. Chem.*, 1934, **106**, 451–456.
69. B. Riegel and V. du Vigneaud, *J. Biol. Chem.*, 1935, **112**, 149–154.
70. H. Jakubowski, *Chem.–Eur. J.*, 2006, **12**, 8039–8043.
71. M. H. Kunzmann, I. Staub, T. Boettcher and S. A. Sieber, *Biochemistry*, 2011, **50**, 910–916.
72. W. J. Leanza, L. S. Chupak, R. L. Tolman and S. Marburg, *Bioconjugate Chem.*, 1992, **3**, 514–518.
73. P. Molina, I. Diaz and A. Tarraga, *Tetrahedron*, 1995, **51**, 5617–5630.
74. J. Momb, P. W. Thomas, R. M. Breece, D. L. Tierney and W. Fast, *Biochemistry*, 2006, **45**, 13385–13393.
75. C. E. McInnis and H. E. Blackwell, *Bioorg. Med. Chem.*, 2011, **19**, 4820–4828.
76. Y. Shinohara, H. Hasegawa, T. Hashimoto and K. Ichida, *J. Labelled Compd. Radiopharm.*, 2010, **53**, 552–555.
77. R. Laliberte, Y. Knobler and M. Frankel, *J. Chem. Soc.*, 1963, 2756–2762.
78. H. Kitano, H. Wolf and N. Ise, *Macromolecules*, 1990, **23**, 1958–1961.
79. K. M. Smith, Y. Bu and H. Suga, *Chem. Biol.*, 2003, **10**, 563–571.
80. D. Ellis, S. E. Norman and H. M. I. Osborn, *Tetrahedron*, 2008, **64**, 2832–2854.
81. A. S. Chubarov, M. M. Shakirov, I. V. Koptyug, R. Z. Sagdeev, D. G. Knorre and T. S. Godovikova, *Bioorg. Med. Chem. Lett.*, 2011, **21**, 4050–4053.
82. K. Okumura, T. Oda, K. Kondo, I. Inoue, T. Danno, H. Yamaguchi and K. Masukawa, *J. Med. Chem.*, 1971, **14**, 226–229.
83. R. Grigg, S. Thianpatanagul and J. Kemp, *Tetrahedron*, 1988, **44**, 7283–7292.
84. R. Grigg and M. A. B. Sarker, *Tetrahedron*, 2006, **62**, 10332–10343.
85. S. Hanessian, N. Moitessier, C. Gauchet and M. Viau, *J. Med. Chem.*, 2001, **44**, 3066–3073.
86. P.A. Gatti, M. Nicola and M. Zacche, *EU Pat.*, 1964838, 2008.

87. P. A. Kendall and E. A. Barnard, *Biochim. Biophys. Acta*, 1969, **188**, 10–24.
88. R. Benesch and R. E. Benesch, *Proc. Natl. Acad. Sci. U. S. A.*, 1958, **44**, 848–853.
89. P. Singh, M. Pirio, D. K. Leung and Y. G. Tsay, *Can. J. Chem.*, 1984, **62**, 2471–2477.
90. R. R. Martodam, D. Y. Twumasi, I. E. Liener, J. C. Powers, N. Nishino and G. Krejcarek, *Proc. Natl. Acad. Sci. U. S. A.*, 1979, **76**, 2128–2132.
91. G. Manecke and H. J. Middeke, *Angew. Makromol. Chem.*, 1980, **91**, 179–201.
92. K. E. Taylor and Y. C. Wu, *Biochem. Int.*, 1980, **1**, 353–358.
93. M. M. Ponpipom and K. M. Rupprecht, *Carbohydr. Res.*, 1983, **113**, 45–56.
94. G. Chassaing, S. Lavielle, S. Julien and A. Marquet, *Tetrahedron Lett.*, 1985, **26**, 623–626.
95. G. Christie, A. M. Breckenridge and B. K. Park, *Biochem. Pharmacol.*, 1989, **38**, 1451–1458.
96. S. C. Kim, N. F. Olson and T. Richardson, *Milk Sci. Int.*, 1990, **45**, 580–583.
97. A. Kumar, S. Advani, H. Dawar and G. P. Talwar, *Nucleic Acids Res.*, 1991, **19**, 4561–4561.
98. O. Blixt and T. Norberg, *J. Org. Chem.*, 1998, **63**, 2705–2710.
99. X. C. Chen, Z. Wen, M. Xian, K. Wang, N. Ramachandran, X. P. Tang, H. B. Schlegel, B. Mutus and P. G. Wang, *J. Org. Chem.*, 2001, **66**, 6064–6073.
100. M. deBarrio, P. Tornero, A. Prieto, T. Sainza, J. M. Zubeldia and T. Herrero, *J. Invest. Allergol. Clin. Immunol.*, 1997, **7**, 193–194.
101. U. Kraatz, H. Wamhoff and F. Korte, *Justus Liebigs Ann. Chem.*, 1972, **758**, 177–184.
102. K. L. Dechant and S. Noble, *Drugs*, 1996, **52**, 875–881.
103. J. Gonella, *US Pat.*, 4411909, 1982.
104. P. Appukkuttan, W. Dehaen, V. V. Fokin and E. Van der Eycken, *Org. Lett.*, 2004, **6**, 4223–4225.
105. H. S. G. Beckmann and V. Wittmann, *Org. Lett.*, 2007, **9**, 1–4.
106. N. M. Smith, M. J. Greaves, R. Jewell, M. W. D. Perry, M. J. Stocks and J. P. Stonehouse, *Synlett*, 2009, 1391–1394.
107. H. M. Kolenbrander, *Can. J. Chem.*, 1969, **47**, 3271–3273.
108. G. L. Anderson, D. L. Bussolotti and J. K. Coward, *J. Med. Chem.*, 1981, **24**, 1271–1277.
109. J. A. Montgomery, H. J. Thomas, M. C. Thorpe and P. K. Chiang, *J. Med. Chem.*, 1981, **24**, 1514–1516.
110. J. A. Montgomery, S. J. Clayton, H. J. Thomas, W. M. Shannon, G. Arnett, A. J. Bodner, I. K. Kion, G. L. Cantoni and P. K. Chiang, *J. Med. Chem.*, 1982, **25**, 626–629.
111. E. Benghiat and P. A. Crooks, *Synthesis*, 1982, 1033–1036.
112. P. D. Halphen and T. C. Owen, *J. Org. Chem.*, 1973, **38**, 3507–3510.

113. W. Hanefeld and H. Schutz, *J. Heterocycl. Chem.*, 1999, **36**, 1167–1174.
114. P. Espeel, F. Goethals and F. E. Du Prez, *J. Am. Chem. Soc.*, 2011, **133**, 1678–1681.
115. M. Uygun, M. A. Tasdelen and Y. Yagci, *Macromol. Chem. Phys.*, 2010, **211**, 103–110.
116. D. Konkolewicz, A. Gray-Weale and S. Perrier, *J. Am. Chem. Soc.*, 2009, **131**, 18075–18077.
117. D. Konkolewicz, C. K. Poon, A. Gray-Weale and S. Perrier, *Chem. Commun.*, 2011, **47**, 239–241.
118. G. W. Coates, P. D. Hustad and S. Reinartz, *Angew. Chem., Int. Ed.*, 2002, **41**, 2237–2257.
119. C. W. Bielawski and R. H. Grubbs, *Prog. Polym. Sci.*, 2007, **32**, 1–29.
120. G. J. Domski, J. M. Rose, G. W. Coates, A. D. Bolig and M. Brookhart, *Prog. Polym. Sci.*, 2007, **32**, 30–92.
121. W. A. Braunecker and K. Matyjaszewski, *Prog. Polym. Sci.*, 2007, **32**, 93–146.
122. T. Yokozawa and A. Yokoyama, *Prog. Polym. Sci.*, 2007, **32**, 147–172.
123. D. Baskaran and A. H. E. Mueller, *Prog. Polym. Sci.*, 2007, **32**, 173–219.
124. S. Penczek, M. Cypryk, A. Duda, P. Kubisa and S. Slomkowski, *Prog. Polym. Sci.*, 2007, **32**, 247–282.
125. J. Liu, Y. Wang, Q. Fu, X. Zhu and W. Shi, *J. Polym. Sci., Part A: Polym. Chem.*, 2008, **46**, 1449–1459.
126. R. L. A. David and J. A. Kornfield, *Macromolecules*, 2008, **41**, 1151–1161.
127. N. ten Brummelhuis, C. Diehl and H. Schlaad, *Macromolecules*, 2008, **41**, 9946–9947.
128. A. Gress, A. Volkel and H. Schlaad, *Macromolecules*, 2007, **40**, 7928–7933.
129. L. Billiet, O. Gok, A. P. Dove, A. Sanyal, L.-T. T. Nguyen and F. E. Du Prez, *Macromolecules*, 2011, **44**, 7874–7878.
130. C. Barner-Kowollik, *Handbook of RAFT Polymerization*, Wiley-VCH, 2008.
131. M. M. Stamenovic, P. Espeel, W. Van Camp and F. E. Du Prez, *Macromolecules*, 2011, **44**, 5619–5630.
132. M. Li, P. De, S. R. Gondi and B. S. Sumerlin, *J. Polym. Sci., Part A: Polym. Chem.*, 2008, **46**, 5093–5100.
133. H. Willcock and R. K. O'Reilly, *Polym. Chem.*, 2010, **1**, 149–157.
134. G. Moad, E. Rizzardo and S. H. Thang, *Polym. Int.*, 2011, **60**, 9–25.
135. J. Xu, L. Tao, C. Boyer, A. B. Lowe and T. P. Davis, *Macromolecules*, 2010, **43**, 20–24.
136. X. P. Qiu and F. M. Winnik, *Macromol. Rapid Commun.*, 2006, **27**, 1648–1653.
137. J. W. Chan, B. Yu, C. E. Hoyle and A. B. Lowe, *Chem. Commun.*, 2008, 4959–4961.
138. C. Boyer, A. Granville, T. P. Davis and V. Bulmus, *J. Polym. Sci., Part A: Polym. Chem.*, 2009, **47**, 3773–3794.

139. S. P. S. Koo, M. M. Stamenovic, R. A. Prasath, A. J. Inglis, F. E. D. Prez, C. Barner-Kowollik, W. V. Camp and T. Junkers, *J. Polym. Sci., Part A: Polym. Chem.*, 2010, **48**, 1699–1713.
140. P. Veld, P. J. Dijkstra and J. Feijen, *Macromol. Chem. Phys.*, 1992, **193**, 2713–2730.
141. T. Ouchi, H. Seike, H. Miyazaki, F. Tasaka and Y. Ohya, *Des. Monomers Polym.*, 2000, **3**, 279–287.
142. S. Cesana, A. Kurek, M. A. Baur, J. Auernheirner and O. Nuyken, *Macromol. Rapid Commun.*, 2007, **28**, 608–615.
143. H. Kudo, F. Sanda and T. Endo, *Macromol. Chem. Phys.*, 1999, **200**, 1232–1239.
144. D. Pappalardo, S. Målberg, A. Finne-Winstrad and A.-C. Albertsson, *J. Polym. Sci., Part A: Polym. Chem.*, 2012, **50**, 792–800.
145. K. Yamaguchi, T. Kato, A. Hirao and S. Nakahama, *Makromol. Chem., Rapid Commun.*, 1987, **8**, 203–207.
146. N. Kihara, C. Kanno and T. Fukutomi, *J. Polym. Sci., Part A: Polym. Chem.*, 1997, **35**, 1443–1451.
147. I. S. Alferiev and I. Fishbein, *Biomaterials*, 2002, **23**, 4753–4758.
148. H. Xu, J. K. Jin, Y. Mao, J. Z. Sun, F. Yang, W. Zhang Yuan, Y. Q. Dong, M. Wang and B. Z. Tang, *Macromolecules*, 2008, **41**, 3874–3883.
149. E. Hrsic, H. Keul and M. Möller, *Eur. Polym. J.*, 2012, **48**, 761–768.
150. R. Nicolaÿ, *Macromolecules*, 2012, **45**, 821–827.
151. C. G. Overberger and H. Aschkenasy, *J. Am. Chem. Soc.*, 1960, **82**, 4357–4360.
152. T. Pauloehrl, G. Delaittre, M. Bastmeyer and C. Barner-Kowollik, *Polym. Chem.*, 2012, **3**, 1740–1749.
153. G. Delaittre, T. Pauloehrl, M. Bastmeyer and C. Barner-Kowollik, *Macromolecules*, 2012, **45**, 1792–1802.
154. M. Kato, K. Toshima and S. Matsumura, *Biomacromolecules*, 2009, **10**, 366–373.
155. K. Yamamoto and A. Takasu, *Macromolecules*, 2010, **43**, 8519–8523.
156. J. Groll, S. Singh, K. Albrecht and M. Moeller, *J. Polym. Sci., Part A: Polym. Chem.*, 2009, **47**, 5543–5549.
157. L. Li, X. Jiang and R. Zhuo, *J. Polym. Sci., Part A: Polym. Chem.*, 2009, **47**, 5989–5997.
158. N. Hisano, H. Iwata, Y. Teramura, H. Chen and Y. Ikada, *J. Polym. Sci., Part A: Polym. Chem.*, 2011, **49**, 671–679.
159. B. Nagel, N. Gajovic-Eichelmann, F. W. Scheller and M. Katterle, *Langmuir*, 2012, **26**, 9088–9093.
160. N. V. Tsarevsky and K. Matyjaszewski, *Macromolecules*, 2005, **38**, 3087–3092.
161. L. X. Wang, J. Kristensen and D. E. Ruffner, *Bioconjugate Chem.*, 1998, **9**, 749–757.
162. V. Bulmus, M. Woodward, L. Lin, N. Murthy, P. Stayton and A. Hoffman, *J. Control. Release*, 2003, **93**, 105–120.

163. L. J. Wong, C. Boyer, Z. F. Jia, H. M. Zareie, T. P. Davis and V. Bulmus, *Biomacromolecules*, 2008, **9**, 1934–1944.
164. E. F. Crownover, A. J. Convertine and P. S. Stayton, *Polym. Chem.*, 2012, **2**, 1499–1504.
165. N. Kihara, H. Tochigi and T. Endo, *J. Polym. Sci., Part A: Polym. Chem.*, 1995, **33**, 1005–1010.
166. A. Sudo, H. Morishita and T. Endo, *J. Polym. Sci., Part A: Polym. Chem.*, 2011, **49**, 1097–1103.
167. P. Espeel, F. Goethals, M. M. Stamenovic, L. Petton and F. E. Du Prez, *Polym. Chem.*, 2012, **3**, 1007–1015.
168. C. J. Hawker, A. W. Bosman and E. Harth, *Chem. Rev.*, 2001, **101**, 3661–3688.

CHAPTER 10

# *Thiol-ene Radical Coupling: A Powerful Technique for the Synthesis of Polymer Precursors, Block Copolymers and Graft Copolymers*

BERNARD BOUTEVIN,* REMI AUVERGNE AND GHISLAIN DAVID

Institut Charles Gerhardt UMR CNRS 5253 Laboratoire Ingénierie et Architecture Macromoléculaire, Ecole Nationale Supérieure de Chimie de Montpellier, 8 rue de l'Ecole Normale, 34296 Montpellier Cedex 05, France
*Email: bernard.boutevin@enscm.fr

## 10.1 Introduction

The thiol-ene radical coupling is a well-known reaction that refers to the radical addition of thiols across non-activated carbon-carbon double bonds.[1] Historically, this reaction was first called telomerization, giving rise to telomers,[2,3] *i.e.* oligomers obtained by chain transfer reaction in the presence of thiols. Thus, this powerful technique was primarily used to obtain functional oligomers as precursors of materials but also for making crosslinked networks. The consequent number of reviewing papers gives a sense of the wide range of polymeric structures obtained from this technique.[4–8] Recently, thiol-ene radical coupling was classified as a "click" reaction and regained lots of

RSC Polymer Chemistry Series No. 6
Thiol-X Chemistries in Polymer and Materials Science
Edited by Andrew B. Lowe and Christopher N. Bowman

Published by the Royal Society of Chemistry, www.rsc.org

interests from academic and industrials researchers due to its robustness and versatility. Thiol-ene radical coupling is clearly increasingly used in polymer functionalization and especially in biomedical applications, showing the compatibility of this technique with sensitive groups. Through some examples, this chapter aims to show how to achieve oligomer structures with controlled functionality and molecular weight.

## 10.2 Synthesis of Reactive Precursors by Thiol-ene Radical Coupling

Thiol-ene radical coupling enables synthesis of reactive precursors of either block or graft copolymers. These reactive precursors can be monofunctional (reactions 1 and 2, scheme 11.1) or telechelic (reactions 3–5, scheme 11.1), referring to a functionality of one and two, respectively. The conventional associated reactive groups (quoted G in scheme 10.1) may be alcohols, acids, amines, epoxy *etc.*, thus leading to block or graft copolymers by poly-condensation reaction. In another way, the precursor functionality may be a vinyl group, *i.e.* acrylate, styrenic *etc.* In this case, the reactive precursor may be polymerized by the controlled radical polymerization techniques. The synthetic strategies to afford either monofunctional or telechelic precursors are given in Scheme 10.1, where G refers to the desired functionality.

To illustrate these synthetic pathways, some examples are given below.

### 10.2.1 Monofunctional Precursors

Monofunctional reactive precursors can be readily obtained from thiol-ene radical coupling either by using a functional thiol compound (reaction 1, scheme 10.1) or by using a vinyl monomer bearing the desired functionality (reaction 2, scheme 10.1). Thus, this versatility allows reaching precursors with a wide range of functional groups, which may enable a further condensation

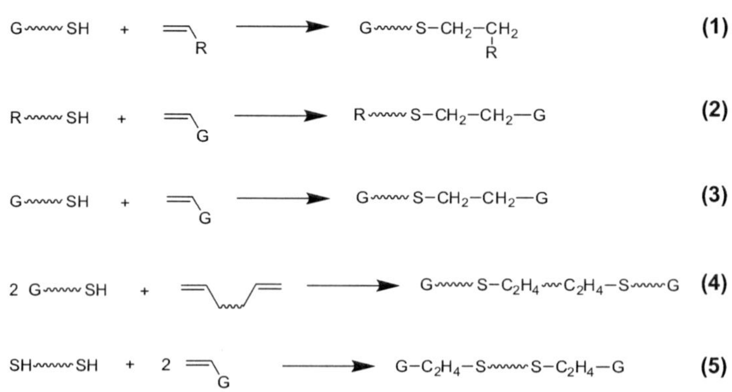


**Scheme 10.1** Synthetic strategies for the syntheses of either monofunctional or telechelicprecursors by thiol-ene radical coupling.

reaction. More specifically, the functional groups can also provide a targeted property to the precursor, such as hydrophilic, lipophobic, adhesion and so on. For instance, Pelaprat *et al.*[9] performed the thiol-ene radical coupling of allyl dialkyl phosphonate monomer in the presence of alkane thiol (Scheme 10.2, reaction a). The phosphonate moieties promote adhesion towards specific surfaces such as metallic surfaces, which then become highly hydrophobic, thanks to alkane moieties. Similarly, David *et al.*[10] carried out the thiol-ene radical coupling of vinyl phosphonic acid in the presence of mercapto acetic acid in water at 70 °C (Scheme 10.2, reaction b). They showed that a low amount of thiol compound (*i.e.* [thiol]/[monomer] about 1 to 3) was required, according to its very high chain transfer constant, and this reaction returned good yields. Another example was presented by Boutevin *et al.*,[11] who carried out the thiol-ene radical coupling of allyl dialkyl phosphonate monomer with mercaptoethanol (Scheme 10.2, reaction c); in the second step this precursor led to acrylation with acryloyl chloride, and provides a new acrylate monomer bearing a phosphonate moiety, which can be readily polymerized by radical polymerization.

Going even further, it is possible to combine, through thiol-ene radical coupling, the functional groups from both the thiol and from the monomer. For instance, several authors[12–15] carried out the synthesis of fluorophosphonate compounds by thiol-ene radical coupling of dialkyl vinyl phosphonate with thiol bearing a fluoroalkyl group (Scheme 10.3, reaction a). These compounds were considered for uses as surfactants, additives for lubricants, impregnation agents for increasing oleophoby, flame-retardant materials, defoaming agents for pesticide or insecticide formulations, mold-release agents and biological chelating agents.

Similarly, fluorinated silanes were prepared[16,17] by reaction of the siliciated telogen thiol $(OCH_3)_2(CH_3)Si(CH_2)_3SH$ upon various fluorinated

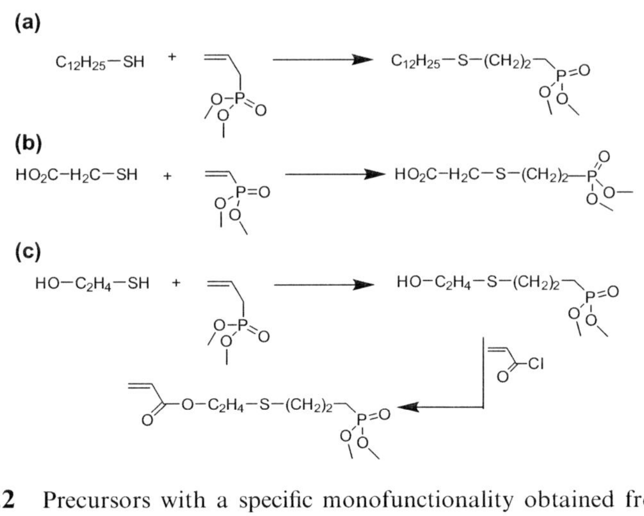


**Scheme 10.2** Precursors with a specific monofunctionality obtained from thiol-ene radical coupling.

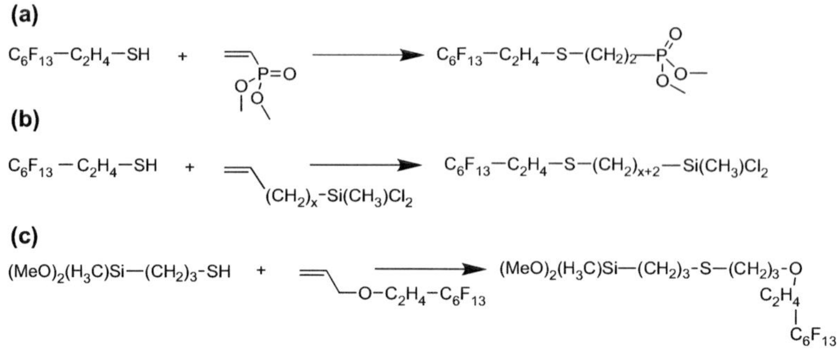


**Scheme 10.3** Precursors with functional groups brought both from the thiol and from the monomer, obtained from thiol-ene radical coupling.

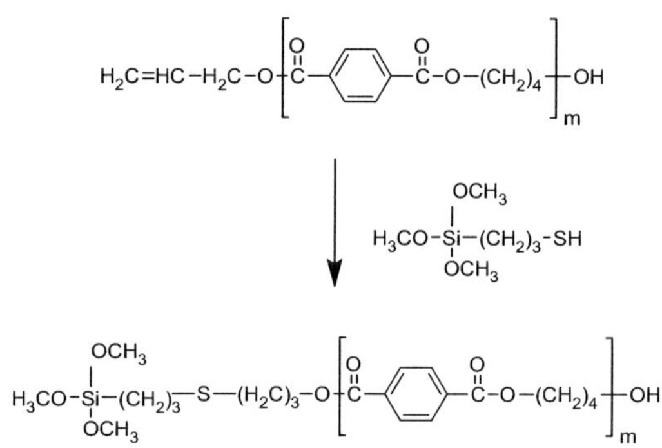


**Scheme 10.4** Poly(butylene terephthalate) end-capped with organosilane group from thiol-ene radical coupling.

olefins. In particular, it was shown that the reaction with the ether $C_6F_{13}(CH_2)_2OCH_2CH{=}CH_2$ gives the monoaddition compound, whereas the reaction of the thiol upon the pentafluorostyrene and the methacrylate gives polyaddition compounds as well.

Silane-coupling agents are of great interest for composite materials such as glass fiber/poly(alkylene terephthalate). They react both with glass-surface silanol groups and with reactive sites of the polymer, thus creating a link between the two phases.

We have also prepared poly(butylene terephthalate) end-capped with organosilane group. α-Allyl oligomers of poly(butylene terephthalate) were first synthesized by a condensation reaction between bis(4-hydroxybutyl) terephthalate and allyl *p*-chlorocarbonylbenzoate, followed by the thiol-ene radical coupling in the presence of 3-(trimethoxysilyl)propanethiol (Scheme 10.4).[18]

## 10.2.2 Telechelic Precursors

Thiol-ene radical coupling is also a powerful technique for synthesizing isomolecular telechelic precursors.[19,20] Basically, three main synthetic strategies can lead to these materials: addition of a functional thiol onto a functional vinyl group (reaction 3, scheme 1); addition of two functional thiols onto a non-functional diene (reaction 4, scheme 1); addition of a non-functional dithiol onto two functional vinyl groups (reaction 5, scheme 1). These pathways were first applied to the synthesis of both aliphatic and aromatic hydroxyl-telechelic compounds as precursors of materials such as polyurethanes.

### *10.2.2.1 Aliphatic and Aromatic Hydroxyl-telechelic Compounds*

The first strategy, through addition of a functional thiol onto a functional vinyl group, obtained mostly diols of various alkyl chains. These diols may contain different groups such as ester, amide, thio and ether bonds. Table 10.1 summarizes the structure of obtained diols from the first strategy, corresponding to the addition of a functional thiol onto a functional vinyl group.

**Table 10.1** Functional aliphatic and aromatic telechelic diols obtained from thiol-ene radical coupling.

| *Hydroxy-telechelic compound* | $M_n$/g $mol^{-1}$ | $T_g$/°C | $T_m$/°C |
|---|---|---|---|
| $HO{-}C_2H_4{-}S{-}(CH_2)_{10}{-}OH$ | 248 | — | 48.3 |
| $HO{-}C_2H_4{-}S{-}(CH_2)_{10}{-}C(O)NH{-}C_2H_4{-}OH$ | 305 | — | 64.2 |
| $HO{-}C_2H_4{-}S{-}(CH_2)_{10}{-}C(=O){-}O{-}C_2H_4{-}OH$ | 306 | — | 15.9 |
| $HO{-}(CH_2)_{12}{-}C(O)O{-}(CH_2)_{10}{-}S{-}(C_2H_4O)_2{-}C_2H_4S{-}(CH_2)_{10}{-}C(O)O{-}(CH_2)_{12}OH$ | 918 | — | 50 |
| $HO{-}(CH_2)_{11}{-}S{-}(CH_2)_n{-}S{-}(CH_2)_{11}{-}OH$ $n = 2$ | 434 | | 93 |
| $n = 5$ | 476 | 12 | 95 |
| $n = 6$ | 490 | | 91 |
| $n = 9$ | 532 | | 89 |
| $HO{-}(CH_2)_{11}{-}S{-}C_2H_4{-}O{-}C_2H_4{-}S{-}(CH_2)_x{-}S{-}C_2H_4{-}O{-}C_2H_4{-}S{-}(CH_2)_{11}{-}OH$ | | | |
| $x = 6$ | 698 | 12 | 64 |
| $x = 10$ | 754 | 10 | 34 |
| 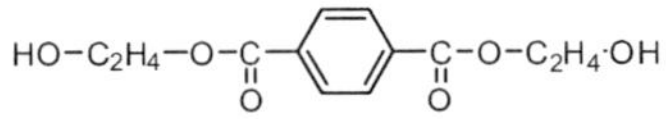 | 254 | — | 109 |
| 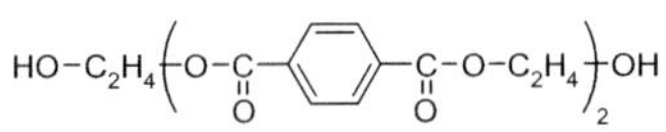 | 446 | — | 169 |
| 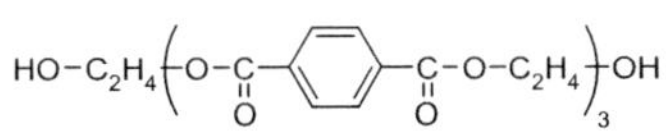 | 639 | — | 200 |
| 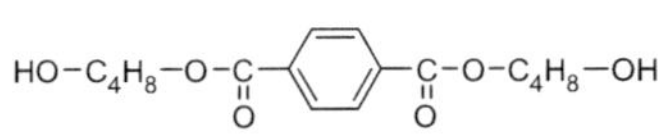 | 310 | — | 72 |

These different diols show melting temperatures, $T_m$, ranging from 16 to 65 °C, as well as molecular weight ($M_w$) values from 250 to 900 g mol$^{-1}$. Both $T_m$ and $M_w$ values were directly connected to the diol chemical structures, *i.e.* chain length and functional groups.[21]

The second strategy relies on the addition of two functional thiols onto a non-functional diene. Ameduri *et al.*[22–24] performed the thiol-ene radical coupling by using thiols of different chain lengths (Scheme 10.5).

The chemical structure of such diols comprises either two or four thio-ether bonds, according to the thiol structure. Commercially available diols do not possess any thio-ether bonds. Thus, Ameduri[25] showed that the synthesized diols exhibit better thermal resistance as compared to commercial ones for the same molecular weight values, thanks to their thio-ether bonds.

The last strategy corresponds to the addition of a non-functional dithiol onto two functional vinyl groups. Following this strategy, aromatics and aliphatic diols have been synthesized. A simple method allows the preparation of monodisperse diols terephtalate oligomers in two steps.[26] From aromatic α,ω-diols, an esterification reaction was carried out with mercapto acetic acid followed by thiol-ene radical addition onto 10-undecen-1-ol (Scheme 10.6).

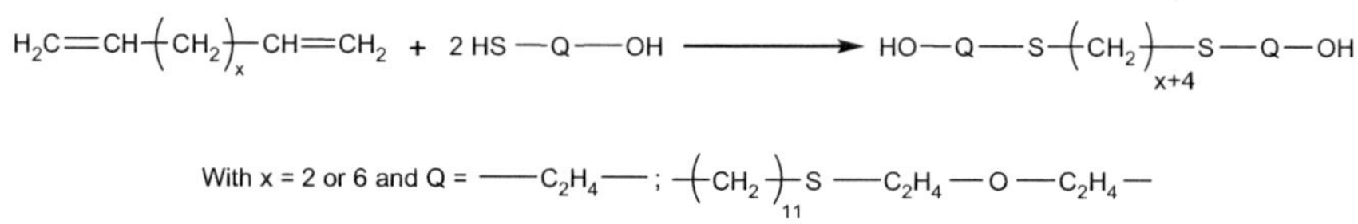


**Scheme 10.5** Synthesis of diol compounds from thiol-ene radical coupling onto dienes.

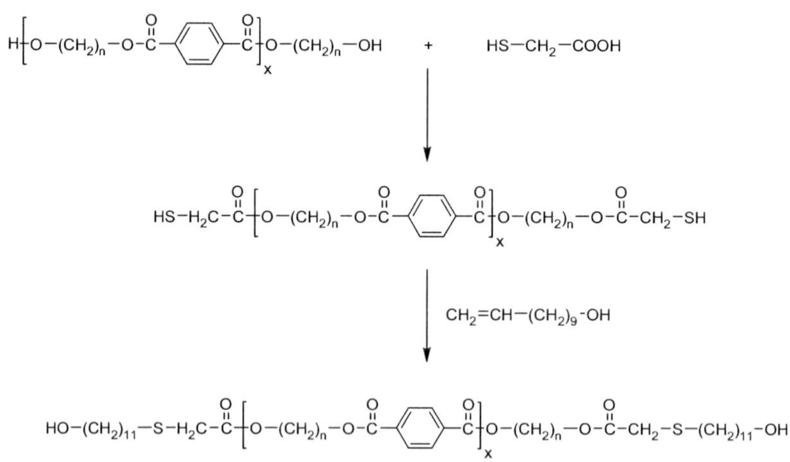


**Scheme 10.6** Synthesis of hydroxyl-telechelic alkylene terephthalate oligomers from thiol-ene radical coupling.

This step led to hydroxyl-telechelic oligo(tetramethylene terephthalate)s and oligo(ethylene terephtalate)s, showing different melting and glass temperatures. Most of the linear diols have been prepared from thiol-ene radical coupling. As compared to aromatic diols, the variety of aliphatic diols is more important: they have different chain lengths; they may possess thio-ether bonds as well as functional groups such as ester, amide or ether oxide. Their melting temperatures are rather low (from 15 to 95 °C), depending on their chemical structure. Interestingly, both aromatic and aliphatic diols provided hard or soft polyurethane resins, respectively.[22–24,27]

More recently, the thiol-ene radical coupling was used as a powerful tool to functionalize the double bonds of fatty acids.[28] Turunk *et al.*[29] performed the thiol-ene addition of fatty acids with functional thiols, *i.e.* alcohols and esters, to prepare linear as well as hyperbranched polyesters containing thio-ether linkages. Recently, we have reported the model study of the radical addition of 2-mercaptoethanol onto oleic acid under mild conditions (*i.e.* initiator free addition under UV light).[30] In particular, we demonstrated that the ratio between the thiol and double bond is a crucial parameter, leading either to efficient thiol-ene addition or to side reactions, such as esterification between the carboxylic acid function of oleic acid and the hydroxyl function of 2-mercaptoethanol. Thus, we have reported the synthesis of pseudo-telechelic diols of vegetable oils containing either esters or amide groups (Schemes 10.7 and 10.8) by using thiol-ene radical coupling. Their reaction with methylene diisocyanate (MDI) afforded bio-based polyurethanes and we especially showed that the amide groups influenced both the curing behavior and the glass transition temperatures of the resulting polyurethanes.[31]

### 10.2.2.2 Other Telechelic Compounds

These above mentioned synthetic pathways were not only used to make diol compounds, but also a wide range of telechelic compounds with different functionalities were able to be obtained. For instance, monodisperse macromolecular disocyanate synthesis was performed by following pathway 3, *i.e.* addition of non-functional dithiol onto a functional vinyl group (Scheme 10.9).[32] In this case, the thiol-ene radical coupling leads to a diacid compound, which requires modification into a diacid chloride by reaction with thionyl chloride. Finally, the reaction with sodium azide generates the corresponding diisocyanate compound. These diisocynanate prepolymers were investigated in the preparation of soft polyurethanes. It is well known that isocyanate groups are very sensitive to hydrolysis. The same authors proposed the synthesis of prepolymers with blocked isocyanate groups from thiol-ene radical coupling.[33] It is also interesting to note that the reaction of dithiol with a non-conjugated diene will provide either a monodispersed dithiol or a telechelic non-conjugated diene, depending on the stoichiometry of both reactants (Scheme 10.9). In both cases, the authors mentioned that batch reaction always leads to a mixture of oligomers, whereas sequential

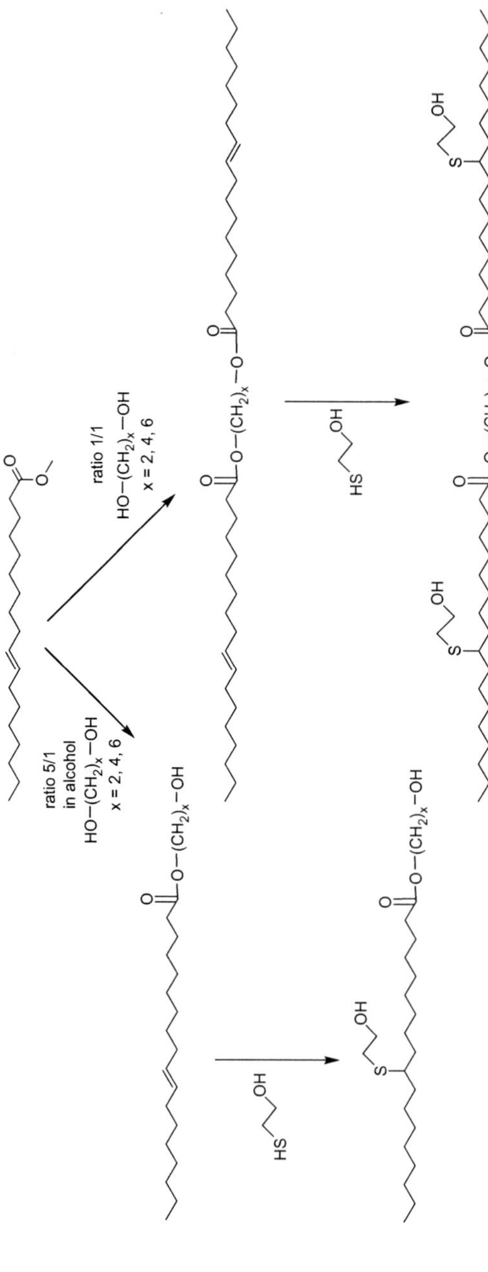


**Scheme 10.7** Synthesis of pseudo-telechelic diol fatty esters containing one or two ester groups.

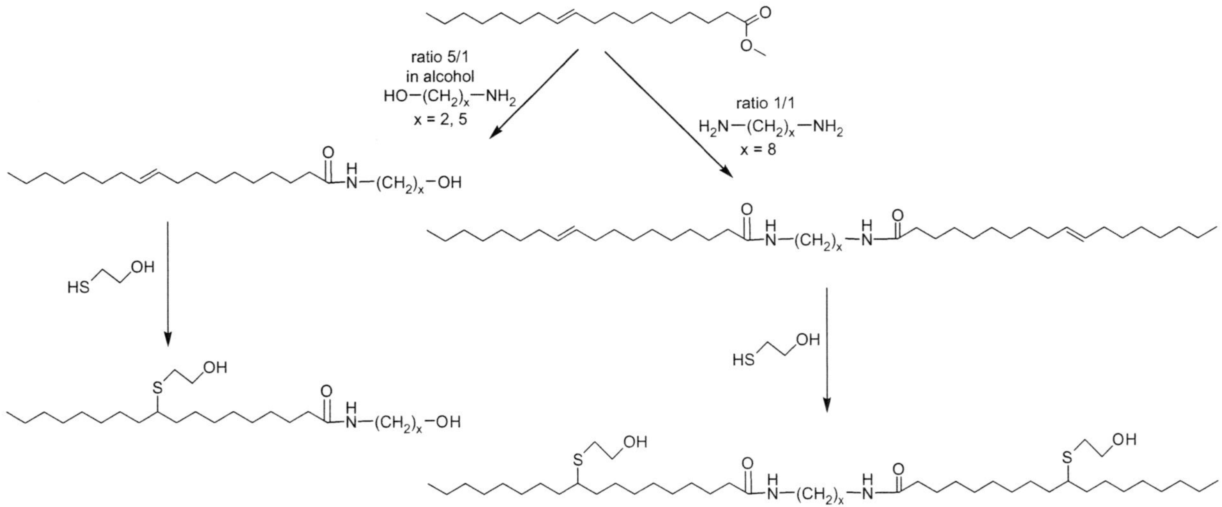


**Scheme 10.8** Synthesis of pseudo-telechelic diol fatty esters containing one or two amide groups.

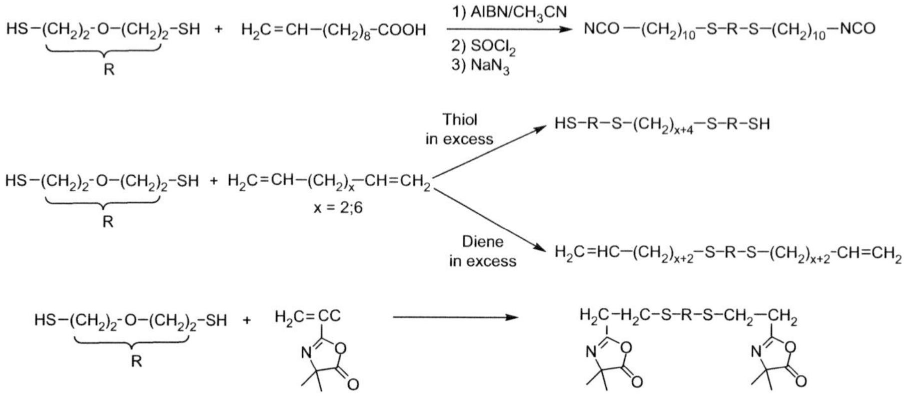

**Scheme 10.9** Synthesis of telechelic compounds from thiol-ene radical coupling.

additions of the reactant by default results in monodispersed telechelic compounds with high yields.[34–36]

More recently, Saint-Loup *et al.*[37] performed the synthesis of bisazlactone-telechelic compounds by thiol-ene radical coupling of a dithiol onto vinylazlactone (Scheme 10.9). The authors showed that this reaction could be clearly improved by using UV initiation. The resulting bislactone compound could be further used as a chain extender for hydroxyl or amino-telechelic oligomers in order to synthesize multi-block copolymers.

Recently, thiol-ene reaction was used to synthesize precursors of isocyanate free polyhydroxyurethanes. The synthesis of thio-ether five and six-membered dicyclocarbonates by radical addition of ethane-1,2-dithiol with 4-(3-butenyl)-1,3-dioxolan-2-one (Scheme 10.10(a))[38] or by reaction between 4-ethenyl-1,3-dioxolan-2-one (AC) with several thiols (Scheme 10.10(b))[39] has been reported. These dicyclocarbonates allow polyhydroxyurethanes without an ester bond to be obtained.

The synthesis of telechelic cyclocarbonates, by using synthesized 4-[(prop-2-en-1-yloxy)methyl]-1,3-dioxolan-2-one (AGC) and commercial 4-ethenyl-1,3-dioxolan-2-one (AC), was described by Boutevin *et al.*[40] Bis-AGC and bis-AC were synthesized by UV thiol-ene coupling with 2,2′-oxydiethanethiol (Scheme 10.10 (c)), and bis-AGC and bis-AC were polymerized with 1,10-diaminedecane (DA10) leading to polyhydroxyurethane materials.

## 10.3 Thiol-ene Radical Coupling onto Vinyl-containing Polymers

Thiol-ene radical coupling performed onto pre-made polymers is of great interest since it can lead to partial modification of the properties of the polymeric backbone. Depending on the type of thiol functional group, it can also bring new interesting properties to the polymer. To this aim, the polymer

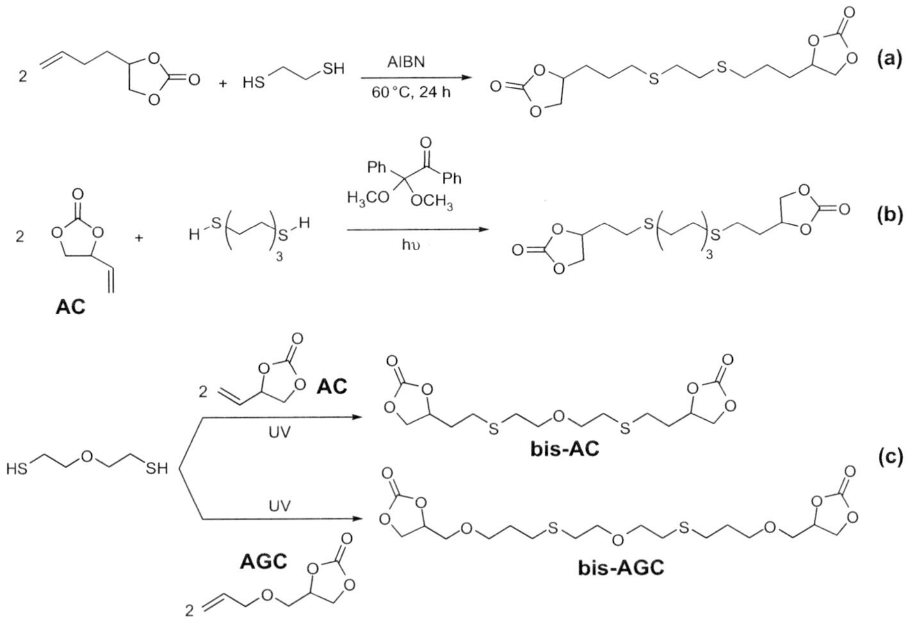


**Scheme 10.10** Synthesis of telechelic cyclocarbonate from thiol-ene radical coupling.

must contain vinyl groups attached alongside the backbone. According to its 1-2/1-4 double bonds, polybutadiene is a good candidate to perform thiol-ene radical coupling. We have indeed carried out thiol-ene radical onto hydroxyl-telechelic polybutadiene (HTPBD) with a series of thiols carrying different functional groups (Scheme 10.11). Indeed, HTPBD is a very attractive prepolymer as it exhibits exceptional softness (very low glass transition temperature of about −75 °C). For instance, thiol-ene radical coupling was performed onto HTPBD in the presence of 2-mercaptoethanol, leading to polybutadiene with a hydroxyl functionality of about 3.[41] Notably, it was demonstrated that thiol-ene radical coupling was achieved by photo-chemical addition towards only 1-2 double bonds. This hydroxyl saturated polybutadiene was then used as a precursor for cross-linked polyurethanes resins. In particular, it was shown that, in the presence of hexamethylene diisocyanate, the gel times were decreased by increasing the hydroxyl functionality.

Similarly, we have performed the photo-chemical addition of a thiol carrying a perfluorinated group onto HTPBD (Scheme 10.11). By using this technique we were able to incorporate a very high fluorine content, up to 50% (w/w), which induced an increase of the glass transition temperature up to −52 °C.[42] As mentioned above, HTPBD is a prepolymer used in the synthesis of polyurethanes (PUs). Furthermore, improving the fire resistance of PUs still remains an important challenge. To this aim, we have proposed the chemical incorporation of phosphonate moieties into HTPBD. Indeed,

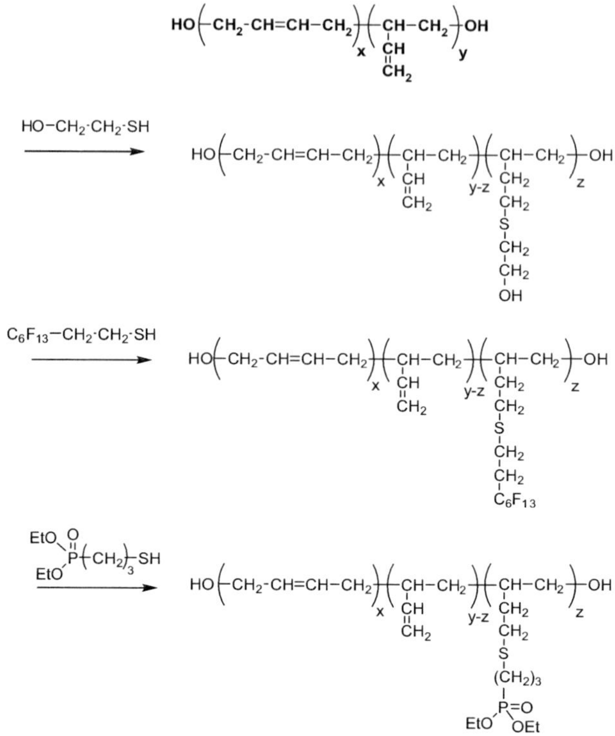


**Scheme 10.11** Modification of HTPBD by thiol-ene radical coupling.

phosphorus compounds are conventionally used as flame retardant additives of polymeric materials. Thus, the thiol-ene radical grafting was applied onto HTPBD (Scheme 10.11) by using a thiol bearing phosphonate moieties.[43] The thiol compound was firstly synthesized *via* a three step reaction: an Arbusov reaction onto allyl bromide, followed by thioacetate addition, then the thiol recovery was possible thanks to the catalytic activity of KCN.[44] From the thiol-ene radical coupling it was shown that about 5 wt% of phosphorus could be efficiently grafted onto HTPBD, mostly by reaction towards the 1-2 double bonds.

Cross-linked elastomers are also of great interest as they allow soft coatings to be obtained. We have been interested in incorporating acrylic groups into polybutadiene since the acrylate enables UV photocross-linking. To obtain such compounds, thiol-ene radical coupling was very useful. Indeed, from carboxy-telechelic polybutadiene (CTPBD), we have performed the thiol-ene radical coupling in the presence of mercapto-acetic acid in order to increase the acid group content (Scheme 10.12).[45] Once again, the radical addition was efficient onto 1-2 double bonds. From CTPBD, the $x$ value was equal to 16 (for $M_n$ of about 4400 g mol$^{-1}$). From NMR investigation, we have shown that the $z$ value (Scheme 10.12) was equal to 10, which denoted an acid functionality

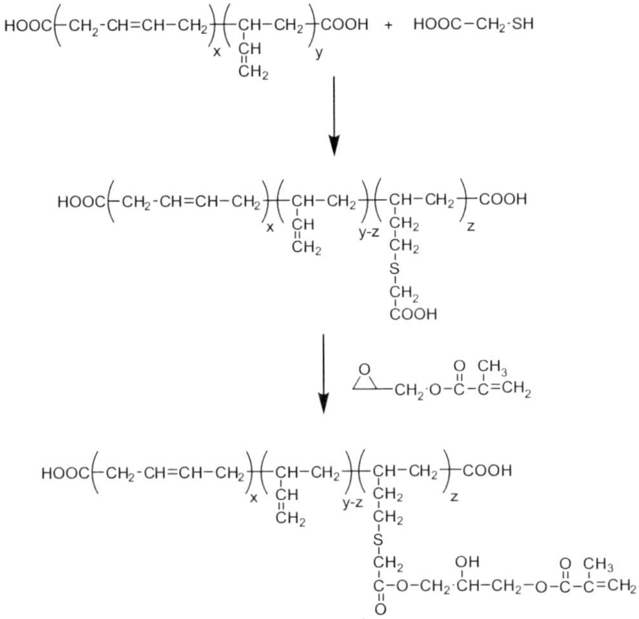

**Scheme 10.12** Modification of CTPBD by thiol-ene radical coupling.

**Table 10.2** Functionalization of poly(butadiene) by using different thiols.

| *Nature of Thiol* | *Initiation* | *References* |
|---|---|---|
| $C_4H_9$–SH | Thermal | 46 |
| HO–$(CH_2)_2$–SH | Photochemical | 41, 47 |
| HOOC–$(CH_2)_n$–SH | Photochemical | 45, 48 |
| $C_nF_{2n+1}$–$(CH_2)_2$–SH | Photochemical, thermal | 48–50 |
| $(EtO)_3Si$–$(CH_2)_3$–SH | Photochemical, thermal | 48, 51 |
| $(EtO)_2P(O)$–$(CH_2)_3$–SH | Photochemical, thermal | 43 |
| $H_2N$–$C_2H_4$–SH | Thermal | 50, 52 |
| HOOC–$(CH_2)_n$–SH | Thermal | 50, 52 |
| $HOCH_2$–CH(OH)–$CH_2$–SH | Thermal | 50, 52 |
| $CH_3$–CO–NH–CH($CO_2$Me)–$CH_2$–SH | Thermal | 50, 52 |

of 12 per chain. Then, the acrylate groups could be introduced by addition with 2,3-epoxy propyl methacrylate onto the acid groups. These functional oligomers were processed to be cross-linked under UV light and a large improvement of flexibility properties was obtained.

Poly(butadiene) was also functionalized by using different thiols compounds through thiol-ene radical coupling. Table 10.2 gathers the overall thiol-ene reactions performed onto poly(butadiene).

The thiol-ene reaction of poly(butadiene) with grafted L-amino acidic side chains (L-cysteine derivatives), allows new materials to be obtained.[53] Most

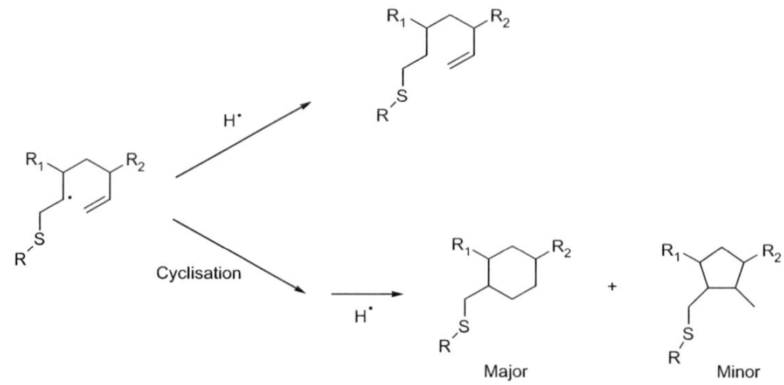


**Scheme 10.13** Polydiene intra-cyclisation from 1,2-double bonds.

importantly, all the modified polymers obtained are chiral polymers starting from a non-chiral polymer, confirming the thiol-ene reaction as a viable approach to obtain optically active polymers by simply 'transferring' this property from a chemical (the L-cysteine derivative).

Schlaad *et al.*[54] worked on polydienes with a high content of 1,2-double bonds and showed that different reactions occurred, based on the two neighboring double bonds cyclization (Scheme 10.13).

This cyclization content depends on the steric hindrance of the thiol and tends to reduce the functionality of polymer from 18 to 48%. MALDI-TOF, NMR and elemental analysis techniques allowed this secondary reaction to be assessed. This side reaction was already described by Hillmyer and Lodge[55] in 2001, who studied the addition of $C_6F_{13}I$ onto 1,2-poly(butadiene) with boranes as initiators. This cyclization reaction has been demonstrated from $^{13}C$ NMR ($CH_2I$ shifts) technique.

As a last example of polymer modification by thiol-ene radical coupling, the synthesis of acrylate-containing poly(dimethylsiloxane) is related. Abdellah *et al.*[56] performed the thiol-ene radical coupling of poly(dimethylsiloxane) (PDMS) containing vinyl groups with mercaptoethanol. Then the hydroxyl groups of PDMS allowed reaction with isocyanate-containing acrylates (Scheme 10.14), leading to the desired acrylate-containing poly(dimethylsiloxane),s with urethane bond distributed alongside the PDMS chains.

## 10.4 Block Copolymers and Ionomers by Thiol-ene Radical Coupling

### 10.4.1 Block Copolymers from Thiol-ene Radical Coupling

Our group[57] also performed the synthesis of triblock co-oligomers of poly(methyl methacrylate)-b-PDMS-b-poly(methylmethacrylate) by thiol-ene radical coupling (Scheme 10.15). The first step consisted of the synthesis of a PMMA macromonomer using γ-mercaptopropyltrimethoxysilane as chain

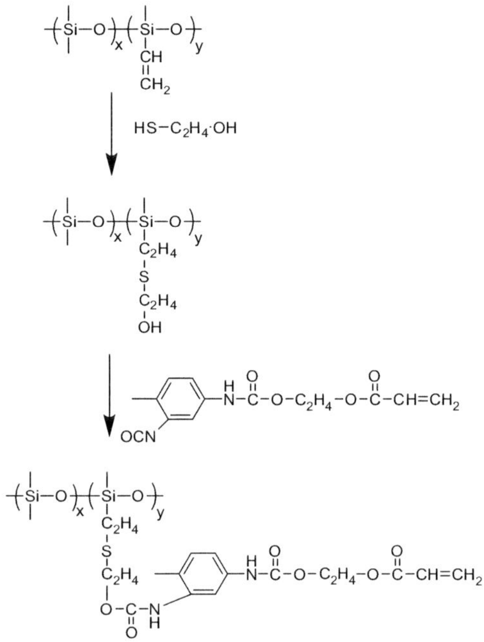


**Scheme 10.14** PDMS containing acrylate groups obtained from thiol-ene radical coupling.

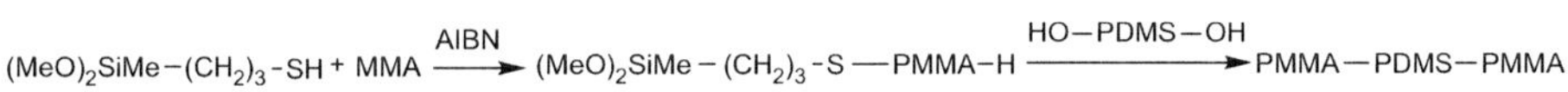


**Scheme 10.15** Synthesis of triblock PMMA-b-PDMS-b-PMMA from thiol-ene radical coupling.

transfer agent, MMA as monomer, and AIBN as initiator. The second step was the condensation reaction between the macromonomer and the silanol-terminated PDMS at 100 °C for 20 h in the presence of a catalyst (2-ethylhexanoic acid/tetramethylguanidine complex). The presence of small amounts of diadduct (reaction of the second methoxy silyl group with a silanol-terminated PDMS) has been observed.

## 10.4.2 Ionomers from Thiol-ene Radical Coupling

Another example of the thiol-ene reaction concerns the synthesis of ionomers. The functionalization of PEG-based molecules, namely a difunctional PEG 400 molecule, a monofunctional derivative of PEG, a non-ionic surfactant Brij 700, Terathane® and aromatic polyester diols obtained by PET glycolysis, with phosphonated groups was performed. Thioglycolic acid (TA) was first grafted onto the hydroxylated chain-end(s) of the polymers before carrying out the thiol-ene radical coupling of vinyl phosphonic acid onto the sulfide moiety.

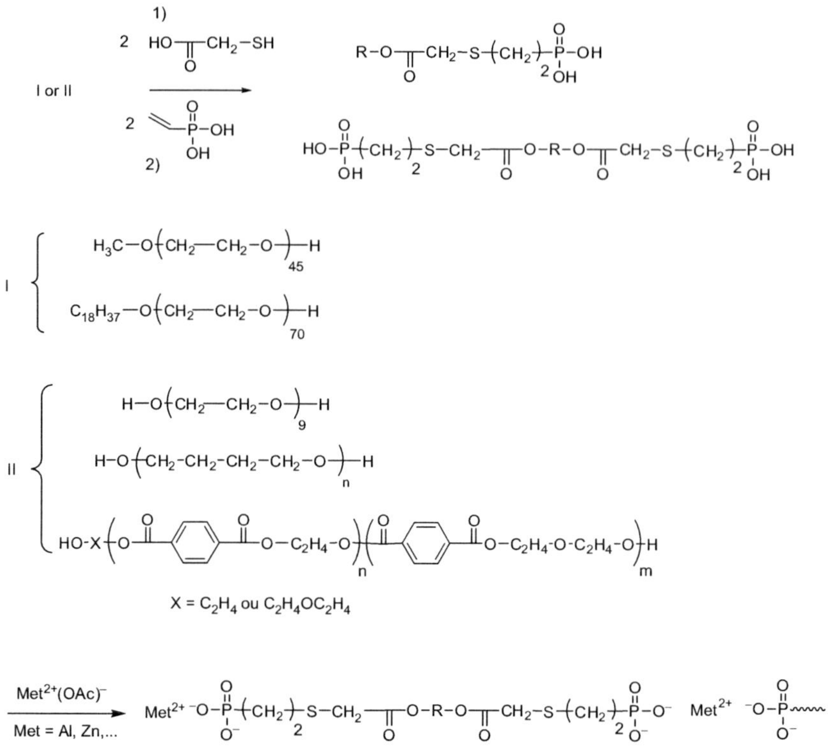

Scheme 10.16 Synthesis of ionomers from thiol-ene radical coupling.

This strategy of mono or diol functionalization allowed preparation of hydrophilic ionomers (Scheme 10.16).[58,59]

## 10.5 Conclusion

We have demonstrated that thiol-ene radical coupling is a powerful and versatile technique that affords oligomers or polymers with a wide range of controlled functionalities. Thiol-ene radical coupling was mostly used to provide telechelic oligomers, used as prepolymers for polycondensation reactions. Interestingly, the versatility of this technique allowed us to functionalize vegetable oils in order to synthesize hydroxyl-telechelic fatty esters, which were then used as precursors for the polyurethane synthesis. Similarly, we used thiol-ene radical coupling to perform the synthesis of telechelic cyclocarbonate, leading to hydroxypolyurethane materials, *i.e.* without any isocyanate.

## References

1. C. E. Hoyle, T. Y. Lee and T. Roper, *J. Polym. Sci., Part A: Polym. Chem.*, 2004, **42**, 5301–5338.

2. G. David, C. Boyer and B. Boutevin, *Adv. Polym. Sci.*, 2007, **206**, 31.
3. B. Boutevin, *J. Polym. Sci., Part A: Polym. Chem.*, 2000, **38**, 3235–3243.
4. M. J. Kade, D. J. Burke and C. J. Hawker, *J. Polym. Sci., Part A: Polym. Chem.*, **48**, 743–750.
5. A. B. Lowe and M. A. Harvison, *Aust. J. Chem.*, **63**, 1251–1266.
6. A. B. Lowe, *Polym. Chem.*, **1**, 17–36.
7. P. J. Roth, C. Boyer, A. B. Lowe and T. P. Davis, *Macromol. Rapid Commun.*, 2011, **32**, 1123–1143.
8. A. Dondoni, *Angew. Chem., Int. Ed.*, 2008, **47**, 8995–8997.
9. N. Pelaprat, G. Rigal, B. Boutevin, A. Manseri and M. Belbachir, *Eur. Polym. J.*, 1996, **32**, 1189–1197.
10. G. David, B. Boutevin, S. Seabrook, M. Destarac, G. Woodward and G. Otter, *Macromol. Chem. Phys.*, 2007, **208**, 635–642.
11. B. Boutevin, B. Hamoui and J. P. Parisi, *J. App. Polym. Sci.*, 1994, **52**, 449–456.
12. N. O. Brace, *J. Chem. Soc.*, 1961, **3**, 3197.
13. T. Jeanmaire, C. Brondino, Y. Hervaud and B. Boutevin, *Phosphorus, Sulfur Silicon Relat. Elem.*, 2002, **177**, 2331–2343.
14. C. Brondino, B. Boutevin, Y. Hervaud, N. Pelaprat and A. Manseri, *J. Fluor. Chem.*, 1996, **76**, 193–200.
15. R. N. Haszeldine, D. L. Hobson and D. R. Taylor, *J. Fluor. Chem.*, 1976, **8**, 115.
16. B. Boutevin, E. Fleury, Y. Pietrasanta and L. Sarraf, *J. Fluor. Chem.*, 1986, **31**, 437–450.
17. B. Boutevin, Y. Pietrasanta and L. Sarraf, *J. Fluor. Chem.*, 1986, **31**, 425–435.
18. B. Boutevin, M. Khamlichi and J.-J. Robin, *Macromol. Chem. Phys.*, 1996, **197**, 1325–1336.
19. B. Boutevin, *Adv, Polym. Sci.*, 1990, **94**, 69–105.
20. B. Boutevin, G. David and C. Boyer, *Adv. Polym. Sci.*, 2007, **206**, 31–135.
21. B. Boutevin, A. El Idrissi and J. P. Parisi, *Makromol. Chem.*, 1990, **191**, 445–456.
22. B. Ameduri, K. Berrada, B. Boutevin and R. D. Bowden, *Polym. Bull.*, 1992, **28**, 497–503.
23. B. Ameduri, K. Berrada, B. Boutevin and R. D. Bowden, *Polym. Bull.*, 1992, **28**, 389–394.
24. B. Ameduri, K. Berrada, B. Boutevin and R. D. Bowden, *Phosphorus, Sulfur Silicon Relat. Elem.*, 1993, **74**, 477–478.
25. B. Ameduri, K. Berrada, B. Boutevin and R. D. Bowden, *Phosphorus Sulfur Relat. Elem.*, 1993, **83**, 39–47.
26. B. Ameduri, B. Boutevin, M. Khamlichi and J. J. Robin, *Macromol. Chem. Phys.*, 1994, **195**, 3425–3443.
27. B. Ameduri, K. Berrada, B. Boutevin, M. Melas and R. D. Bowden, *Makromol. Chem.*, 1993, **194**, 3001–3014.
28. G. B. Bantchev, J. A. Kenar, G. Biresaw and M. G. Han, *J. Agric. Food Chem.*, 2009, **57**, 1282–1290.

29. O. Turunc and A. R. Meier Michael, *Macromol. Rapid Commun.*, 2010, **31**, 1822–1826.
30. M. Desroches, S. Caillol, V. Lapinte, R. Auvergne and B. Boutevin, *Macromolecules*, 2011, **44**, 2489–2500.
31. M. Desroches, S. Caillol, R. Auvergne, B. Boutevin and G. David, *Polym. Chem.*, 2012, **3**, 450–457.
32. B. Boutevin, M. Chaib and J. J. Robin, *Polym. Bull.*, 1991, **26**, 177–179.
33. B. Boutevin, M. Chaib and J. J. Robin, *Polym. Bull.*, 1992, **29**, 613–616.
34. B. Ameduri, K. Berrada, B. Boutevin and R. D. Bowden, *Polym. Bull.*, 1993, **28**, 531–537.
35. B. Ameduri, K. Berrada, B. Boutevin, R. D. Bowden and L. Pereira, *Polym. Bull.*, 1991, **26**, 377–382.
36. B. Ameduri, B. Boutevin, M. Khamlichi and J. J. Robin, *Macromol. Chem. Phys.*, 1994, **195**, 3425.
37. R. Saint-Loup, J.-J. Robin and B. Boutevin, *Macromol. Chem. Phys.*, 2002, **203**, 199–211.
38. H. Tomita, F. Sanda and T. Endo, *J. Polym. Sci., Part A: Polym. Chem.*, 2001, **39**, 4091–4100.
39. C. N. Tang, H. B. Nulwala, K. Damodaran, P. Kaur and D. R. Luebke, *J. Polym. Sci., Part A: Polym. Chem.*, 2011, **49**, 2024–2032.
40. S. Benyahya, M. Desroches, R. Auvergne, S. Carlotti, S. Caillol and B. Boutevin, *Polym. Chem.*, 2011, **2**, 2661–2667.
41. G. Boutevin, B. Ameduri, B. Boutevin and J.-P. Joubert, *J. Appl. Polym. Sci.*, 2000, **75**, 1655–1666.
42. B. Boutevin, Y. Hervaud and M. Nouiri, *Eur. Polym. J.*, 1990, **26**, 877–882.
43. B. Boutevin, Y. Hervaud and G. Mouledous, *Polym. Bull. (Berlin)*, 1998, **41**, 145–151.
44. B. Boutevin, Y. Hervaud, G. Mouledous and N. Pelaprat, *Phosphorus, Sulfur Silicon Relat. Elem.*, 1998, **140**, 125–133.
45. B. Boutevin, E. Fleury, J. P. Parisi and Y. Pietrasanta, *Makromol. Chem.*, 1989, **190**, 2363–2372.
46. P. Claudy, J. M. Letoffe, Y. Camberlain and J. P. Pascault, *Polym. Bull.*, 1983, **9**, 206.
47. A. Priola, F. Ferrero, G. Gozzelino and M. Panetti, *Chem. Ind.*, 1984, **66**, 471.
48. T. D. Wickard, E. Nelsen, N. Madaan, N. Brummelhuis, C. Diehl, H. Schlaad, R. C. Davis and M. R. Linford, *Langmuir*, 2010, **26**, 1923–1928.
49. B. Ameduri, B. Boutevin and M. Nouiri, *J. Polym. Sci., Part A: Polym. Chem.*, 1993, **31**, 2069.
50. J. Justynska and H. Schlaad, *Macromol. Rapid Commun.*, 2004, **25**, 1478–1481.
51. F. Schapman, J. P. Couvercelle and C. Bunel, *Polymer*, 1998, **39**, 4955–4962.
52. J. Justynska, Z. Mordyjewick and H. Schlaad, *Polymer*, 2005, **4**, 12057–12064.

53. L. Lotti, S. Coiai, F. Ciardelli, M. Galimberti and E. Passaglia, *Macromol. Chem. Phys.*, 2009, **210**, 1471–1483.
54. N. T. Brummelhuis, C. Diehl and H. Schlaad, *Macromolecules*, 2008, **41**, 9946–9947.
55. Y. Ren, T. Lodge and H. A. Hillmyer, *Macromolecules*, 2001, **34**, 4780–4787.
56. L. Abdellah, B. Boutevin and B. Youssef, *Eur. Polym. J.*, 1991, **27**, 821–825.
57. Y. Pietrasanta, E. Fleury, B. Boutevin and L. Sarraf, *Polym. Bull.*, 1986, **15**, 107.
58. M. Essahli, F. Ganachaud, M. In and B. Boutevin, *J. Appl. Polym. Sci.*, 2008, **108**, 483–490.
59. M. Essahli, G. Colomines, S. Monge, J. J. Robin, A. Collet and B. Boutevin, *Polymer*, 2008, **49**, 4510–4518.

CHAPTER 11

# *Hybrids of Synthetic Polymers and Natural Building Blocks Using Thio-click*

MARTINA H. STENZEL

Centre for Advanced Macromolecular Design School of Chemical Engineering, University of New South Wales, Sydney, Australia
Email: m.stenzel@unsw.edu.au

## 11.1 Hybrids of Polymers and Nature's Building Block

Hybrids of polymers with biopolymers or other natural building blocks have been the focus of attention for many years now. The aim is to combine the best properties of both worlds to create new materials. Synthetic polymers have the advantage that they can take on many different properties, which can be easily fine-tuned. Ranging from water- to oil-soluble, stimuli-responsive, brittle, elastomeric, and conductive; there are not many properties that cannot be created using synthetic polymers. Biopolymers such as proteins, oligonucleotide, polysaccharides or smaller naturally occurring building blocks such as amino acids and sugar are often biocompatible and degradable, but most importantly they are frequently bioactive. Combining both materials allows the design of polymers that can take on the properties of synthetic polymers while being bioactive. The versatility of these materials with their various facets makes them highly attractive for biomaterial design.

In the following, the different types of thiol-X reactions used to create these hybrids of synthetic polymers and nature's building blocks are discussed.

RSC Polymer Chemistry Series No. 6
Thiol-X Chemistries in Polymer and Materials Science
Edited by Andrew B. Lowe and Christopher N. Bowman

Published by the Royal Society of Chemistry, www.rsc.org

The focus is the functionalization of soluble polymer chains only, while surface functionalization has been omitted.

## 11.2 Glycopolymers

Carbohydrates are involved in a myriad of biological communication events, including cellular recognition, inflammation, signal transmission and infection of pathogens displayed by them.[1–5] Moreover, many studies on drug targeting in the biological system have shown that lectins on cell surfaces mediate cell–cell interactions by interacting with their complementary carbohydrates.[6–10] The interaction between one carbohydrate molecule with the corresponding lectin is rather weak, but the simultaneous interaction of many sugar molecules arranged along a polymer change can lead to strong binding. This effect is coined the "glycocluster effect".[11] The interest in the synthesis of glycopolymers, which are synthetic polymers with pendant sugar groups, is therefore not surprising.[12] Glycopolymers have initially been prepared using established procedures such as free radical polymerization. Wulff *et al.*[13] highlighted, in an early feature article, possible avenues to generate glycomonomers, which can be utilized in a radical process. Synthetic routes to glycopolymers were later expanded to living anionic polymerization,[14] cationic polymerization,[15] ring-opening polymerization[16] and ring-opening metathesis polymerization.[17] A significant impact made later was the extensive use of living radical polymerization techniques (nitroxide mediated polymerization (NMP), atom transfer radical polymerization (ATRP) and reversible addition fragmentation polymerization (RAFT)) to generate well-defined glycopolymers with various polymer architectures.[12,18]

Although the direct polymerization of sugar-containing monomers is now a well-established technique, post-modification of preformed polymers with carbohydrates offers an excellent alternative synthetic route. The post-functionalization approach allows the production of libraries of glycopolymers with the same macromolecular architecture by attaching different sugar moieties to pre-formed polymer scaffolds. It also generally provides a simpler procedure, as some sugar-containing monomers have the tendency to self-polymerize during the purification procedures. Furthermore, some reaction conditions, such as high temperatures in NMP, are not suited for sensitive monomers such as glycomonomers. Initially, the focus of post-modification was heavily on amino sugars as an active carbohydrate resulting in amide linkages between the polymer backbone and sugar. Due to the good nucleophilicity of amines compared to other functional groups such as alcohols, selectivity is offered without the need for protecting groups.[19,20] The higher nucleophilicity of thiols[21] resulted in the domination of thiol-X chemistry to post-modify reactive polymers with pendant carbohydrates.

### 11.2.1 Thio-halogeno Reaction

Probably the first report on postfunctionalization of a polymer with a thiolated sugar employed the nucleophilic substitution of a chloride. The commercially available polylysine was modified with chloroacetic anhydride (and bromoacetic anhydride) to yield polymers with chloroacetamide functions (Scheme 11.1.1).[22] Four different thiolated sugars were used, but only thioglucose with its secondary thiol functionality was able to reach complete conversion when using triethyl amine ($NEt_3$) as catalyst. Other carbohydrates, which carried primary thiols, required the use of the stronger base 1,8-diazabicyclo[5.4.0]-undec-7-ene (DBU) to achieve complete functionalization.[23] A thiolated derivative of sialic acid was clicked to backbones with various molecular weights using DBU. Remaining functional groups were then reacted with thioglycerol in the presence of 3 eq of $NEt_3$ to yield a range of glycopolymers with an array of epitope densities, which were then further tested towards their ability to inhibit E-selectin.[23] The procedure was then later extended to dendritic structures.[24]

The described approach required a two-step post-modification procedure, which includes the conversion of the amine to the chloroacetamide functional group. The direct synthesis of polymers with halogenides is in fact only suitable for some selected polymerization procedures. One example is the palladium catalyzed Sonogashira coupling polymerization between 2,7-diiodo-9,9-bis-(6′-bromohexyl)fluorine and 1,4-phenyldiboronic acid, which led to a conjugated fluorene based polymer (Scheme 11.1.2). Reaction of the polymer in THF using $K_2CO_3$ and acetylated 1-thio-α-D-glucose sodium salt for 24 hours led to 98% completion, but also the direct route using the deprotected sugar in DMF was successful.[25]

This approach was also tested using other conjugated systems. Poly(*p*-phenylene)s were prepared using a bromide bearing diiodoaryl monomer and 1,4-phenyldiboronic acid in the palladium-catalyzed Suzuki coupling reaction (Scheme 11.1.3).[26] Thiol-X reactions with 1-thiolethyl-β-D-mannose tetraacetate or 1-thiol-β-D-glucose tetraacetate in THF solution in the presence of $K_2CO_3$ resulted in fluorescent glycopolymers. The mannose containing polymer showed significant fluorescent quenching after interaction with ConA. The conjugated fluorene based polymer (Scheme 11.1.2) was further modified to introduce polyethylene glycol spacers between the stiff conjugated backbone and the bromide to enhance hydrophilicity of the polymer. Subsequent reaction with 1-thiol-β-D-glucose tetraacetrate or 1-thiolethyl-α-D-mannose afforded a highly water-soluble glycopolymer. Incubation of the mannose bearing polymer with *E. coli* resulted in the formation of fluorescent bacteria clusters. Meanwhile, the UV/Vis absorption underwent a significant red-shift.[27]

Another technique that is reasonably robust in the presence of chloride-containing monomers is RAFT polymerization, albeit that possible chain transfer to the chloride needs to be considered. Two types of polymer architectures were prepared using chloromethyl styrene (Scheme 11.1.4), a statistical polymer with *t*-butyl acrylate[28] and a starpolymer.[29] The statistical polymer was functionalized with 1-thio-β-D-glucose sodium salt in DMSO/acetone with an excess of $NEt_3$. The pink colour of the RAFT polymer disappeared probably due to the alkaline reaction environment.[28] Interestingly, it was

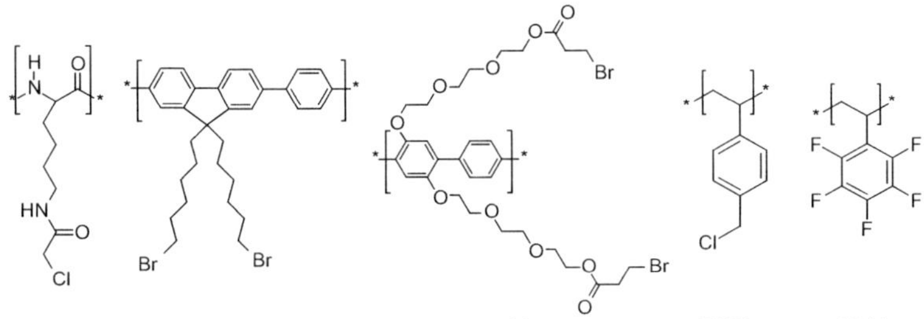


**Scheme 11.1** Polymers for the nucleophilic substitution of halogenides by thiolated carbohydrates.

observed that an excess of 1-thio-β-D-glucose and base is not required to achieve the same outcome. Various poly(chloromethyl styrene) 4-arm star polymers were reacted with 1-thio-β-D-glucose sodium salt in DMSO in the absence of any base. As a result, the only side product was NaCl and the polymer could be used directly without any further purification. However, this could only be achieved after a lengthy reaction time of 120 hours according to on-line monitoring using $^{1}$H-NMR.[29] Binding with ConA showed that a medium molecular weight of the glycopolymer star is most suitable for fast and efficient binding.

## 11.2.2 Thio-para Fluoro Reaction

Thiols are, compared to amines, softer nucleophiles and they are more reactive in undergoing nucleophilic substitution. A typical example here is the reaction of poly(pentafluoro styrene) with thiols (Scheme 11.1.4). Pentafluorostyrenes can be polymerized with a range of techniques including free radical polymerization,[32] ATRP,[33] RAFT[34] and NMP.[35] The latter is considered unsuitable for the direct polymerization of glycomonomers due to the high reaction temperature although glycopolymers have been prepared directly using a low temperature approach.[36] NMP has been employed as a metal-free polymerization technique to generate well-defined poly(pentafluorstyrene) for post-functionalization. 2,3,4,6-tetra-*O*-acetyl-1-thio-β-D-glucopyranose was mixed with the polymer in DMF and a three times excess of $NEt_3$ at 40 °C to yield complete conversion after only a few hours. The overall procedure, polymerization and click reaction was completely metal-free and was therefore considered a good avenue for biological polymers. The clear advantage of this pathway is that it allows direct monitoring of the progress of the reaction using $^{19}$F-NMR spectroscopy. A detailed investigation revealed that the reaction is rather rapid and more than 90% of functional groups have reacted after 30 minutes.[31] This reaction was not only successful with homopolymers, but also with block copolymers. A polystyrene-β-poly(pentafluorostyrene) was clicked with two different kind of thiolated carbohydrates, 1-thio-β-D-glucopyranose and 1-thio-β-D-galactopyranose. These polymers were subsequently employed as a biocompatible coating for adhesion of 3T3 and MC3T3-E1 cells.[30]

### 11.2.3 Thiol-ene Reaction

The most common pathway to create glycopolymers with post-functionalization is the reaction between thiols and pendant double bonds. Prerequisite is the synthesis of polymers with pendant double bonds, which limits the approach to polymerization techniques that are orthogonal to the presence of double bonds. Glycoconjugation using thiol-ene reaction to create low molecular weight compounds has been a well-established technique for many years,[37] but it took several years more before polymer science discovered the simplicity of this pathway.

The first report on the design of glycopolymers using thiol-ene chemistry described the use of poly[2-(3-isobutenyl)-2-oxaline] as reactive substrate (Scheme 11.2.6). The polymer (around 4 wt%) was mixed with a small excess of various thiols, among them 2,3,4,6-tetra-*O*-acetyl-1-thio-β-D-glucopyranose, and the degassed samples were irradiated under light for 24 hours. The reaction was found to have all the features of a click reaction, with side products such as cyclic units in the backbone being absent. In addition, solely the anti-Markovnikov addition product was formed.[38] Unique materials were created when copolymerizing 2-isopropyl-2-oxaline with 2-(3-isobutenyl)-2-oxaline *via* cationic ring-opening polymerization (Scheme 11.2.7). Poly(2-isopropyl-2-oxazoline) (PIPOX) is a structural isomer to poly(*N*-isopropyl acrylamide) (PNIPAAm) and exhibits a LCST similar to PNIPAAm at 37 °C. The two monomers were copolymerized at different ratios, followed by the thiol-ene click reaction with 2,3,4,6-tetra-*O*-acetyl-1-thio-β-D-glucopyranose, which was subsequently deprotected. The LCST could be fine-tuned with the amount of carbohydrate present along the polymer chain.[39]

The thiol-ene reactions were successful in the absence of any catalyst or initiator. Irradiation of thiol and substrate in a degassed solution resulted in complete conversion, albeit reaction times of 24 hours were required. To accelerate the reaction, photoinitiators such as 2,2-dimethoxy-2-phenylacetophenone (DMPA) have been added. The presence of DMPA accelerates the thiol-ene click reaction leading to complete conversion in less than 2 hours.[46] This pathway was then used to modify block copolymers creating temperature-responsive micelles. Binding with the lectin ConA was observed to be enhanced with micelle formation (Scheme 11.3).

Photoinitiators were later also employed for the modification of poly(2-oxazoline) using a slightly different substrate (Scheme 11.2.8). Copolymers of 2-ethyl-2-oxazoline and 2-(dec-9-enyl)-2-oxazoline were

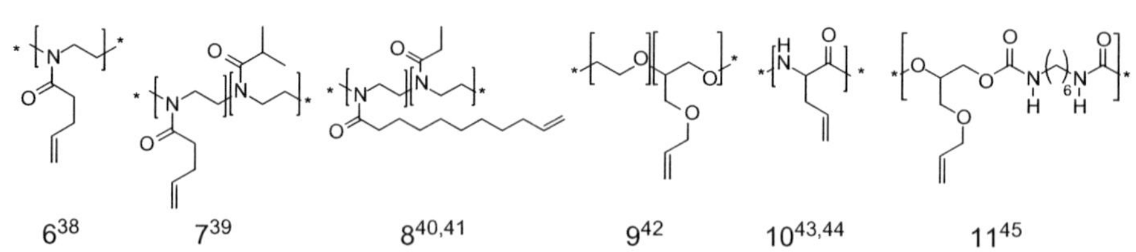


**Scheme 11.2** Reactive polymers used for the thiol-ene reaction with thiolated carbohydrates.

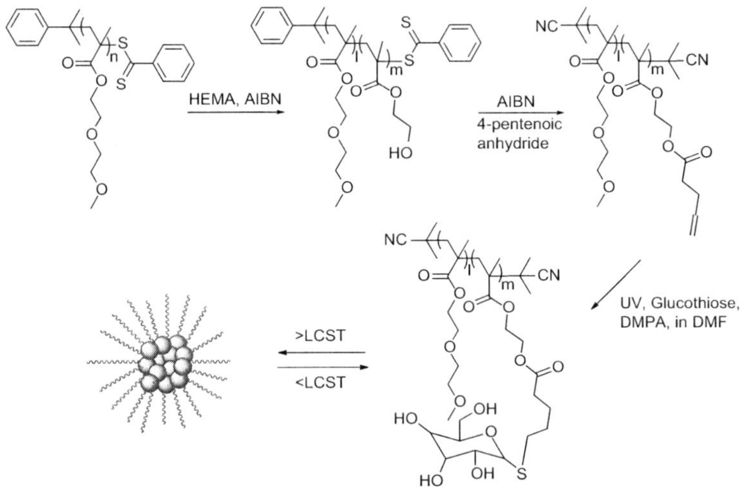


**Scheme 11.3** Synthetic strategies for the preparation of glucose functionalised block copolymers using thiol-ene chemistry.[46]

prepared and reacted with a slight excess of 2,3,4,6-tetra-*O*-acetyl-1-thio-β-D-glycopyranose in THF initiated using DMPA under UV light. In contrast to expectations,[39] the cloud point (LCST) of the deprotected glycopolymer now decreased with increasing sugar content. This has been explained by the higher flexibility of the alkyl spacer allowing the carbohydrate to diffuse close to the backbone for intramolecular H-bonding, exposing the alkyl chain to the aqueous environment.[40] Subsequently, cationic ring-opening was employed to generate a terpolymer from 2-ethyl-2-oxazoline, 2-(dec-9-enyl)-2-oxazoline and a new glycomonomer, which was obtain by reacting 2-(4-trimethyl-silylethynylphenyl)-2-oxazoline with (2-azidoethyl)-1-β-D-glucopyranose tetraacetate *via* copper(I)-catalyzed azide-alkyne cycloaddition (CuAAC) click reaction. The terpolymer was reacted with a range of thiols including 2,3,4,6-tetra-*O*-acetyl-1-thio-β-D-glucopyranose.[41]

Living anionic ring-opening polymerization was employed to create another polymer substrate with pendant double bonds.[42] The block copolymer poly(ethylene oxide)-β-poly(allyl glycidyl ether) (Scheme 11.2.9) was subsequently reacted with 2,3,4,6-tetra-*O*-acetyl-1-thio-β-D-glucopyranose using again the radical approach. Instead of using UV-irradition, a thermal radical initiator, AIBN, was employed. The reaction was carried out at 70 °C for 24 hours. $^1$H NMR analysis showed the expected product with a small impurity indicating the formation of some Markovnikov product. A small signal at 1.23 ppm was assigned to thiol-ene coupling to the α-carbon of the double bonds, which resulted in the formation of a methyl group.[42]

Radical initiation with AIBN at 70 °C was the pathway of choice to functionalize polypeptides with thiolated sugars (Scheme 11.2.10).[43] The yield was found to be below 50% and switching to photointiation did not improve the outcome. The low solubility of the polypeptide in most solvents was identified as a possible reason for the incomplete reaction. Complete conversion was achieved with a shorter synthetic peptide with the same allylglycine repeating units. The overall peptide structure was slightly different, but the ten-fold excess of 2,3,4,6-tetra-*O*-acetyl-1-thio-β-D-glucopyranose facilitated the reaction.[44]

A very different type of substrate was created by reacting a diol with a pendant double bond with 1,6 hexamethylene diisocyanate leading to reactive polyurethanes (Scheme 11.2.11). The following thiol-ene step, initiated with a photoinitiator, resulted in glucose and galactose-modified polyurethanes.[45]

The fast reaction between thiols and double bonds is especially attractive for the synthesis of dendrimers where the multitude of steps can lead to a lengthy synthesis procedure. Less than one hour reaction time was required to glycosylate all 48 double bonds of a dendrimer using DMPA as photoinitiator.[47]

## 11.2.4 Thiol-yne Reaction

Although the thiol-ene reaction is highly efficient, synthesis of the reactive polymer is limited to polymerizations that are indifferent to the presence of vinyl groups. The reactive polymer displayed in Scheme 11.3 was obtained by postmodification of a polymer with vinyl groups by reacting PHEMA with pentenoic anhydride.[46] Some efforts have been made to synthesize polymers with vinyl functionalities directly *via* radical polymerization techniques by carefully adjusting the reactivity ratios of an asymmetric divinyl monomer. Examples of asymmetric divinyl compounds are 2-vinyloxyethyl methacrylate,[48] allyl methacylate,[49] 4-(3′-buten-1′-oxy)-2,3,5,6-tetrafluorostyrene,[50,51] and 2-(5-norbornene)methyl methacrylate,[52] which were successfully polymerized *via* RAFT polymerization while gelation was noticeably delayed but not fully avoided.

More suitable for radical polymerization techniques is in contrast the thiol-yne approach since the trimethylsiliyl protecting groups prevents unwanted crosslinking events, but can also easily be removed. Hyperbranched poly(3-trimethylsilanyl-prop-2-ynyl)acrylate was synthesized in the presence of a RAFT agent and a small amount of dimethacrylate as crosslinker. Deprotection using tetra-*n*-butylammonium fluoride (TBAF) and acetic acid led to the reactive propinyl group, which was subsequently reacted with two 1-thio-β-D-glucose equivalents initiated using DMPA and UV-irradiation for 8 hours (Scheme 11.4).[53]

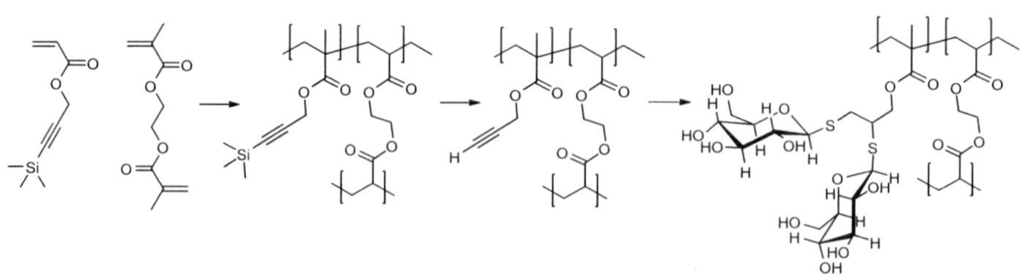

**Scheme 11.4** Hyperbranched polymers prepared *via* RAFT polymerization, functionalized with glucose using thiol-yne reaction.[53,54]

The fact that two thiols can react with one triple bond makes this an excellent avenue to create dendritic structures.[55] With only a few synthesis steps, polymers with glycodendrimers as terminal units can be created.[56]

## 11.3 Polymer-peptide Conjugates

Peptides are reasonably large biological molecules built up from a selection of amino acids. The amount of amino acids in a chain is significantly lower than that in proteins with approximately 100 aminoacids in a chain considered the cut-off between peptide and protein. Similar to carbohydrates, peptides take part in many biological recognition events. Hormones, which are involved in signaling events between cells, typically have peptide structures.[57] Neuropeptides are another class of peptides with an important function in neural tissue.[58] Peptides can also play a crucial role in the defense of an organism. Alkaloids, which can be found in plants, fungi and some animals like shellfish can create adverse effects in their predators.[59] More deadly for other organisms can be antibiotics or toxins, although not all of them are based on a peptide structure.[60] The involvement of peptides in these various naturally occurring events have led to the development of synthetic peptides, which can act as new drugs.[61] The interest in peptides does not only stem from their bioactivity, but also their capability to build up new materials. Strong H-bonding allows the creation of stable self-assembled structures.[62]

The preparation of polymer-peptide conjugates tries to address slightly different problems compared to glycopolymers. The polymer chain should enhance solubility of peptides, but also an enhanced bioactivity is envisaged by changing the mode of delivery. The combination of peptides and polymers also aims at the design of new polymer hybrids with the advantages of synthetic polymers being combined with the strong H-bonding of peptides.[63] Since the multivalent effect, which is typical for glycopolymers, is less important, the types of polymer-peptide conjugates are slightly different. It seems that there is a stronger focus on endfunctional polymers, which are then coupled to only one peptide. The coupling chemistry is extensive, but the centre of attention here are chemistries that involve thiols. Thiols are widely abundant in peptides. Cysteine, which carries a thiol group, plays a vital part in protein chemistry, since its ability to form disulfide bridges can assist the stabilization of these biological molecules. From a practical point of view, the strong tendency of these molecules to form disulfide bridges in an oxidative environment needs to be considered when attempting to conjugate peptides to polymers *via* their cysteine building block.

In the following, different types of chemistries are discussed. It becomes evident that although some conjugation reactions are similar to glycopolymers, the reaction with peptides is dominated by different approaches.

### 11.3.1 Thio-halogeno Reaction

The nucleophilic substitution of halogenides by a thiolate was a favourite in the synthesis of glycopolymers. This reaction has also been employed in the

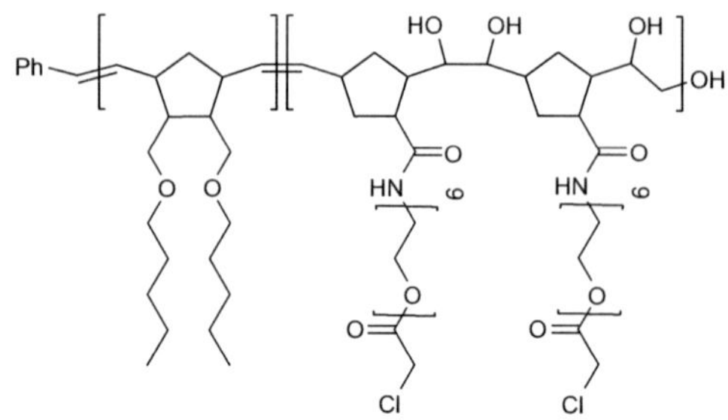


**Scheme 11.5** Thiol-reactive block copolymers for peptide conjugation prepared *via* ROMP with Grubbs' catalyst, coupling of hexaethylene glycol side chains to the second block, chloroacetylation of terminal hydroxyl group on hexaethylene glycol chains, and dihydroxylation of the double bonds along the polymer backbone.[54]

synthesis of polymers with pendant peptides. ROMP of a norbornene monomer with an amine reactive functionality, bicyclo[2.2.1]hept-5-ene-exo-2-carboxylic acid *N*-hydroxysuccinimide ester, led to homo-[64] and blockcopolymers,[54] which were subsequently modified in two steps to generate a reactive polymer scaffold with chloroacetyl groups (Scheme 11.5). The homopolymer was reacted with the cysteine of the peptide Ac-HTSTYWWLDGAPC-Am in DMSO,[64] while the block copolymer was first self-assembled into micelles and then reacted at pH = 8 in water.[54] Conjugation was evident by the increased water-solubility[64] and increase in nanoparticle size.[54] Both approaches lead to a bioactive structure that can inhibit anthrax toxin.

## 11.3.2 Thiol-ene Reaction

The radical approach to conjugate peptides *via* a thiol group to polymers is more prevalent in the literature, but also the Michael-type addition has been reported. A cyclic carbonate monomer with acrylate functionality was polymerized using ring-opening polymerization (Scheme 11.6, top). The polymer was then used as a substrate for the subsequent reaction with various thiols in DMF using pyridine as base. An excess of thiol has been tried but even the equimolar amounts of vinyl and thiol groups led to complete conversion as demonstrated using cysteine alone. The tripeptide RGD however only led to around 60% conversion under the same conditions. Even an excess of RGD did not improve the outcome, indicating that steric hindrance may prevent further reaction.[65]

Similar observations were made when the peptide was conjugated to the polymer using a radical approach. Pentafluorostyrene was first polymerized *via* ATRP and then modified with allyl amine in order to avoid common side reaction when directly polymerizing allyl-containing monomers (Scheme 11.6, bottom). The radical thiol-ene reaction was initiated with AIBN at 70 °C using

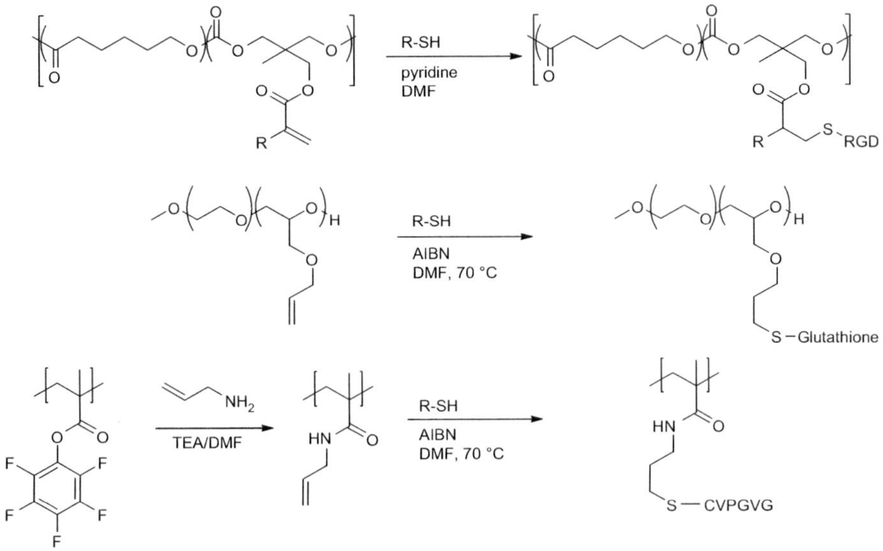


**Scheme 11.6** Polymer for thiol-ene reaction with peptides synthesized by ROP of a cyclic carbonate monomer with an acrylate and ε-caprolactone (top),[65] anionic ROP of allylglycidyl ether with ethylene oxide (middle)[67] and two-step postmodification of poly(pentafluorostyrene), which was prepared using ATRP (bottom).[66]

CVPGVG, which carries the cysteine functionality as the terminal unit. Maximum conversion of 50% suggests again steric hindrance.[66]

Steric hindrance can play a more influential role than other factors preventing complete peptide conjugation, which has been demonstrated by the use of a statistical polymer. The distance between two allyl groups was increased by copolymerization of allylglycidyl ether (AGE) with ethylene oxide (EO) (Scheme 11.6, middle). The polymer P($EO_{132}$-co-$AGE_{13}$) was then reacted with various commercial thiols with glutathione, a tripeptide, among them. Overnight reaction with ten-time excess of thiol was used and the reaction, initiated with AIBN at 70 °C, resulted in complete conversion.[67] As a conclusion, it was shown that the radical thiol-ene reaction is in principle a powerful tool to make polymer-peptide conjugates, although potential side reactions such as partial dimerization at low thiol concentrations were observed.[67]

An almost complete conjugation can also be achieved with one a single terminal allyl unit. An α-glutathione, ω-biotin functionalized poly(*N*-isopropylacrylamide) (PNIPAAm) was synthesized *via* RAFT polymerization using a RAFT agent with an allyl functionality.[68] The reaction between the allyl functional group and the thiol group was carried out with a 20-times excess of glutathione for two hours under UV light (2,2-dimethoxy-2-phenylacetophenone (DMPA) as photoinitiator) to reach almost full conversion. Endgroup analysis of polymers using techniques such as NMR can be subject to large errors.

Therefore, an elegant colorimetric technique was applied to determine the exact amount of conjugated glutathione by reacting 2,4,6-trinitrobenzene sulfonic acid (TNBSA) with the amines present in glutathione.[68]

Photoinitiation was also a popular route to generate hydrogels with peptides as crosslinker or with peptides as bioactive groups. Peptides as crosslinker can offer a pathway to degradable networks and in particular to networks that degrade in the presence of certain enzymes. The reaction between the two cysteine units of KCGGYRGCK, a peptide that can be cleaved by matrix metalloproteinase (MMP), and a four-arm PEG-star polymer with norbornene functional groups at each chain end led to the formation of hydrogels within a matter of a few minutes (Scheme 11.7). Initiated by 4-(2-hydroxyethoxy)phenyl-(2-hydroxy-2-propyl)ketone (Irgacure 2959) and UV-light, the reaction was fast and efficient enough to allow survival of human mesenchymal stem cells (hMSCs), which were mixed into the reaction mixture. Most of the living cells survived the treatment for 24 hours and the process was therefore termed cytocompatible.[69] The hMSCs were able to proliferate even when the degradability was reduced by mixing with a non-degradable PEG-dithiol during hydrogel formation.[70] A related, but slightly different approach used strain-promoted [3 + 2] Huisgen cycloaddition to generate hydrogels around living cells. The hydrogel was not prepared *via* thiol-ene reaction, but the click reaction was carried out in the presence of peptide crosslinkers with a pendant allylester group. This allowed further modification of gels with peptide using the photoinitiated thiol-ene coupling reaction. The added advantage of this approach that it allows patterning of the gel with different peptides by applying selective UV irradiation with a photomask.[71] Irradiation time and concentration of photoinitiator can control the conversion of the thiol-ene reaction offering an avenue to create a gradient of peptide functionalization.[72] A modified version of this approach allowed the

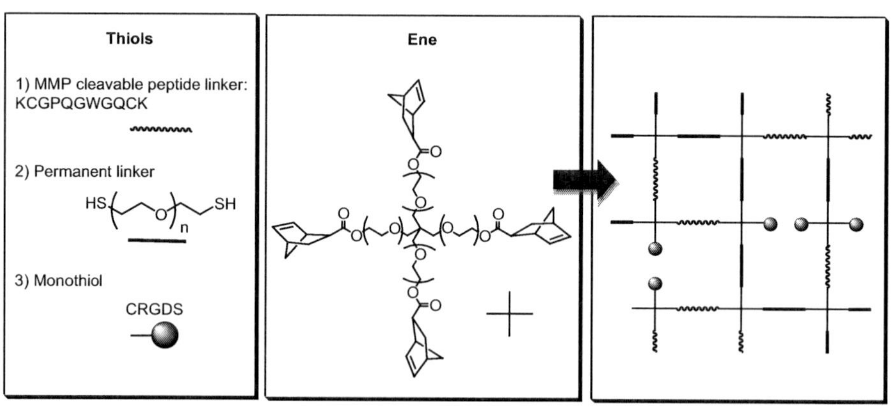


**Scheme 11.7** Synthesis of hydrogels using degradable and permanent crosslinker in addition to a peptide with only one cysteine, which is a integrin binding peptide for cell-adhesivity.[70]

incorporation of various biomolecules, including thiolated immunoglobulin G (IgG).[73]

This efficient approach was found to be suitable even for living cells. Nevertheless, there are concerns that the presence of radicals may damage certain sensitive compounds. A detailed study has been undertaken to test if it is possible to achieve functionalization of allyl groups on the surface of nanoparticles even when a radical initiator or strong base are absent. Initially benzylthiol was used as a model compound using different solvents and temperatures ranging from 25–35 °C. Up to 62% of allyl groups on the surface of the nanoparticle had reacted with thiols after a reaction time of 72 hours. Although the reaction time seems lengthy, it needs to be considered that only an equimolar amount of thiol was employed and any other reactants such as catalysts or base were absent, making this a soft and comparatively efficient technique. In fact, the increase of thiols did benefit the reaction only to a limited extent and more disulfide formation was observed.[74] Subsequent functionalization of the nanoparticles with peptides resulted in the formation of drug delivery carriers with targeting units.[74,75]

## 11.3.3 Thiol-vinylsulfone Reaction

Another technique that does not require any added catalyst is the reaction between thiols and vinylsulfone groups. Stirring of two reactants in either aqueous solution (preferably slightly basic) or organic solvent for up to one day is sufficient to achieve complete conversion. The β-thiosulfonyl linkage is stable in aqueous solution and cannot be cleaved. Polymers with pendant vinyl sulfone groups could directly be prepared by ring-opening copolymerization of a cyclic carbonate monomer with vinyl sulfone functionalities with ε-caprolactone. The reactive groups were indifferent to the polymerization conditions. Reaction with the peptide GRGDC in DMF led to complete conversion after 24 hours. It was also noted that the Michael reaction was selective and robust in the presence of other functional groups.[76]

The absence of any catalyst makes this process also attractive for the functionalization of self-assembled block copolymer aggregates, where added catalyst or initiator could potentially accumulate in the hydrophobic compartments. Polymersomes[77] and micelles[78] based on vinyl sulfone-terminated poly(ethylene oxide)-block-poly(γ-methyl-ε-caprolacton)[77] or poly-ethylene glycol-poly(L-lactide) diblock copolymer,[78] respectively, were coupled to peptides in aqueous solution. Reaction rate analysis showed that after 1 hour at pH 8 a significant amount of thiol has already reacted. L-Cystein and CREKA had similar reaction rates, while the TAT peptide was slightly slower. The cyclic peptide c(RGDfC) led to a significantly delayed reaction. Detailed investigation revealed that during the preparation process according to the manufacturer's recommendation the pH value dropped to a lower value. This highlighted the fact that the reaction indeed required slightly alkaline condition for success.[78]

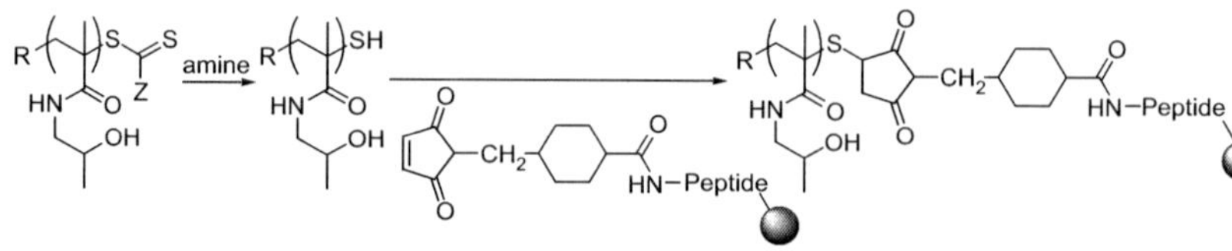


**Scheme 11.8** Synthesis of PHPMA-peptide block copolymers by reacting thiol-terminated PHPMA with resin-bound peptides.[79]

## 11.3.4 Thiol-maleimide Reaction

Another reaction that does not require heat or a catalyst is the nucleophilic addition of thiols to maleimide. Stirring the two reactants at room temperature is often enough to achieve complete conversion. This avenue was utilized to conjugate PHPMA to a peptide. The polymer was prepared *via* RAFT polymerization, which allows easy access to polymers with a thiol endfunctionality after aminolysis. Addition of TCEP was necessary to prevent disulfide formation, but the reaction between the thiol group of PHPMA and maleimide group of the resin-bound peptide $CH_3CO$–QQRFQWQFEQQ–$NH_2$, a β-sheet peptide, proceeds in DMF at room temperature without any catalysis, although the yields were found to be rather low (Scheme 11.8).[79]

## 11.3.5 Thiol-pyridyl Disulfide (PDS) Reaction

A very popular route to polymer-peptide conjugates is the rapid reaction between thiols and pyridyl disulfides (PDS).[80] The pyridyl disulfide will undergo rapid exchange with free thiol groups to result in mixed disulfides. The driving force of the reaction is the formation of the coloured pyridine-2-thione, which can also be used as an indication for the proceeding reaction. The popularity in this route to create polymer-peptide conjugates lies not only in the ease of monitoring the reaction by visible inspection, but also in the possibility to create disulfide bridges, which can be cleaved again in the reductive cell environment. In contrast to the techniques discussed above, the thiol-pyridyl disulfide reaction provides the opportunity to release the peptide again from the polymer scaffold, a possibility that is attractive for the delivery of peptide drugs.

The reaction has been used to generate polymers with pendant peptide groups, and also with a peptide as terminal group only. Monomers with pendant PDS groups were prepared either by radical polymerization techniques[81–83] or by step-growth polymerizations.[84,85] RAFT polymerization of monomers with pendant PDS groups such as pyridyldisulfide ethylmethacrylate proceeded in a well-controlled manner.[81] The linear homopolymers were then reacted with glutathione, as a model peptide, and other thiols. The amount of reacted thiols could be easily monitored *via* UV/Vis, confirming efficient and complete reaction with small thiols. The tripeptide glutathione was, however, subject to steric hindrance and full functionalization could not be achieved.[81] A statistical copolymer of PDS monomer and PEGMEMA, which was obtained again by RAFT polymerization, was processed into nanogels by converting a small fraction of PDS groups to disulfide bridges. Excess PDS groups on the surface of the nanogel were then decorated with thiolated TAT peptides.[83]

The PDS group was also found to be robust during the Michael addition of amines to acrylate groups, leading to reactive polyesters for RGDC conjugation (Scheme 11.9).[84] A detailed investigation on the effect of reaction time and thiol concentration led to a detailed reaction map. Complete conversion could for example be achieved with ratio of 1.5 between thiol and PDS and a reaction time of 44 hours.

A related pathway utilized, instead of a PDS monomer, oxidized cystamine as crosslinker. The gel was then reacted with 2,2′-dithiodipyridine, leading to the linear polymers with PDS side groups. Addition of equimolar amounts glutathione to the PBS group in TRIS buffer resulted in an intensely yellow solution within minutes, indicating the successful reaction with the peptide.[85]

Polymers with endfunctional PDS units can be designed either by employing an initiator or chain transfer agent with a PDS group or by post functionalizing an existing polymer. RAFT agents with PDS groups were found to be a versatile option for the synthesis of a polymer with one PDS endfunctionality[86] or telechelic polymer, which carried a PDS group at each end.[87] RAFT polymerization provides, in addition, an alternative route that enables simple postfunctionalization. Polymers prepared *via* RAFT polymerization can be aminolyzed using hexylamine, leading to thiol endfunctionalities that are in danger to oxidize quickly. Addition of 2,2′-dithiodipyridine was shown to suppress this side reaction by forming PDS functional groups, which can then react with peptides and other thiols.[88] Another established procedure to convert endfunctional hydroxyl groups such as in pluronics (PEO-PPO-PEO) to introduce a 2-pyridyl disulfide group is depicted in Scheme 11.10.[89]

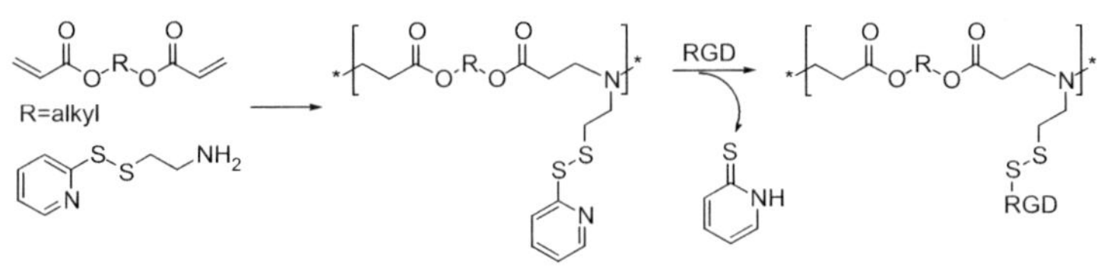


**Scheme 11.9** Reaction of peptides with polymers with pendant PDS group, which creates cleavable disulfide bridges.[84]

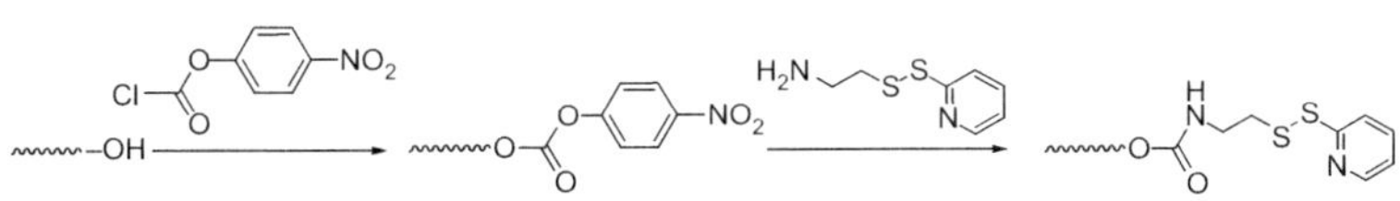

**Scheme 11.10** Chain end transformation of an hydroxyl-terminated polymer (PEO-PPO-PEO) to introduce a PDS functionality.[89]

Endfunctional block copolymers with PDS groups were often coupled with peptides after they have been self-assembled into aggregates such as micelles. Accessibility of the PDS group in the shell by the peptide was found to be as high as 80%. In this case, a ten-times excess of glutathione was reacted for 2 hours.[87] However, concerns have been raised that some peptides could potentially be physically encapsulated in the core and conjugation below the critical micelle concentration was considered to be the preferred route.[90]

## 11.4 Conjugates with Other Biopolymers

Thio-X reactions have also been used to conjugate polymers to a range of biopolymers, such as proteins, DNA or monoclonal antibodies. The conjugation chemistry is similar to the techniques employed to couple peptides to a polymer although the progress is now often hampered by the large size of the biomolecule. In the following, the techniques employed that related to thiol-X chemistries for conjugation are briefly summarized.

### 11.4.1 Polymer-protein Conjugates

The concept of "protein pegylation", the attachment of a PEG chain to a protein to increase stability and solubility, has been a very vibrant research area for years. Moreover, polymer-protein conjugates have led to a range of FDA approved products.[91] The high activity in this field is evidenced by the large number of review articles,[63,92–95] which either give a general overview or focus on a specific aspect such as the activity of the modified protein.[96] Owing to the large amount of review articles, this section is kept rather brief only highlighting thiol-X approaches.

The target for bioconjugation to proteins *via* thiol-X chemistry is similar to peptide conjugation. Cysteine is a vital building block in protein design. The disulfide bridges ensure that the three-dimensional structure of the protein is maintained. As a result, many cysteine units are not available for conjugation. Reduction of cysteine units can increase the number of thiols accessible, but only with a loss of structural integrity of the protein. Conjugation to the thiol of cysteine is, however, still one of the most commonly used avenues. The number of accessible thiol groups then determines the number of attached polymer chains.

There are basically two approaches to the polymer-protein conjugates: the synthesis of a reactive endfunctional polymer followed by conjugation to the polymer, and the modification of a protein with an initiator or chain transfer agent followed by grafting from the protein. While the first pathway is the more traditional approach, the latter was found to circumvent the problem that may arise when trying to react a long polymer chain to a bulky protein, which may lead to a low coupling efficiency.

The reaction between protein thiols and maleimides is one of the most popular pathways to polymer-protein conjugates. Key is the reaction between polymers with maleimide endfunctionality, which were either introduced as part of the ATRP initiator[97,98] or as part of the chain transfer agent in

RAFT polymerization.[99–101] The only consideration is that the maleimide functionality needs to be protected with furan prior to the polymerization to avoid crosslinking.

Pyridyl disulfide (PDS) as an active end group for the conjugation to thiol containing proteins, such as bovine serum albumin (BSA), is emerging now as a more popular approach when conjugating a polymer to cysteine.[102] The first report described an ATRP initiator, which was functionalized with PDS.[103] Subsequently, 2-hydroxyethyl methacrylate was polymerized *via* ATRP followed by attachment of the polymer chain to BSA. This success has been complemented by the development of a variety of PDS containing RAFT agents.[104,105] In addition, RAFT agents and ATRP initiators with PDS functionalities were reacted with proteins, which were then used as initiator or chain transfer agents to grow polymers from the protein.[106–108]

Also, the thiol-vinyl sulfone reaction, which has been discussed in the peptide section, is a versatile route, especially when RAFT polymers are employed. Polymers made *via* RAFT polymerization need to undergo only a simple aminolysis step to create polymers with a thiol as endfunctionality. Reactions with excess divinyl sulfone create reactive semitelechelic polymers, which were then conjugated to BSA.[109,110]

Thiosulfonates are another reactive functionality that is an emerging force in polymer-protein conjugate design.[111] Thiolsulfonates can be easily introduced to the polymer chain by nucleophilic substitution of halogenides (Scheme 11.11).[112] The application of this reaction for protein conjugation was demonstrated using modified PEG. Papain and recombinant human granulocyte colony-stimulating factor (rhG-CSF) were used as model proteins for the subsequent conjugation, resulting in the formation of a disulfide bridge between protein and polymer.[112]

Thiol-ene reactions now make a subtle appearance in polymer-protein conjugation. Disulfide bridges in a polymer were reduced to two thiol groups using water-soluble organic phosphines. Subsequent reaction with poly(monomethoxy ethylene glycol)-(meth)acrylates was carried out in one pot without the need of any added catalysis.[113] The resulting product carried two polymer chains connected to the two parts of the original disulfide bridge.

Another avenue that can specifically target disulfide bridges is the reaction with bisulfone. Polymer conjugation to interferon-α2a (IFN) using the amine functionality was observed to be difficult. Instead, the disulfide could be opened

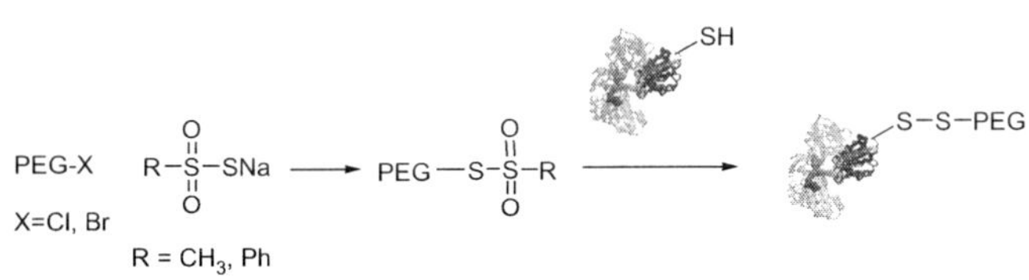


**Scheme 11.11** Alternative route to disulfide bridges between polymer and protein using thiosulfonates.[112]

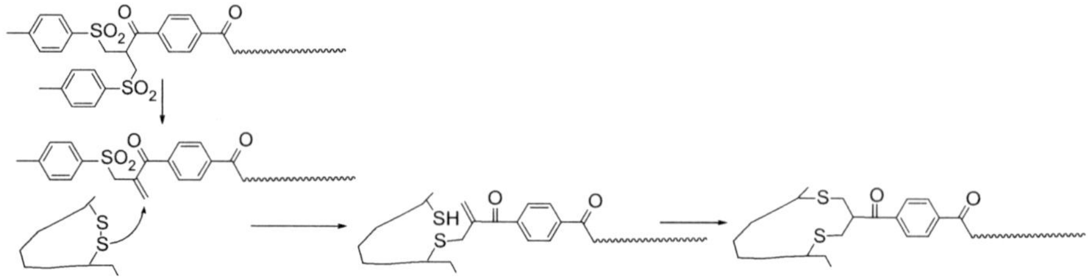


**Scheme 11.12** Bisulfones as a pathway to conjugate one polymer chain to a disulfide bridge.[114]

with the linker shown in Scheme 11.12. The interesting part is that instead of attaching two chains and permanently cleaving the bridge, only one polymer chain has been attached, while the functionality acts now as the junction point.[114]

### 11.4.2 Polymer-DNA Conjugates

Conjugation of polymers has recently been proposed as an alternative way to deliver oligonucleotides. In contrast to the formation of a complex between the negatively charged oligonucleotide and positively charged polymers, polymer-DNA conjugates aim as well at increasing the therapeutic efficiency of the oligonucleotide but without the use of potentially toxic cationic polymers.[115] Conjugation of polymers to siRNA was carried out by employing siRNA carrying an endfunctional thiol group. Polymers prepared by RAFT polymerization with PDS as a terminal unit were then coupled to siRNA.[116] Reaction with PEG containing polymers could indeed improve serum stability and nuclease resistance compared to free siRNA.[117] Instead of using polymers with one endfunctional PDS reactive group, which leads to the formation of block copolymers consisting of a synthetic polymer and siRNA, polymers with multiple PDS groups along the polymer chain were prepared by RAFT polymerization. Several replicas of siRNAs were conjugated to the polymer scaffold, resulting in polymer-siRNA conjugates with the drug bound *via* disulfide bridges. The release of siRNA from the polymer in a reductive environment was then successfully demonstrated.[118]

The modification of siRNA was not limited to the thiol-PDS reaction. Reaction with poly(PEG methyl ether acrylate) *via* thiol-ene reaction[119] or the reaction between thiol-terminated PNIPAAm, which was obtained by aminolysis of RAFT made PNIPAAm, to 5′-maleimide-modified DNA *via* Michael addition complete the picture.[120]

### 11.4.3 Polymer-monoclonal Antibodies Conjugates

Binding of a thiol group of monoclonal antibodies to maleimide functionalized polymers *via* Michael addition is a well-established technique with numerous examples in literature, thus only a few selected references are added here.[121–124]

## 11.5 Conclusions

Thiol-X chemistry is now a commonly used approach to form hybrids between synthetic polymers and nature's building blocks, such as sugars, peptides and proteins. The most prevalent type of thiol-X reactions depends on the kind of material that is being conjoined. Glycopolymers are often prepared by thiol-halogenide nucleophilic substitution, while this reaction is rare for modification with peptides and it is completely absent in the preparation of polymer-protein conjugates. The reason for this may be the need for alkaline conditions to accelerate this reaction. Thiol-PDS conjugation is, in contrast, the preferred route to react polymers with DNA. The most significant aspect might be the formation of the disulfide bridge, which means that DNA can be released again. The choice of conjugation chemistry depends on the stability of the reactants in the presence of catalysts, heat and solvents, but also on the stability of the final product. While most approaches lead to stable thioethers, thiol-PDS and thio-thiosulfate result in the formation of disulfide bonds, which can be cleaved again under reductive conditions. In summary, conjugation chemistry based on thiols has now reached a mature state and allows the researcher to choose from a variety of suitable reactions.

## References


1. R. A. Dwek, *Chem. Rev.*, 1996, **96**, 683–720.
2. H. Lis and N. Sharon, *Chem. Rev.*, 1998, **98**, 637–674.
3. J. J. Lundquist and E. J. Toone, *Chem. Rev.*, 2002, **102**, 555–578.
4. Y. Miura, *J. Polym. Sci., Part A: Polym. Chem.*, 2007, **45**, 5031–5036.
5. M. Mammen, S.-K. Chio and G. M. Whitesides, *Angew. Chem., Int. Ed.*, 1998, **37**, 2755–2794.
6. A. David, P. Kopeckova, J. Kopecek and A. Rubinstein, *Pharm. Res.*, 2002, **19**, 1114–1122.
7. A. David, P. Kopeckova, A. Rubinstein and J. Kopecek, *Bioconjugate Chem.*, 2001, **12**, 890–899.
8. X. Montet, M. Funovics, K. Montet-Abou, R. Weissleder and L. Josephson, *J. Med. Chem.*, 2006, **49**, 6087–6093.
9. Y. Shamay, D. Paulin, G. Ashkenasy and A. David, *J. Med. Chem.*, 2009, **52**, 5906–5915.
10. L. A. J. M. Sliedregt, P. C. N. Rensen, E. T. Rump, P. J. Van Santbrink, M. K. Bijsterbosch, A. R. P. M. Valentijn, G. A. Van der Marel, J. H. Van Boom, T. J. C. Van Berkel and E. A. L. Biessen, *J. Med. Chem.*, 1999, **42**, 609–618.
11. R. T. Lee and Y. C. Lee, *Glycoconjugate J.*, 2001, **17**, 543–551.
12. S. Pearson, G. Chen and M. H. Stenzel, in *Engineered Carbohydrate-Based Materials for Biomedical Applications: Polymers, Surfaces, Dendrimers, Nanoparticles, and Hydrogels*, ed. R. Narain, John Wiley & Sons, Inc., Hoboken, NJ, USA, 2011 pp. 1–119.

13. G. Wulff, J. Schmid and T. Venhoff, *Macromol. Rapid Commun.*, 1996, **197**, 259–274.
14. S. Loykulnant, M. Hayashi and A. Hirao, *Macromolecules*, 1998, **31**, 9121–9126.
15. K. Yamada, M. Minoda and T. Miyamoto, *J. Polym. Sci., Part A: Polym. Chem.*, 1997, **35**, 751–757.
16. T. Uryu, C. Yamaguchi, K. Morikawa, K. Terui, T. Kanai and K. Matsuzaki, *Macromolecules*, 1985, **18**, 599–605.
17. K. H. Mortell, M. Gingras and L. L. Kiessling, *J. Am. Chem. Soc.*, 1994, **116**, 12053–12054.
18. S. R. S. Ting, G. J. Chen and M. H. Stenzel, *Polym. Chem.*, 2010, **1**, 1392–1412.
19. G. Martinez, M. Fernandez-Garcia and M. Sanchez-Chaves, *J. Polym. Sci., Part A: Polym. Chem.*, 2005, **43**, 18–27.
20. R. Auzely-Velty, M. Cristea and M. Rinaudo, *Biomacromolecules*, 2002, **3**, 998–1005.
21. B. D. Mather, K. Viswanathan, K. M. Miller and T. E. Long, *Prog. Polym. Sci.*, 2006, **31**, 487–531.
22. G. Thoma, J. T. Patton, J. L. Magnani, B. Ernst, R. Ohrlein and R. O. Duthaler, *J. Am. Chem. Soc.*, 1999, **121**, 5919–5929.
23. G. Thoma, R. O. Duthaler, J. L. Magnani and J. T. Patton, *J. Am. Chem. Soc.*, 2001, **123**, 10113–10114.
24. G. Thoma, M. B. Streiff, A. G. Katopodis, R. O. Duthaler, N. H. Voelcker, C. Ehrhardt and C. Masson, *Chem.–Eur. J.*, 2006, **12**, 99–117.
25. C. H. Xue, V. R. R. Donuru and H. Y. Liu, *Macromolecules*, 2006, **39**, 5747–5752.
26. C. H. Xue, S. P. Jog, P. Murthy and H. Y. Liu, *Biomacromolecules*, 2006, **7**, 2470–2474.
27. C. Xue, S. Velayudham, S. Johnson, R. Saha, A. Smith, W. Brewer, P. Murthy, S. T. Bagley and H. Liu, *Chem.–Eur. J.*, 2009, **15**, 2289–2295.
28. C. Boyer, A. Bousquet, J. Rondolo, M. R. Whittaker, M. H. Stenzel and T. P. Davis, *Macromolecules*, 2010, **43**, 3775–3784.
29. Y. Chen, G. J. Chen and M. H. Stenzel, *Macromolecules*, 2010, **43**, 8109–8114.
30. K. Babiuch, C. R. Becer, M. Gottschaldt, J. T. Delaney, J. Weisser, B. Beer, R. Wyrwa, M. Schnabelrauch and U. S. Schubert, *Macromol. Bioscience*, 2011, **11**, 535–548.
31. C. R. Becer, K. Babiuch, D. Pilz, S. Hornig, T. Heinze, M. Gottschaldt and U. S. Schubert, *Macromolecules*, 2009, **42**, 2387–2394.
32. W. A. Pryor and T.-L. Huang, *Macromolecules*, 1969, **2**, 70–77.
33. K. T. Powell, C. Cheng and K. L. Wooley, *Macromolecules*, 2007, **40**, 4509–4515.
34. C. S. Gudipati, M. B. H. Tan, H. Hussain, Y. Liu, C. He and T. P. Davis, *Macromol. Rapid Commun.*, 2008, **29**, 1902–1907.

35. C. Ott, R. Hoogenboom and U. S. Schubert, *Chem. Commun.*, 2008, 3516–3518.
36. S. R. S. Ting, E. H. Min, P. Escale, M. Save, L. Billon and M. H. Stenzel, *Macromolecules*, 2009, **42**, 9422–9434.
37. A. Dondoni and A. Marra, *Chem. Soc. Rev.*, 2012, **41**, 573–586.
38. A. Gress, A. Volkel and H. Schlaad, *Macromolecules*, 2007, **40**, 7928–7933.
39. C. Diehl and H. Schlaad, *Macromol. Biosci.*, 2009, **9**, 157–161.
40. K. Kempe, T. Neuwirth, J. Czaplewska, M. Gottschaldt, R. Hoogenboom and U. S. Schubert, *Polym. Chem.*, 2011, **2**, 1737–1743.
41. K. Kempe, C. Weber, K. Babiuch, M. Gottschaldt, R. Hoogenboom and U. S. Schubert, *Biomacromolecules*, 2011, **12**, 2591–2600.
42. Z. G. Hu, X. S. Fan, J. X. Wang, S. P. Zhao and X. L. Han, *J. Polym. Res.*, 2010, **17**, 815–820.
43. J. Sun and H. Schlaad, *Macromolecules*, 2010, **43**, 4445–4448.
44. N. Franz, L. Menin and H. A. Klok, *Org. Biomol. Chem.*, 2009, **7**, 5207–5218.
45. C. Ott, C. D. Easton, T. R. Gengenbach, S. L. McArthur and P. A. Gunatillake, *Polym. Chem.*, 2011, **2**, 2782–2784.
46. G. J. Chen, S. Amajjahe and M. H. Stenzel, *Chem. Commun.*, 2009, 1198–1200.
47. M. L. Conte, M. J. Robb, Y. Hed, A. Marra, M. Malkoch, C. J. Hawker and A. Dondoni, *J. Polym. Sci., Part A: Polym. Chem.*, 2011, **49**, 4468–4475.
48. Z. Jia, J. Liu, T. P. Davis and V. Bulmus, *Polymer*, 2009, **50**, 5928–5932.
49. D. Valade, C. Boyer, T. P. Davis and V. Bulmus, *Aust. J. Chem.*, 2009, **62**, 1344–1350.
50. J. Ma, C. Cheng, G. Sun and K. L. Wooley, *Macromolecules*, 2008, **41**, 9080–9089.
51. J. Ma, J. W. Bartels, Z. Li, K. Zhang, C. Cheng and K. L. Wooley, *Aust. J. Chem.*, 2010, **63**, 1159–1163.
52. Z. M. Dong, X. H. Liu, X. L. Tang and Y. S. Li, *Macromolecules*, 2009, **42**, 4596–4603.
53. M. Semsarilar, V. Ladmiral and S. Perrier, *Macromolecules*, 2010, **43**, 1438–1443.
54. A. Carrillo, M. J. Yanjarappa, K. V. Gujraty and R. S. Kane, *J. Polym. Sci., Part A: Polym. Chem.*, 2006, **44**, 928–939.
55. G. J. Chen, J. Kumar, A. Gregory and M. H. Stenzel, *Chem. Commun.*, 2009, 6291–6293.
56. J. Kumar, A. Bousquet and M. H. Stenzel, *Macromol. Rapid Commun.*, 2011, **32**, 1620–1626.
57. G. Gimpl and F. Fahrenholz, *Phys. Rev.*, 2001, **81**, 629–683.
58. B. A. Eipper, D. A. Stoffers and R. E. Mains, *Ann. Rev. Neurosci.*, 1992, **15**, 57–85.
59. M. M. Cowan, *Clin. Microbiol. Rev.*, 1999, **12**, 564.

60. E. C. Hobbs, F. Fontaine, X. F. Yin and G. Storz, *Curr. Opin. Microbiol.*, 2011, **14**, 167–173.
61. L. A. Thompson and J. A. Ellman, *Chem. Rev.*, 1996, **96**, 555–600.
62. J. D. Hartgerink, E. Beniash and S. I. Stupp, *Science*, 2001, **294**, 1684–1688.
63. O. D. Krishna and K. L. Kiick, *Peptide Sci.*, 2010, **94**, 32–48.
64. A. Carrillo, K. V. Gujraty, P. R. Rai and R. S. Kane, *Nanotechnology*, 2005, **16**, S416–S421.
65. W. Chen, H. C. Yang, R. Wang, R. Cheng, F. H. Meng, W. X. Wei and Z. Y. Zhong, *Macromolecules*, 2010, **43**, 201–207.
66. N. K. Singha, M. I. Gibson, B. P. Koiry, M. Danial and H. A. Klok, *Biomacromolecules*, 2011, **12**, 2908–2913.
67. B. Obermeier and H. Frey, *Bioconjugate Chem.*, 2011, **22**, 436–444.
68. X. Huang, C. Boyer, T. P. Davis and V. Bulmus, *Polym. Chem.*, 2011, **2**, 1505–1512.
69. B. D. Fairbanks, M. P. Schwartz, A. E. Halevi, C. R. Nuttelman, C. N. Bowman and K. S. Anseth, *Adv. Mater.*, 2009, **21**, 5005.
70. S. B. Anderson, C. C. Lin, D. V. Kuntzler and K. S. Anseth, *Biomaterials*, 2011, **32**, 3564–3574.
71. C. A. DeForest, B. D. Polizzotti and K. S. Anseth, *Nat. Mater.*, 2009, **8**, 659–664.
72. C. A. DeForest, E. A. Sims and K. S. Anseth, *Chem. Mater.*, 2010, **22**, 4783–4790.
73. P. S. Hume, C. N. Bowman and K. S. Anseth, *Biomaterials*, 2011, **32**, 6204–6212.
74. A. van der Ende, T. Croce, S. Hamilton, V. Sathiyakumar and E. Harth, *Soft Matter*, 2009, **5**, 1417–1425.
75. A. E. van der Ende, V. Sathiyakumar, R. Diaz, D. E. Hallahan and E. Harth, *Polym. Chem.*, 2010, **1**, 93–96.
76. R. Wang, W. Chen, F. H. Meng, R. Cheng, C. Deng, J. Feijen and Z. Y. Zhong, *Macromolecules*, 2011, **44**, 6009–6016.
77. M. A. Petersen, L. G. Yin, E. Kokkoli and M. A. Hillmyer, *Polym. Chem.*, 2010, **1**, 1281–1290.
78. J. W. Bae, E. Lee, K. M. Park and K. D. Park, *Macromolecules*, 2009, **42**, 3437–3442.
79. L. C. Radu, J. Y. Yang and J. Kopecek, *Macromol. Bioscience*, 2009, **9**, 36–44.
80. D. R. Grassetti and J. F. Murray Jr, *Arch. Biochem. Biophys.*, 1967, **119**, 41–49.
81. L. J. Wong, C. Boyer, Z. F. Jia, H. M. Zareie, T. P. Davis and V. Bulmus, *Biomacromolecules*, 2008, **9**, 1934–1944.
82. J.-H. Ryu, R. T. Chacko, S. Jiwpanich, S. Bickerton, R. P. Babu and S. Thayumanavan, *J. Am. Chem. Soc.*, 2010, **132**, 17227–17235.
83. J.-H. Ryu, S. Jiwpanich, R. Chacko, S. Bickerton and S. Thayumanavan, *J. Am. Chem. Soc.*, 2010, **132**, 8246–8247.
84. G. T. Zugates, D. G. Anderson, S. R. Little, I. E. B. Lawhorn and R. Langer, *J. Am. Chem. Soc.*, 2006, **128**, 12726–12734.

85. E. Ranucci, P. Ferruti, M. A. Suardi and A. Manfredi, *Macromol. Rapid Commun.*, 2007, **28**, 1243–1250.
86. C. L. Duvall, A. J. Convertine, D. S. W. Benoit, A. S. Hoffman and P. S. Stayton, *Mol. Pharm.*, 2010, **7**, 468–476.
87. J. Q. Liu, H. Y. Liu, C. Boyer, V. Bulmus and T. P. Davis, *J. Polym. Sci., Part A: Polym. Chem.*, 2009, **47**, 899–912.
88. C. Boyer, J. Q. Liu, V. Bulmus and T. P. Davis, *Aust. J. Chem.*, 2009, **62**, 830–847.
89. J.-T. Li, J. Carlsson, J.-N. Lin and K. D. Caldwell, *Bioconjugate Chem.*, 1996, **7**, 592–599.
90. P. R. Joshi, J. McGuire and J. A. Neff, *J. Biomed. Mater. Res. Part B*, 2009, **91B**, 128–134.
91. S. N. S. Alconcel, A. S. Baas and H. D. Maynard, *Polym. Chem.*, 2011, **2**, 1442–1448.
92. G. N. Grover and H. D. Maynard, *Curr. Opin. Chem. Bio.*, 2010, **14**, 818–827.
93. A. K. Shakya, H. Sami, A. Srivastava and A. Kumar, *Prog. Polym. Sci.*, 2010, **35**, 459–486.
94. C. Boyer, X. Huang, M. R. Whittaker, V. Bulmus and T. P. Davis, *Soft Matter*, 2011, **7**, 1599–1614.
95. B. Le Droumaguet and J. Nicolas, *Polym. Chem.*, 2010, **1**, 563–598.
96. M. A. Gauthier and H.-A. Klok, *Polym. Chem.*, 2010, **1**, 1352–1373.
97. G. Mantovani, F. Lecolley, L. Tao, D. M. Haddleton, J. Clerx, J. J. L. M. Cornelissen and K. Velonia, *J. Am. Chem. Soc.*, 2005, **127**, 2966–2973.
98. B. Le Droumaguet, G. Mantovani, D. M. Haddleton and K. Velonia, *J. Mater. Chem.*, 2007, **17**, 1916–1922.
99. P. De, M. Li, S. R. Gondi and B. S. Sumerlin, *J. Am. Chem. Soc.*, 2008, **130**, 11288.
100. M. Li, H. M. Li, P. De and B. S. Sumerlin, *Macromol. Rapid Commun.*, 2011, **32**, 354–359.
101. L. Tao, C. S. Kaddis, R. R. O. Loo, G. N. Grover, J. A. Loo and H. D. Maynard, *Macromolecules*, 2009, **42**, 8028–8033.
102. C. Boyer, J. Liu, V. Bulmus, T. P. Davis, C. Barner-Kowollik and M. H. Stenzel, *Macromolecules*, 2008, **41**, 5641–5650.
103. D. Bontempo, K. L. Heredia, B. A. Fish and H. D. Maynard, *J. Am. Chem. Soc.*, 2004, **126**, 15372–15373.
104. L. Tao, J. T. Xu, D. Cell and T. P. Davis, *Macromolecules*, 2010, **43**, 3721–3727.
105. J. Q. Liu, H. Y. Liu, V. Bulmus, L. Tao, C. Boyer and T. P. Davis, *J. Polym. Sci., Part A: Polym. Chem.*, 2010, **48**, 1399–1405.
106. J. Liu, V. Bulmus, D. Herlambang, C. Barner-Kowollik, M. Stenzel and T. Davis, *Angew. Chem., Int. Ed.*, 2007, **46**, 3099–3103.
107. C. Boyer, V. Bulmus, J. Q. Liu, T. P. Davis, M. H. Stenzel and C. Barner-Kowollik, *J. Am. Chem. Soc.*, 2007, **129**, 7145–7154.
108. K. L. Heredia, D. Bontempo, T. Ly, J. T. Byers, S. Halstenberg and H. D. Maynard, *J. Am. Chem. Soc.*, 2005, **127**, 16955–16960.

109. G. N. Grover, S. N. S. Alconcel, N. M. Matsumoto and H. D. Maynard, *Macromolecules*, 2009, **42**, 7657–7663.
110. S. N. S. Alconcel, G. N. Grover, N. M. Matsumoto and H. D. Maynard, *Aust. J. Chem.*, 2009, **62**, 1496–1500.
111. P. J. Roth, D. Kessler, R. Zentel and P. Theato, *J. Polym. Sci., Part A: Polym. Chem.*, 2009, **47**, 3118–3130.
112. Y. J. Zhao, Y. Q. Zhai, Z. G. Su and G. H. Ma, *Polym. Adv. Technol.*, 2010, **21**, 867–873.
113. M. W. Jones, G. Mantovani, S. M. Ryan, X. X. Wang, D. J. Brayden and D. M. Haddleton, *Chem. Commun.*, 2009, 5272–5274.
114. A. Lewis, Y. Tang, S. Brocchini, J.-w. Choi and A. Godwin, *Bioconjugate Chem.*, 2008, **19**, 2144–2155.
115. D. Smith, A. C. Holley and C. L. McCormick, *Polym. Chem.*, 2011, **2**, 1428–1441.
116. K. L. Heredia, T. H. Nguyen, C.-W. Chang, V. Bulmus, T. P. Davis and H. D. Maynard, *Chem. Commun.*, 2008, 3245–3247.
117. K. Gunasekaran, T. H. Nguyen, H. D. Maynard, T. P. Davis and V. Bulmus, *Macromol. Rapid Commun.*, 2011, **32**, 654–659.
118. A. W. York, F. Huang and C. L. McCormick, *Biomacromolecules*, 2010, **11**, 505–514.
119. S. C. Kow, J. McCarroll, D. Valade, C. Boyer, T. Dwarte, T. P. Davis, M. Kavallaris and V. Bulmus, *Biomacromolecules*, 2011, **12**, 4301–4310.
120. K. Isoda, N. Kanayama, D. Miyamoto, T. Takarada and M. Maeda, *Reactive Funct. Polym.*, 2011, **71**, 367–371.
121. R. N. Johnson, P. Kopeckova and J. Kopecek, *Bioconjugate Chem.*, 2009, **20**, 129–137.
122. R. N. Johnson, P. Kopeckova and J. Kopecek, *Biomacromolecules*, 2012, **13**, 727–735.
123. J. Hongrapipat, P. Kopeckova, J. Liu, S. Prakongpan and J. Kopecek, *Mol. Pharm.*, 2008, **5**, 696–709.
124. J. H. Liu, P. Kopeckova, P. Buhler, P. Wolf, H. Z. Pan, H. Bauer, U. Elsasser-Beile and J. Kopecek, *Mol. Pharm.*, 2009, **6**, 959–970.

CHAPTER 12

# *Surface Engineering with Thiol-click Chemistry*

RYAN M. HENSARLING AND DEREK L. PATTON*

School of Polymers and High Performance Materials, University of Southern Mississippi, 118 College Drive #5050, Hattiesburg, MS 39406, USA
*Email: derek.patton@usm.edu

## 12.1 Overview of Thiol-click Surface Engineering

Surface engineering collectively describes a set of processes – both physical and chemical – for the design and modification of surfaces, or the near-surface regions of a material, enabling the surface to perform functions that are distinct from those functions demanded from the bulk of the material.[1] As surfaces are ubiquitous in nature, surface engineering has played an important role in enabling technologies for everyday applications, such as corrosion inhibition, wear resistance, antifouling, biomedical implants, and drug delivery systems among many others. Tailoring the chemical composition of the material interface is a major impetus of surface engineering. In recent years, this impetus has been particularly driven by the desire to endow surfaces with well-defined chemical functionality, with high specificity and conversion, and with fine control over the spatial arrangement of the surface chemical composition. Among several strategies reported for surface modification, those that exploit "click" reactions, such as the archetypical copper catalyzed azide-alkyne Huisgen 1,3-dipolar cycloaddition (CuAAC),[2,3] are particularly attractive for transforming the aforementioned desires into the realities of precisely engineering the chemical composition of surfaces. The click approach to engineer surfaces circumvents several limitations often associated with conventional

RSC Polymer Chemistry Series No. 6
Thiol-X Chemistries in Polymer and Materials Science
Edited by Andrew B. Lowe and Christopher N. Bowman

Published by the Royal Society of Chemistry, www.rsc.org

conjugation chemistries, including poor conversions on the surface, harsh reaction conditions, side reactions, and the need for highly active reactants which often require extensive synthetic preparation. A recent review by Nebhani and Barner-Kowollik[4] highlighted the admirable characteristics of CuAAC for engineering a variety of functional surfaces, hence few details regarding CuAAC will be provided here. However, concern over the presence of residual metal impurities following copper-catalyzed click reactions has motivated the development of alternate, metal-free surface modification strategies. Consequently, a wide variety of metal-free click reactions – such as strain promoted azide-alkyne cycloadditions,[5–9] and Diels–Alder cycloadditions[10–14] – are increasingly becoming methods of choice to synthesize functional surfaces particularly in the realm of biomaterials.

In a complementary direction to those metal-free click reactions previously mentioned, thiol-based click reactions provide a powerful approach for engineering multifunctional surfaces in a modular fashion. Specifically, thiols readily react with electron rich alkenes (radical-mediated), alkynes (radical-mediated), electron poor alkenes (amine or phosphine catalyzed), isocyanates (amine catalyzed), epoxies (amine catalyzed $S_N2$ ring opening), and halogens (amine catalyzed $S_N2$ nucleophilic substitution), creating a diverse thiol-click toolbox which exploits a large library of functional moieties for rapid manipulation of surface properties (Scheme 12.1). Thiol-click reactions are advantageous for this purpose in that they proceed at room temperature with

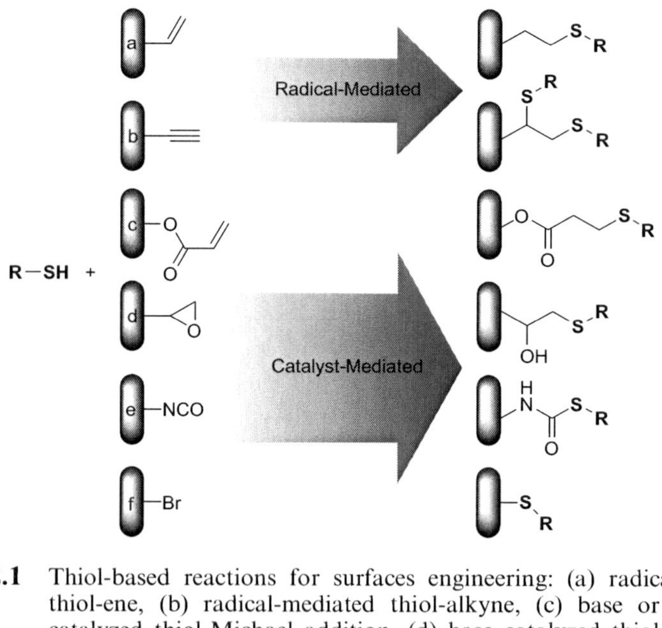


**Scheme 12.1** Thiol-based reactions for surfaces engineering: (a) radical-mediated thiol-ene, (b) radical-mediated thiol-alkyne, (c) base or phosphine catalyzed thiol-Michael addition, (d) base catalyzed thiol-epoxy ring opening, (e) base catalyzed thiol-isocyanate, and (f) base catalyzed thiol-halogen substitution. The reverse configuration of reactants (*i.e.* thiol-functionalize surface) is also possible and is often exploited.

high efficiency and rapid kinetics, in the presence of oxygen/water, without expensive and potentially toxic catalysts, and are highly tolerant of a wide range of functional groups.[15–18] In many cases, these thiol-click transformations fulfill the robust-efficient-orthogonal (REO) strategy by proceeding orthogonally with one another (*i.e.* radical-mediated thiol-yne and base-catalyzed thiol-ene[19]) and with other click reactions such as the CuAAC and Diels-Alder cycloadditions, providing a powerful approach to engineer surfaces with complex architectures and multicomponent chemistries.[20]

Presently, the utility of thiol-click reactions for the purpose of manipulating the chemistry of surfaces has not been reviewed in a comprehensive manner. In the current chapter, we aim to provide a comprehensive review of works illustrating simple, versatile, and modular synthetic methods for modifying surfaces based on thiol-click reactions. Rather than the specific type of thiol-click reaction, the chapter is organized based on the type of substrate surface being modified, including (i) monolayers and other ultrathin films, (ii) polymer surfaces, (iii) microporous membranes, (iv) nano- and microparticle surfaces, and (v) biological surfaces. Given that several publications have recently reviewed the modification of soft material assemblies (*i.e.* polymer micelles, vesicles, etc.) *via* thiol-based conjugations,[16,20] this chapter does not cover this topic.

## 12.2 Monolayers and Other Ultrathin Films

Modifying the terminal group of self-assembled monolayers (SAMs) *via* thiol-click reactions offers a viable way to add two-dimensional functionality to either planar or curved surfaces enabling control over interfacial properties such as wetting and adhesion, chemical resistance, bio-compatibility, sensitization, and molecular recognition for sensors and nano-fabrication.[21] To date, Michael-type thiol-ene[22–24] as well as radical-mediated thiol-ene[25–28] and thiol-yne[28] reactions have been utilized to modify the terminal group of SAMs, creating complex functional surfaces used for microarrays to biochips. Inherent to SAMs, the surface-attached molecules have the ability to fully extend maximizing packing density and Van der Waals interactions allowing for a dense distribution of desirable functionality at the air interface upon surface modification with thiol-click reactions.[29]

As an approach to engineer functional SAMs, Michael-type thiol-ene conjugations involving the reaction between thiols and maleimides have perhaps received the most attention in recent years – particularly for immobilization of biomolecules. The thiol-maleimide reaction exhibits several salient features that are particularly attractive for bio-immobilizations, including selectivity in the presence of multiple functional groups, rapid and quantitative conversions at low concentrations, and minimal synthetic modification of ligands prior to immobilization. In an early example, Schreiber *et al.*[30] exploited these features for the preparation of printed microarrays to probe protein–ligand interactions using maleimide-terminated SAMs to conjugate several cysteine-containing bio-ligands (*i.e.* biotin) to the surface. Similarly,

Mrksich and coworkers fabricated peptide and carbohydrate biochips *via* Michael-type thiol-ene reactions of thiol-functionalized glycosides and peptides on maleimide-functionalized SAMs,[22] while Corn *et al.*[31] used thiol-terminated SAMs as a platform for the immobilization of maleimide-terminated nucleic acids. The key for successful biomolecule immobilization for microarray applications is preventing nonspecific interactions with biomolecules, while enabling quantitative analysis of specific binding events. To improve deoxyribonucleic acid (DNA) hybridization from complex media for DNA microarray and biosensor applications, Castner and co-workers investigated the surface structure and performance of thiolated single-strand DNA (HS-ssDNA) immobilized on maleimide-ethylene glycol disulfide (MEG) monolayers on gold.[23] The MEG thiolate monolayers were formed *via* a single-step solution self-assembly process, as opposed to a conventional approach involving the preparation of an amino-terminated layer followed by reaction with a bifunctional linker molecule that contains both an amino-reactive site and a maleimide group. The combination of surface-exposed maleimide functional moieties within a background of inert ethylene glycol units aided in selective and efficient reactivity toward HS-ssDNA end groups while providing resistance to nonspecific binding of DNA and proteins. Michael-type thiol-ene additions between maleimides and thiols are ideal due to being stable in aqueous environments and maleimide selectivity toward thiols.[23] In a similar approach, Magnusson and co-workers immobilized cysteine-functionalized chemoattractant peptides onto mixed SAMs composed of maleimide- and hydroxyl-terminated oligo(ethylene glycol) alkyldisulfides (Figure 12.1).[24] The mixed monolayer approach allowed for site-specific immobilization of peptides *via* Michael-type thiol-ene reactions along with spatial and temporal control of chemoattractant distribution to develop defined circumstances for recruitment and activation of cells.

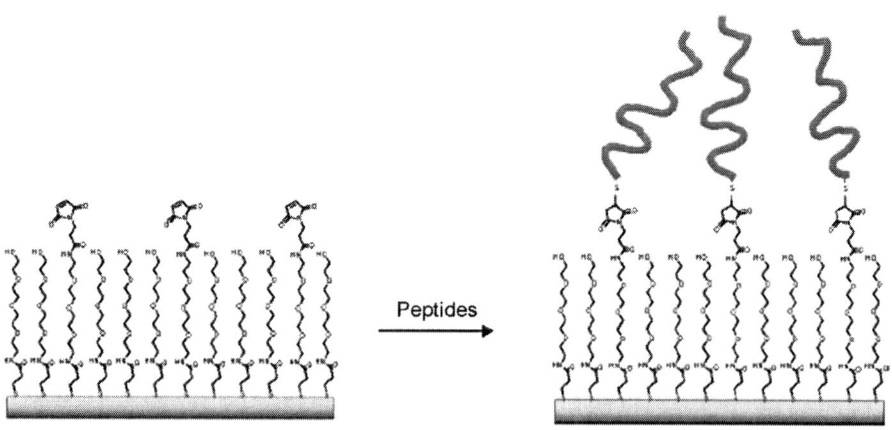


**Figure 12.1** Peptide immobilization on maleimide-terminated SAMs. Reprinted from ref. 24 with permission from the American Chemical Society.

Radical-mediated thiol-ene and thiol-yne reactions have also been demonstrated as versatile approaches to engineer the chemistry of SAM surfaces. In an early example, a radical-mediated thiol-ene reaction was employed to graft linseed oil onto a thiol functionalized aluminium surface to investigate thin vegetable oil films for friction reduction.[26,32] The ability to initiate thiol-ene and thiol-yne reactions with light is particularly attractive as a way to fabricate patterned functional surfaces using photopatterning techniques.[25,27,28] Zuilhof and coworkers[27] developed radical-mediated thiol-ene click chemistry as an efficient, facile route to create patternable biofunctional monolayers on silicon hydride substrates – an approach advantageous for oxide-free silicon substrates useful for biofunctional electronic applications. H-terminated Si(111) surfaces were reacted with neat 1,13-tetradecadiene at 80 °C under an inert atmosphere for 16 h to facilitate the covalent attachment of an alkene-terminated monolayer. Under ambient atmosphere, the alkene-terminated monolayer was then exposed to 365 nm light in the presence of various thiols, such as thioglycolic acid, 3-mercaptopropyl tetraacetate, 9-fluorenylmethoxy-carbonyl cysteine, and a typical photoinitiator, 2,2-dimethoxy-2-phenylacetophenone (DMPA) at room temperature employing minimal amounts of solvent to solubilize the thiols. High surface coverage of the attached thiols was achieved (45–75%) for a wide variety of functional thiols, while maintaining oxide-free surfaces. Zuilhof and coworkers[27] extended their approach by combining photo-induced thiol-ene reactions with microcontact printing (μCP) to prepare chemical patterns on alkene-terminated oxide-free surfaces. In this approach, a PDMS stamp with pillar features was inked with thioglycolic acid and brought into conformal contact with the alkene-terminated monolayer. UV irradiation through the backside of the transparent stamp facilitated the thiol-ene reaction in the contacted area, resulting in circular patterns of carboxylic acid groups.

Ravoo and colleagues[33] extended the scope of photochemical μCP to include thiol-ene and thiol-yne reactions on alkene and alkyne-terminated alkoxysilane monolayers (Figure 12.2a shows the process for thiol-ene μCP), and additionally demonstrated the patterning concept with a broader range of functional thiols (Figure 12.2b). In 2011, Ravoo *et al.*[28] further probed structure–reactivity relationships for a range of fluorescent thiols undergoing photoinduced thiol-ene or thiol-yne reactions with alkene or alkyne terminated SAMs, respectively, within the nanoconfined environment between the PDMS stamp and the substrate. Using fluorescence intensity of the modified SAM, the authors probed the reactivity of various thiol/ene and thiol/yne pairs (*i.e.* aromatic thiol *versus* aliphatic thiol) and found the structure–reactivity relationships for confined reactions were similar to those observed in solution (*i.e.* electron-poor alkynes react with thiols faster that electron-rich alkynes). The dependence of confined thiol-click reactions on irradiation conditions (*i.e.* UV *vs.* visible *vs.* dark) was also developed.[28] Figure 12.3 correlates the fluorescence intensity of the fluorescent thiol-modified alkene surfaces to printing time and light conditions. As expected, under irradiation with UV light the reaction rate was the highest, while under ambient light conditions, the yield was still remarkably high. However, the reactions proceeded very slowly in the

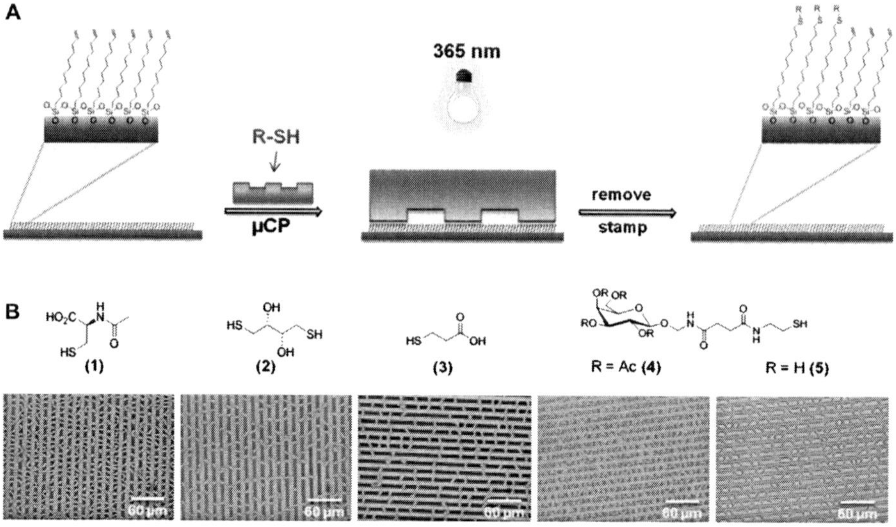


**Figure 12.2** (A) Schematic illustration of photochemical μCP by thiol-ene chemistry: an oxidized PDMS stamp inked with a thiol is placed on an alkene-terminated SAM and irradiated with UV light (365 nm). Immobilization of the thiol occurs exclusively in the area of contact. (B) Light microscopy of water condensation experiments on surfaces patterned by printing *N*-acetyl-L-cysteine (1), D,L-dithiothreitol (2), 3-mercaptopropionic acid (3), tetraacetylgalactoside–thiol conjugate (4), and galactoside–thiol conjugate (5).

Reprinted from ref. 33 with permission from the American Chemical Society.

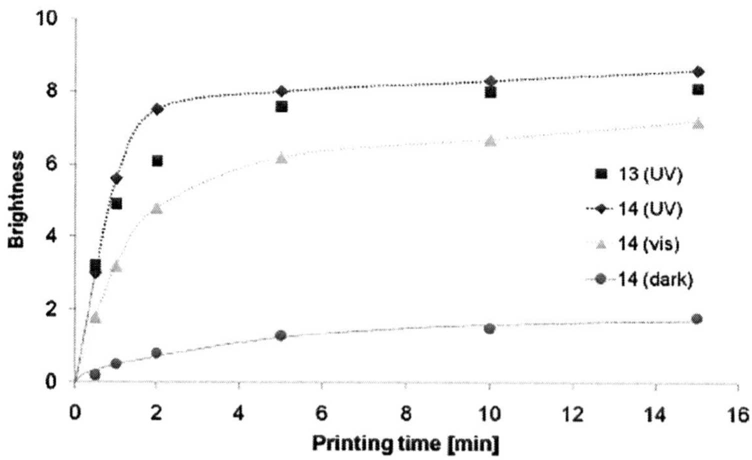


**Figure 12.3** Thiol inks aminoethanethiol (13) and aminothiophenol (14) printed onto alkene SAM (C, Figure 13.2). Printing time dependence of the fluorescence intensity under varying light conditions. Trend lines serve to guide the eye. Reprinted from ref. 28 with permission from the Royal Society of Chemistry.

dark suggesting that the confinement plays little role in mediating the thiol-click surface reactions.

Direct photomasking and writing strategies with radical-mediated thiol-ene reactions have also been demonstrated on other ultrathin film platforms. For example, Waldmann and coworkers used the thiol-ene reaction to pattern proteins and other biomolecule arrays onto surfaces using a biotin/streptavidin approach.[34,35] Silicon wafers were first modified with a thin film of COOH-terminated dendrimers followed by conversion to thiol or ene-modified surfaces, as shown in Figure 12.4. The thiol-terminated surface was coated with an olefin-modified biotin derivative (or *vice versa* for the ene-terminated surface) and then exposed to UV light through a photomask, yielding the thioether linkage and biotin micropatterns. Exposure of the biotin-modified surfaces to Cy5-labeled streptavidin provided homogeneous protein patterns over centimetre-wide areas. The patterns were also used to demonstrate protein–protein interactions, where the results indicated the thiol-ene based

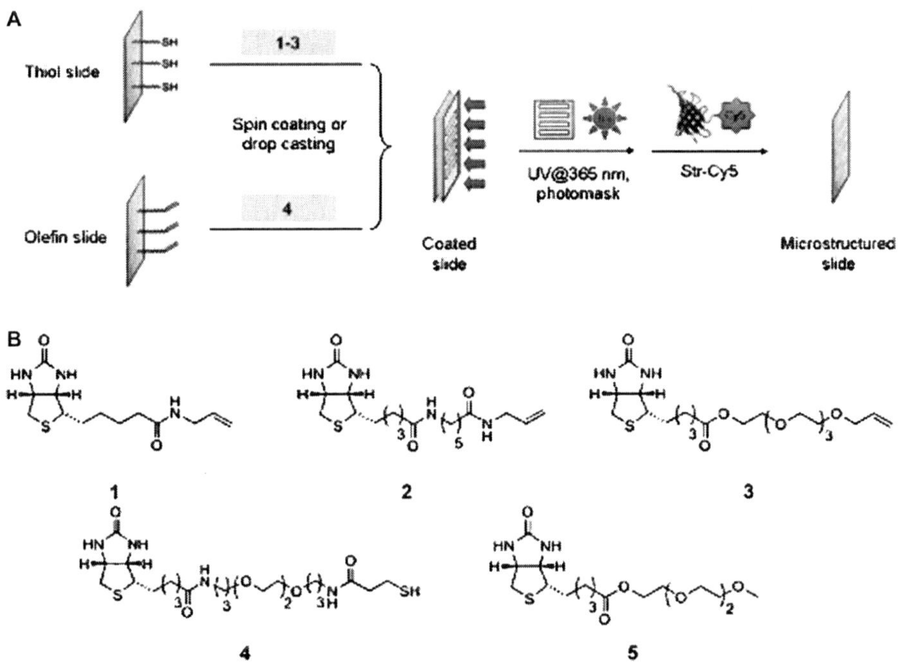


**Figure 12.4** Photopatterning with the thiol-ene reaction. (A) Thiol- or olefin-functionalized $SiO_x$/Si or glass slides with an optional layer of PAMAM dendrimers were coated with biotin probes 1–4 by using spin coating or drop casting. Coated slides were exposed to UV light at 365 nm through a photomask to obtain biotin microstructures (10 mW cm$^{-2}$). After washing and blocking, slides were incubated with Cy5-labeled streptavidin. Microstructures of biotin-streptavidin-Cy5 were detected with a microarray fluorescence scanner. (B) Biotin derivatives employed in the development of the thiol-ene photoimmobilization.
Reprinted from ref. 35 with permission from Wiley Interscience.

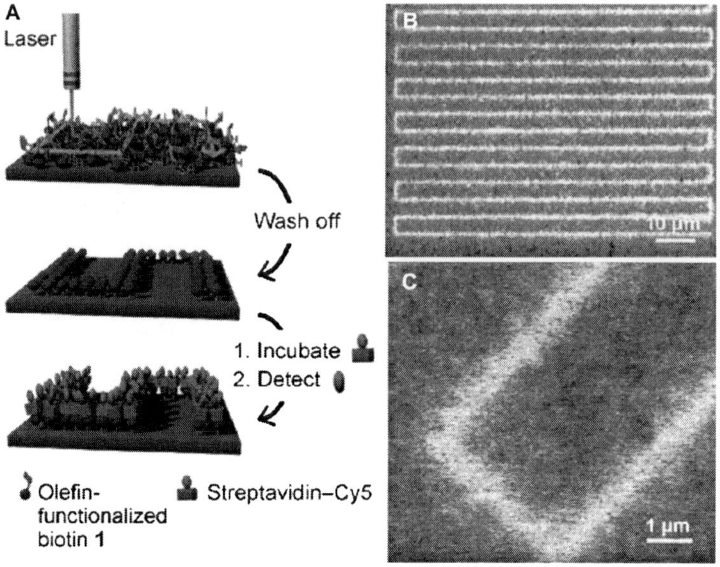


**Figure 12.5** Photochemical nanopatterning. (A) Olefin-functionalized biotin is deposited onto a thiol-modified surface. After laser-assisted ($\lambda = 411$ nm) nanopatterning in square-wave mode, the pattern is visualized by incubating the surface with SAv-Cy5. (B,C) Scanning confocal fluorescence microscopy images of the nanopatterns ($\lambda = 630$ nm).
Reprinted from ref. 34 with permission from Wiley Interscience.

immobilization strategy provided immobilized proteins with retention of their structure and function, as well as a preferred orientation.[36] Waldman *et al.* further demonstrated protein immobilization onto biotin line patterns using a laser scanned over the surface to directly facilitate the thiol-ene reaction between the thiol surface and the olefin-modified biotin derivative (Figure 12.5).[34]

Photolabile protecting groups (PPGs) can be removed upon exposure to light under neutral and reagent-free conditions to yield reactive functional groups (*e.g.* acids, alcohols, amines, and thiols).[37,38] As a stimulus for deprotection, light has the added benefit of spatial and temporal control when used for surface patterning. Smith *et al.*[39] employed the PPG approach to design "caged" thiol monolayers on amorphous carbon surfaces for thiol-click mediated biomolecule immobilization. In this approach, an *o*-nitrobenzyl-protected thiopropyl phosphoramidite moiety was immobilized onto a hydroxyl-terminated oligo(ethylene glycol) layer on ultrathin amorphous carbon (15 nm). Photolytic cleavage of the *o*-nitrobenzyl protecting group was performed using a maskless array synthesizer with programmed illumination locations generating thiol reactive functionality only in the light exposed areas. These reactive thiols enabled the attachment of biomolecules onto the surface pattern *via* thiol-maleimide reactions and disulfide exchange. Figure 12.6 shows a pattern generated by photodeprotection followed by immobilization of

**Figure 12.6** Fluorescence image of an amorphous carbon substrate after biotin immobilization via maleimide-thiol reaction followed by staining with fluorescently tagged avidin.
Reprinted from ref. 39 with permission from the American Chemical Society.

biotin-maleimide and incubation with fluorescently labeled avidin. This strategy enabled facile covalent attachment of biotin, DNA, and proteins while maintaining their biological activity.

## 12.3 Polymer Surfaces

Surface-initiated polymerization (SIP), or the "grafting from" approach, represents one of the most effective and versatile methods for tailoring the physico-chemical properties of surfaces.[40] SIP offers a direct means to control the density, thickness, and functionality of ultrathin films by growing polymer chains directly from surface bound initiators. At high grafting, the macromolecules adopt a highly stretched conformation extending perpendicular to the substrate surface in order to avoid chain overlap, or the so-called "polymer brush" conformation.[41] Using difunctional thiols and difunctional enes, the Bowman research group provided early examples of employing radical-mediated thiol-ene polymerization, base-catalyzed thiol-acrylate polymerization, and combinations thereof to fabricate polymer thin films by "grafting from" either thiol or ene terminated monolayers (Figure 12.7).[42–44] The click and step-growth mechanism of these reactions were exploited to control film thickness in a self-limiting manner by small manipulations in diene to dithiol stoichiometric ratio. Similar approaches have been used to fabricate patterned thiol-ene thin films,[25,45] and films exhibiting gradients in composition and thickness.[46]

The ability to endow a surface with three-dimensional functionality has tremendous advantages for applications where high functional group densities

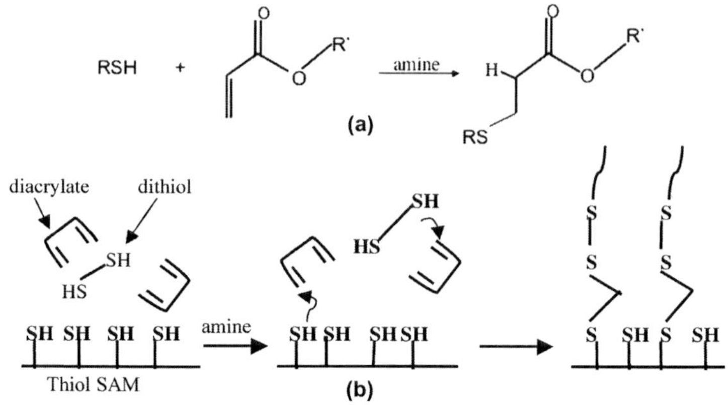


**Figure 12.7** (a) Thiol–acrylate conjugate addition reaction: a thiol molecule adds across the carbon–carbon double bond of the acrylate to form the thiol–acrylate product; (b) Schematic of the thiol–acrylate conjugate addition reaction on a surface.
Reprinted from ref. 44 with permission from the American Chemical Society.

are required, *e.g.* membranes and biosensor chips.[47] The three-dimensional conformation of polymer brushes greatly enhances the functionality of the surface by providing not only an opportunity to present functional groups at the interface, but also throughout the film – as each monomer unit is capable of carrying a functional moiety.[48] However, despite recent advances in the SIP approach, there remains a large number of pendent functional groups that cannot be directly polymerized from the surface due to (i) exorbitant cost of functional monomer synthesis and/or (ii) intolerance of the functional moiety in the polymerization process (*i.e.* reactivity, steric bulk). This often necessitates the development of a modular approach to surface engineering in the form of post-polymerization modification (PPM). The utility of the PPM strategy has been highlighted in several recent approaches implementing thiol-click reactions for the modification of the interfacial properties of materials without sacrificing bulk properties.[49–52] For example, properties such as biocompatibility, wettability, corrosion resistance, friction, affinity to a specific target molecule, and adsorption capacity can be controlled through decorating surfaces with polymer chains bearing specific pendent functional groups amenable to PPM.

Within the last few years, we have demonstrated radical mediated thiol-yne,[49] base-catalyzed thiol-isocyanate,[50] thiol-epoxy, and thiol-bromo chemistries[51] along with Michael-type thiol-ene reactions[52] as modular PPM brush platforms for rapid fabrication of highly functional, multicomponent surfaces. Figure 12.8a outlines the process for the thiol-yne PPM platform, which involves immobilization of a photoinitiator on an $SiO_x$ surface and SIP of propargyl methacrylate, yielding a three-dimensional configuration of reactive "handles" expressed by densely tethered "yne"-containing polymer brushes. The resulting "yne" functionalized brush serves as a "universal" reactive precursor for subsequent thiol-yne reactions, eliminating the synthetic effort associated with the use of multiple functional monomers. The efficacy of the thiol-yne reaction on polymer brushes was demonstrated with a library of commercially available thiols, including 3-mercaptopropionic acid (MPA) of interest for pH responsive surfaces, *N*-acetyl-L-cysteine as a model for the attachment of peptide fragments to brush surfaces and thiocholesterol as a ubiquitous component of biomembrane structures.[53] Thiol-yne reactions were carried out in the presence of DMPA (2 wt% DMPA : thiol) at 365 nm under ambient air, temperature and humidity conditions to afford the functional brushes. To illustrate the modularity of our approach, we conducted sequential and area-selective thiol-yne brush modifications using a simple photopatterning technique (Figure 12.8b) to produce multicomponent hydrophobic (dodecanethiol, DDT)/hydrophilic, MPA) surfaces, both in the lab and outdoors using sunlight as irradiation source. Figures 12.8c (lab) and 12.8d (sunlight) show the optical condensation images for the MPA/DDT patterned surface. As a final point, we demonstrated that homogeneous, pH responsive MPA functionalized brushes can be prepared in a facile manner. These surfaces exhibit reversible wettability upon protonation and deprotonation of the carboxylic acid functionalities as shown in Figure 12.8e. Similarly, sequential/area-selective base-catalyzed thiol-isocyanate brush modifications[50] were performed using an elastomeric microcapillary patterning process;[54] however, the isocyanate group is inert to radical polymerization conditions

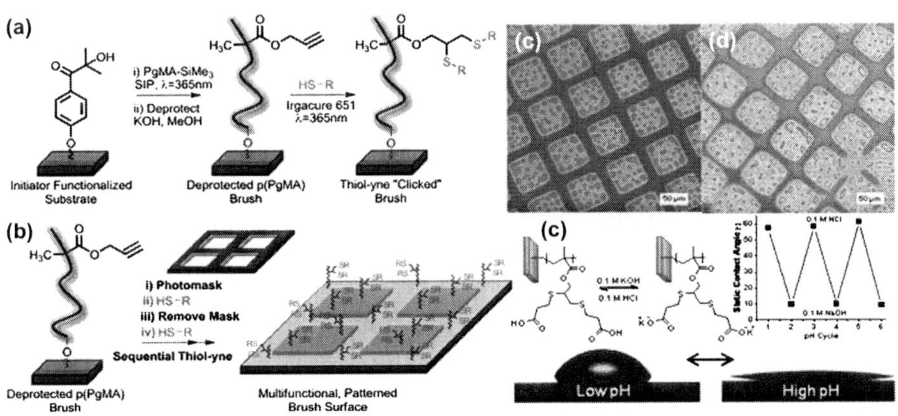


**Figure 12.8** (a) Schematic procedure of surface-initiated photopolymerization of TMS-protected propargyl methacrylate, deprotection, and subsequent thiol-yne functionalization. (b) Schematic procedure for photopatterning "yne"-containing polymer brush surfaces with sequential thiol-yne reactions. Condensation images of sequential thiol-yne micropatterned brushes by (c) UV light and (d) sunlight, (e) pH responsive brush surfaces.
Reprinted from ref. 49 with permission from the American Chemical Society.

eliminating any need and synthetic effort to protect the "clickable" moiety during SIP. This platform is analogous, but orthogonal to the radical-mediated thiol-yne click approach.[49]

Notably, one only has to look within the thiol-click class of reactions to realize a powerful set of orthogonal transformations that enable the installation of multiple chemical functionalities on a surface with high efficiency and modularity. With this in mind, multifunctional polymer brush surfaces have been created *via* combination of SIP and orthogonal thiol-click reactions. One of the principal advantages of the post-modifiable brush platform is that it provides a much larger number of modifiable sites per unit area of substrate as compared to conventional SAMs, while decoupling the polymer synthesis step from the immobilization of sensitive functional groups on the surface, thereby avoiding expensive monomer synthesis and reducing potential side reactions.[55] Specifically, two routes to multifunctional brush surfaces have been explored: in the first approach, alkyne-functionalized homopolymer brushes were modified with multiple thiols *via* a statistical, radical-mediated thiol-yne co-click reaction; and in the second approach, statistical copolymer brushes carrying two distinctly-addressable reactive moieties were sequentially modified *via* orthogonal base-catalzyed thiol-X (where X represents an isocyanate, epoxy, or α-bromoester) and radical-mediated thiol-yne reactions.[51] In both cases, the surface properties, in the form of wettability, were easily tuned over a wide range by judicious choice of brush composition and thiol functionality. One-pot statistical co-click reactions of brush pendent alkyne groups was successfully applied to various mixtures of thiols including a combination of thioglycerol and dodecanethiol, which bestow hydrophilic hydroxyl and hydrophobic aliphatic groups to the polymer brush, respectively, and a combination of *N*-acetyl-L-cysteine and dodecanethiol to form a model polymer coating with a biologically relevant molecule. The one-pot approach was also shown to be easily extendable to a mixture of three thiols. The ability to control the surface properties (hydrophilicity in this work) by facile control over relative concentration of thiols in a thiol mixture also exhibits the robustness and versatility of the one-pot approach. Additionally, sequential and orthogonal thiol-click reactions have been successfully applied to synthesize dual-functional polymer brushes by PPM of statistical copolymer brushes, as demonstrated in Figure 12.9. In this case, the surface properties were tailored by controlling the relative concentrations of monomers in the monomer feed during the SIP, which in turn dictates the composition of the thiol-clicked surface.

With an ongoing theme of biologically relevant surface modifications, peptide-functionalized thin films are of great interest and have been explored for an array of applications including biomedical,[56] anti-fouling,[57] inorganic/peptide hybrids,[58] and stimuli-responsive materials.[59] Polypeptide brushes are a unique class of surface-grafted polymers in that they allow the incorporation of structural motifs typical of those found in proteins, such as α-helices, β-sheets, and random coils. Like their native counterparts in proteins, these surface-grafted motifs can respond dynamically to changes in their environment.[60]

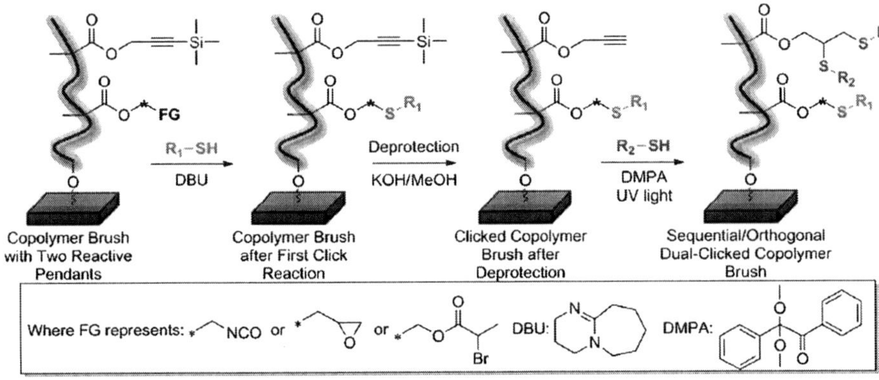


**Figure 12.9** Schematic for synthesis of dual-functional polymer brushes by sequential and orthogonal thiol-based click reactions.
Reprinted from ref. 51 with permission from the Royal Society of Chemistry.

This feature presents an opportunity to control the brush structure, function, and response at a level difficult to achieve with conventional organic polymers. In this direction, we have recently demonstrated nickel-mediated SIP of α-amino *N*-carboxyanhydrides, in particular NCA-*S*-*tert*-butylmercapto-L-cysteine, to facilitate PPM of thiol-clickable polypeptide brushes.[61] This approach demonstrates the versatility of thiol-click reactions by utilizing the pendent thiol of tethered poly(cysteine) brushes upon deprotection of the *tert*-butyl mercapto moieties with dithiothreitol in DMF at 60 °C. Figure 12.10a shows the Michael-type thiol-ene reaction employed to functionalize the poly(cysteine) brush with a fluorine-labeled maleimide. The brush functionalization was characterized by X-ray photoelectron spectroscopy (XPS), which confirmed the attachment of the pendent maleimide as indicated by the peak at 689.5 eV in the high resolution F1s spectrum (Figure 12.10b). Although the maleimide used in this study was convenient for XPS analysis, the modification of the pendent thiol can easily be extended to other "enes" carrying pendent moieties useful for a broad range of applications, *i.e.*, bioconjugation.

In most cases, the creation of functional surfaces *via* self-assembled monolayers or surface-initiated polymerization requires specific surface chemistry as anchoring groups as a first step, *i.e.* thiols on gold or silanes on oxides. The specificity of the anchoring step can potentially limit the types of surfaces that can be facilely modified in a post-modification approach. Chemical vapor deposition (CVD) polymerization decouples the polymerization process from the underlying substrate, significantly broadening the range of materials that may serve as substrates.[62] Additionally, a number of unique attributes arise from the CVD process, such as coating conformity, film homogeneity, solvent-free environments, and absence of initiators and plasticizers.[63–65] In 2012, Chen and co-workers[66] combined CVD with thiol-ene and thiol-yne post-modification processes to demonstrate a versatile, and

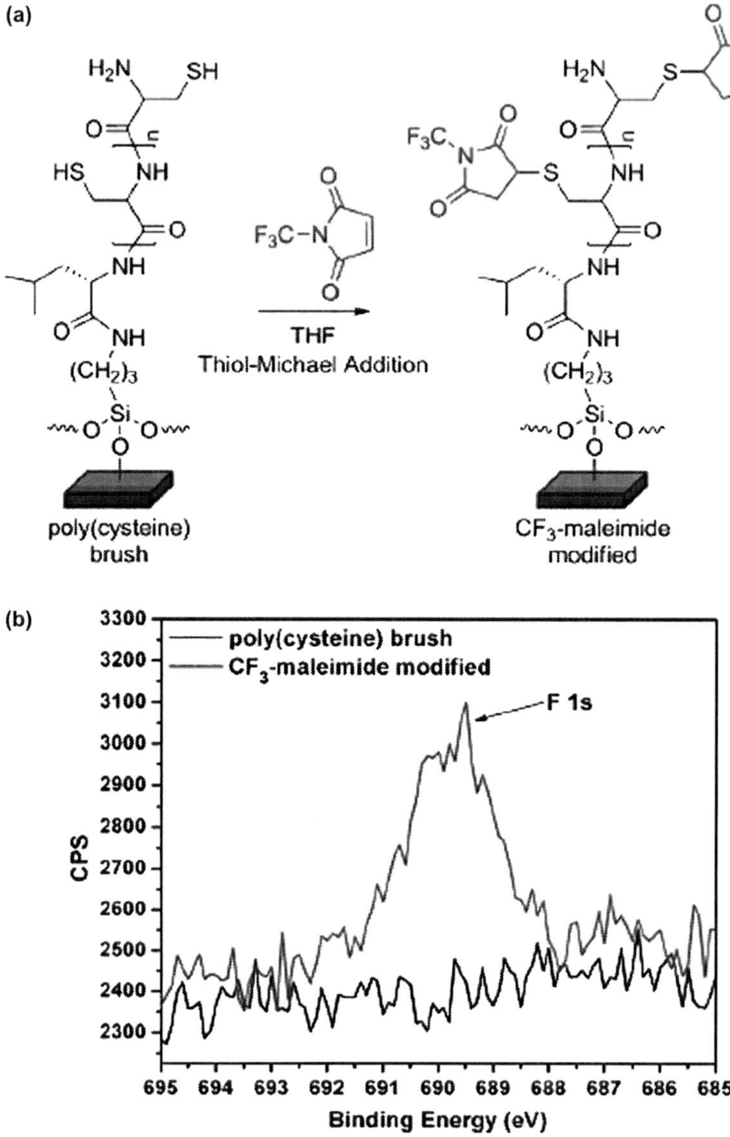


**Figure 12.10** (a) Synthetic scheme for Michael-type thiol-ene modification of a deprotected poly(cysteine) brush with a fluorine-labeled maleimide. (b) XPS data indicates a successful modification as shown by the observation of a peak at 689.5 eV in the high resolution F1s spectrum. Reprinted from ref. 61 with permission from the Royal Society of Chemistry.

substrate independent route to functional polymer films (Figure 12.11). Alkene and alkyne functionalized polymers, poly(4-vinyl-*p*-xylylene-co-*p*-xylylene) and poly(4-ethynyl-*p*-xylylene-co-*p*-xylylene) were prepared *via* CVD

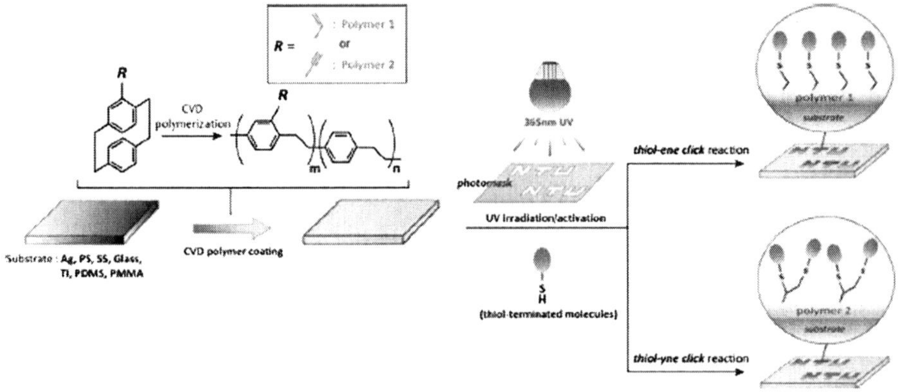


**Figure 12.11** Schematic illustration of thiol-ene/-yne click reactions based on CVD reactive polymer coatings. Poly(4-vinyl-*p*-xylylene-co-*p*-xylylene) (polymer 1) and poly(4-ethynyl-*p*-xylylene-co-*p*-xylylene) (polymer 2) are coated on various substrates *via* CVD polymerization process. Surfaces are conjugated with designed molecules through thiol-ene and thiol-yne click reactions that are activated by UV irradiation. A photomask is assisted for selective registration.
Reprinted from ref. 66 with permission from Wiley Interscience.

polymerization of 4-vinyl[2.2]paracyclophane and ethynyl-[2.2]paracyclophane, respectively, onto various substrates (silver, titanium, stainless steel, polystyrene, poly(methyl methacrylate), silicon, glass, poly(dimethylsiloxane), and poly(tetrafluoroethylene)) to support thiol-ene and thiol-yne click reactions.[66] Thiol-terminated poly(ethylene glycol) molecules as well as cell-adhering peptides containing thiols were used to investigate click reactions for antifouling and protein adsorption applications on various substrates. The specific activation locations were resolved using a high-resolution photomask, enabling activation of the thiol-ene/thiol-yne radical reactions only in light exposed areas.

## 12.4 Microporous Membranes

Microporous polymer membranes have found widespread use for various applications in industry, medicine, pharmacology, and for separation of particles, colloids, proteins, enzymes, and cells. For many membrane applications, it would be desirable to have straightforward surface engineering approaches that enable the simple alteration of surfaces, for example to prevent biofouling, without significantly altering the properties of the bulk membrane material. For example, the hydrophobic nature of common polymers (*i.e.* poly(vinylidene fluoride) (PVDF) and poly(propylene)) used in microporous membrane applications limit their practical use. In this sense, thiol-click reactions, particularly radical-mediated thiol-ene[67] and thiol-yne[68,69] processes, have proven to be a viable approach to improve membrane performance for purposes such as fouling minimization, hydrophilicity/hydrophobicity modulation, biocompatibility, biomembrane mimicry, as well as providing bio-and/or chemical functionalities that differ from the bulk membrane composition.

Kang and co-workers synthesized functional poly(vinylidene fluoride) (PVDF) copolymer membranes *via* thermally-induced graft copolymerization of allyl methacrylate (AMA) from ozone-preactivated PVDF chains (PVDF-g-PAMA). Microporous membranes were prepared from PVDF-g-PAMA by phase inversion from 15 wt% polymer solutions (NMP solvent) in water. The (PVDF-g-PAMA) graft copolymer membranes with active allyl groups served as a platform for surface modification *via* radical-mediated thiol-ene and thiol-Michael reactions (Figure 12.12). PVDF-g-PAMA membranes were modified initially through thermally-induced radical-mediated thiol-ene coupling of 3-mercaptopropionic acid (MPA) or 1,6-hexanedithiol (HDT) in the presence of azobisisobutyronitrile, AIBN, at 70 °C under an argon atmosphere for 24 hours.[67]

Successful immobilization of MPA and HDT onto PVDF-g-PAMA membranes was observed by XPS. The MPA-modified membranes exhibited pH-dependent permeability in aqueous solutions. Changes in the permeability were attributed to changes in the conformation of carboxylic acid functionalized graft chains on the membrane surface and within the pores

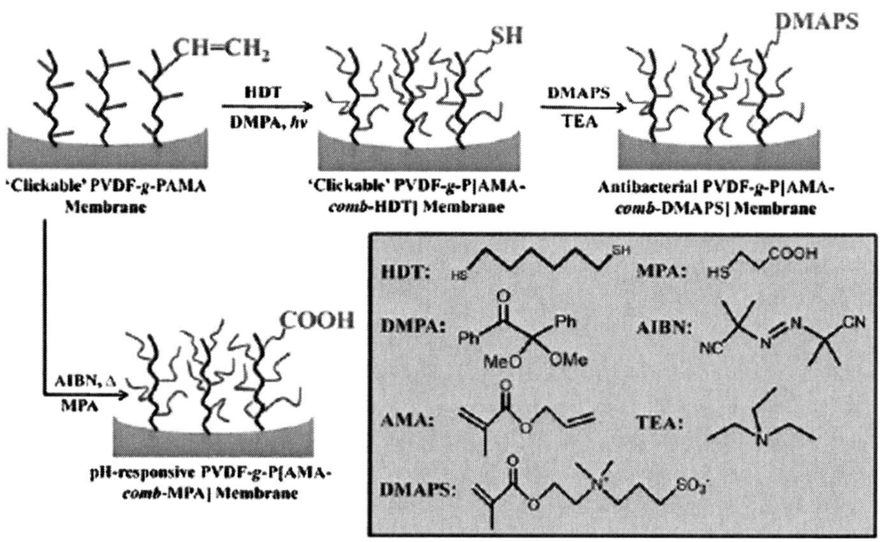


**Figure 12.12** Preparation of pH-responsive PVDF-g-PAMA membranes *via* thermally-initiated surface thiol-ene click reaction of mercaptopropionic acid, and antibacterial PVDF-g-PAMA membranes *via* UV-initiated surface thiol-ene click of hexanedithiol followed by a thio-Michael addition reaction of *N,N′*-dimethyl-(methylmethacryloyl ethyl) ammonium propanesulfonate (DMAPS).
Reprinted from ref. 67 with permission from the Royal Society of Chemistry.

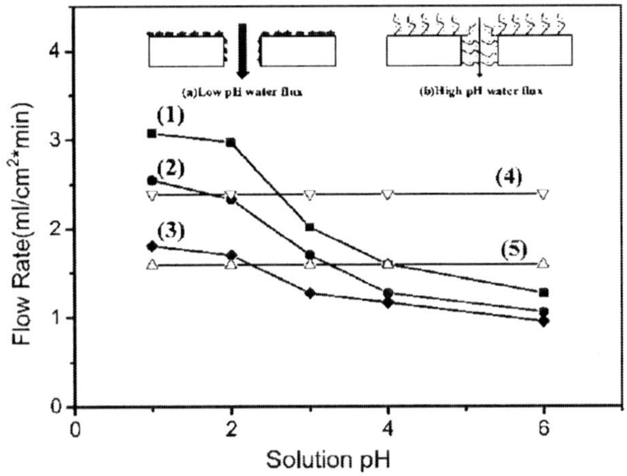


**Figure 12.13** pH-dependent permeability of aqueous solution through the pristine PVDF and PVDF-g-P[AMA-comb-MPA] microporous membrane. Curves 1, 2 and 3 are obtained from three PVDF-g-P[AMAcomb-MPA] microporous membranes with surface graft concentrations ([–COOH]/[–$CH_2CF_2$–])surface of 0.30, 0.22, 0.14, respectively. Curves 4 and 5 are from fluxes through two commercial PVDF membranes (standard pore diameter: d = 0.65 and 0.45 mm, respectively). Reprinted from ref. 67 with permission from the Royal Society of Chemistry.

upon changes in pH (Figure 12.13). Modification of the membrane with HDT resulted in excess thiol groups on the membrane surface, which served as reactive sites for thiol-Michael immobilization of *N*,*N*′-dimethy-(methylmethacryloyl ethyl) ammonium propanesulfonate (DMAPS) – a functional moiety known for excellent antimicrobial properties. DMAPS-modified membranes showed good antibacterial properties (*S. epidermidis*) under 1 mL $min^{-1}$ flow conditions. Successful incorporation of DMAPS was indicated by XPS and SEM, through observing changes in binding energy, as well as antibacterial efficiency when exposing unmodified and modified copolymer membranes to S. *epidermidis* bacteria cells, respectively.

Analogously, Kang and co-workers[68] demonstrated a modular membrane platform with "clickable" alkyne surfaces. The surface-enriched alkyne groups within the poly(vinylidene fluoride)-graft-poly(propargyl methacrylate) (PVDF-g-PPMA) copolymer membranes enabled tailoring of the surface *via* thiol-yne click reactions with thiols such as 3-mercapto-1-propanesulfonic acid (MPS), or alkyne-azide click reactions with azide-functionalized macromolecules such as azido-β-cyclodextrin (β-CD). The PDVF-g-PPMA membranes modified with MPS exhibited electrolyte-dependent permeability for aqueous solutions, while the membranes modified with azido-β-CD showed protein adsorption resistance.[68] Highly functionalized biomaterial surfaces for specific separation and purification of proteins were created by Xu and co-workers by glycosylating microporous polypropylene membranes (MPPMs) *via*

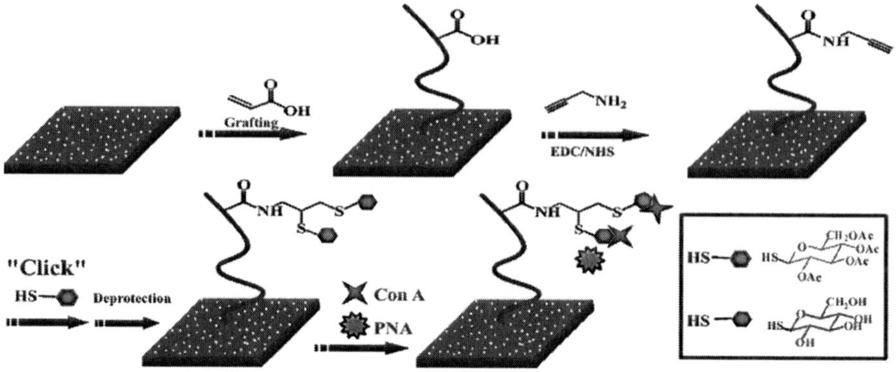


**Figure 12.14** Schematic illustration for the construction of glycosylated surface by thiol -yne click chemistry for lectin recognition.
Reprinted from ref. 69 with permission from the Royal Society of Chemistry.

thiol-yne click chemistry.[69] Carbohydrate–protein interactions are of great interest for mimicking the "glycoside cluster effect" by providing glycosylated or carbohydrate-decorated surfaces with a high density of saccharides for lectin recognition and affinity adsorption. Azide-alkyne click chemistry is a viable method to provide this "glycoside cluster effect" on MPPMs; however, the triazole moieties formed *via* the cycloaddition reaction show an affinity for the amino acid residues of proteins, promoting hydrogen bonding and stacking interactions, which decrease the specific recognition capability of the glycosylated membrane.[2,70] As shown by Xu *et al.*,[69] the use of radical-mediated thiol-yne click chemistry yields membranes with high carbohydrate grafting densities while avoiding undesirable interactions. UV-induced graft polymerization of acrylic acid from MPPMs provided carboxylic acid moieties on the membrane surface (Figure 12.14). The COOH groups were subsequently activated with 1-ethyl-3-[3-dimethylaminopropyl]carbodiimide hydrochloride/ *N*-hydroxysuccinimide and reacted with proparylamine to give an alkyne-modified MPPM susceptible to reaction with 2,3,4,6-tetra-*O*-acetyl-β-D-glucopyranosyl thiol when exposed to UV irradiation in the presence of a photoinitiator.[69,70] The glycosylated membranes showed significant recognition specificity and affinity adsorption towards fluorescently labeled lectins (FITC-Con A).[69] Rapid modification of microporous membranes represents yet another niche where thiol-click reactions have impacted the development of specific bio-related surface functionalization without interference from the route used for immobilization.

## 12.5 Nano- and Microparticle Surfaces

In recent years, considerable effort has been devoted to designing nanomaterials capable of performing a specific function. Such materials are of interest for a broad range of applications including catalysis, therapeutics, diagnostics, and composite materials. The design of functional nanoparticles (NPs) has certainly benefited from fast and efficient coupling reactions (*i.e.* click chemistry), enabling precise control of composition, overall size/shape, and location of functional groups (*i.e.* internal *vs.* surface). The ability to modify the surface chemistry of nanostructured materials, ultimately manipulating interfacial surface energies, greatly extends both function and use of this class of materials. Thiol-click reactions offer a direct pathway to modify surfaces with a diverse range of chemical functionalities from a set of highly efficient chemical transformations on nanoscale components such as single-walled carbon nanotubes and vertically aligned carbon nanofibers,[71] shape amphiphiles,[72,73] micro- and nano-spheres,[74–77] and NPs.[78–81] Although there are numerous excellent examples of modification of soft NPs (*i.e.* polymer assemblies, see the review by Iha *et al.*[20]), this section focuses only on thiol-click modification of inorganic nanomaterials and highly cross-linked polymer microspheres.

One of the initial reports of employing thiol-ene reactions to modify nanoparticle surfaces was that of Hayashi *et al.*[78] This approach demonstrated a one-pot synthesis and biofunctionalization of cysteine-modified iron oxide ($Fe_3O_4$) nanoparticles through an *in situ* hydrolysis-condensation of iron(III) 3-allylacetylacetonate at 80 °C, followed by radical-mediated addition of cysteine to the surface allyl groups. The cysteine-modified NPs showed good biocompatibility, superparamagnetism, and hyperthermic properties making the NPs potentially useful as MRI contrast agents and heating devices for magnetic hyperthermia. Integration of multifunctional molecules into NPs can be quite challenging due to the possibility of undesirable interactions between the various functionalities of interest and the surface. Warner and co-workers[79] demonstrated the synthesis of diphosphonic acid functionalized $Fe_3O_4$ NPs *via* thiol-ene addition of allyl diphosphonic acid onto thiol-containing NPs (Figure 12.15). The rapid rate and efficiency of the thiol-ene reaction circumvented competitive binding of the diphosphonic acid ligand at the $Fe_3O_4$ surface, yielding NPs useful for heavy metal remediation. More recent examples have employed similar thiol-ene modification approaches to modify the surface of $Fe_3O_4$-based NPs with folic acid,[80] glycopolymers,[82] and ferrocene[83] functional moieties for cancer targeting and biosensing applications. Kang's research group made use of thiol-ene and azide-alkyne click reactions to prepare binary brush modified core-shell hybrid NPs by grafting thiol and azide end-functionalized polymers, respectively, onto ene and yne copolymer shells.[74] A similar approach was used to prepare fluorescent and temperature responsive polymer shells on silica cores.[76] Removal of the silica core by dissolution in hydrofluoric acid provided hairy hollow nanospheres with poly(styrene) and poly(ethylene glycol) binary brush surfaces in the former example, and fluorescent poly(*N*-vinylcarbozole)/poly(*N*-isopropoylacrylamide) surfaces in the latter. The same authors recently extended this approach to fabricate hairy hybrid microrattles with metal nanoparticle cores.[84]

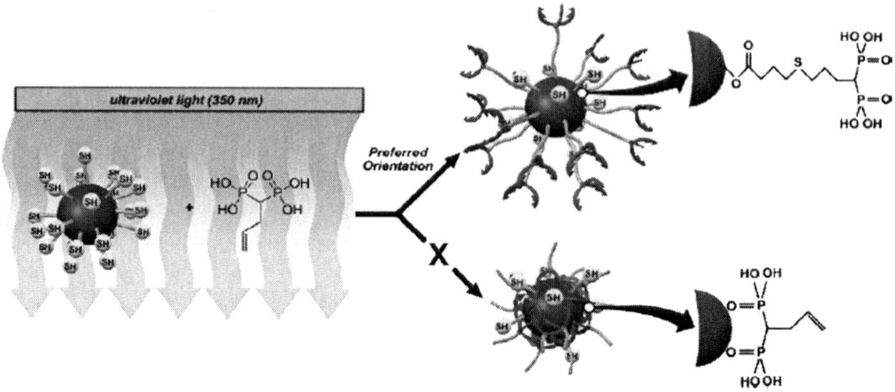


**Figure 12.15** The UV-promoted thiol-ene click reaction allows for the directed functionalization of otherwise troublesome ligands, circumventing such problems as undesirable head group affinity for the nanoparticle and low resultant ligand densities.
Reprinted from ref. 79 with permission from the American Chemical Society.

In 2009, Goldman and colleagues[85] reported the surface functionalization of cross-linked poly(divinylbenzene) microspheres *via* a "grafting to" thiol-ene approach. Thiol end-functionalized poly(*N*-isopropylacrylamide) prepared by reversible addition-fragmentation chain transfer (RAFT) polymerization was clicked onto vinyl-terminated microspheres *via* radical-mediated thiol-ene addition. Likewise, the surface of the microspheres was transformed to azide terminal groups by thiol-ene modification with 1-azido-undecane-11-thiol, enabling further modification of the microspheres by copper catalyzed azide-alkyne cycloaddition. Gokmen *et al.*[86] investigated the nature of thiol-click reactions for solid phase synthesis by fabricating thiol- and alkyne-modified mircobeads *via* a microfluidic approach. A small stoichiometric excess of either a multifunctional thiol or alkyne added during the bead synthesis provided a facile route to achieve beads rich in the excess functional group at the surface-air interface. The so-formed beads were probed as ligation platforms for nine different thiol-mediated reactions with the mobile phase reactants in 2× excess, relative to the surface thiol concentration (Table 12.1). Molecules such as non-activated enes and norbornenes (radical-mediated), acrylates and maleimides (Michael addition), halogens ($S_N2$ nucleophilic substitution), epoxies and aziridines ($S_N2$ ring opening), isocyanates and iso(thio)cyanates (nucleophilic addition) were compared in terms of kinetics and conversion on the bead surface. The radical-mediated thiol-norbornene reaction was found to provide the fastest reaction rate for ligation. Gokmen also showed the radical-mediated thiol-yne reaction occurred faster than the copper catalyzed azide-alkyne click reaction on the bead surface.[86] Using an approach called "sandwich" micro-contact printing, Kaufmann *et al.* extended the functionalization of the previously described beads to fabricate bifunctional Janus particles by printing alkyne inks onto thiol-functionalized beads.[87]

**Table 12.1** Overview of thiol-click reactions performed on thiol beads. Reprinted from ref. 86 with permission from the Royal Society of Chemistry.

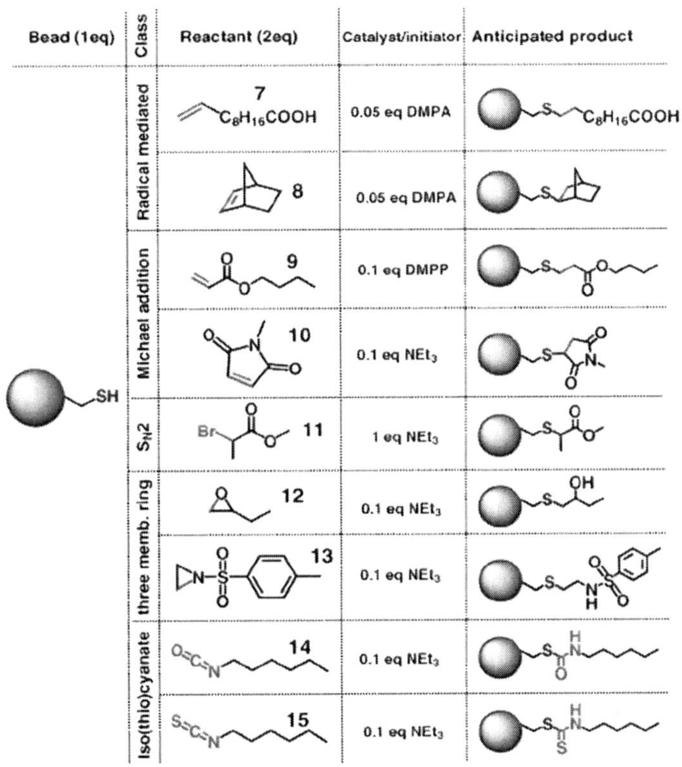

| Bead (1eq) | Class | Reactant (2eq) | Catalyst/initiator | Anticipated product |
|---|---|---|---|---|
| SH | Radical mediated | 7 ($C_8H_{16}COOH$) | 0.05 eq DMPA | S–$C_8H_{16}COOH$ |
| | Radical mediated | 8 | 0.05 eq DMPA | |
| | Michael addition | 9 | 0.1 eq DMPP | |
| | Michael addition | 10 | 0.1 eq $NEt_3$ | |
| | $S_N2$ | 11 | 1 eq $NEt_3$ | |
| | three memb. ring | 12 | 0.1 eq $NEt_3$ | |
| | three memb. ring | 13 | 0.1 eq $NEt_3$ | |
| | Iso(thio)cyanate | 14 | 0.1 eq $NEt_3$ | |
| | Iso(thio)cyanate | 15 | 0.1 eq $NEt_3$ | |

## 12.6 Biological Surfaces

Mediating the interface between synthetic molecules/materials and biological systems – or the so called biointerface – represents one of the most challenging fields in science and technology. The need to mediate such interactions is driven by industrial (antifouling coatings) and clinical applications (biosensors, bioimplants, tissue engineering), but also by the desire to fundamentally understand processes that occur at the biointerface. Engineering specific interactions or installing specific chemical functional groups at the biointerface using thiol-click chemistry has recently gained significant attention. While there are numerous examples of thiol-based surface modifications of biological entities, such as viral constructs,[88–94] this section will only highlight two recent, but exciting, examples employing thiol-click reactions to modify the surfaces of living cells and human hair.

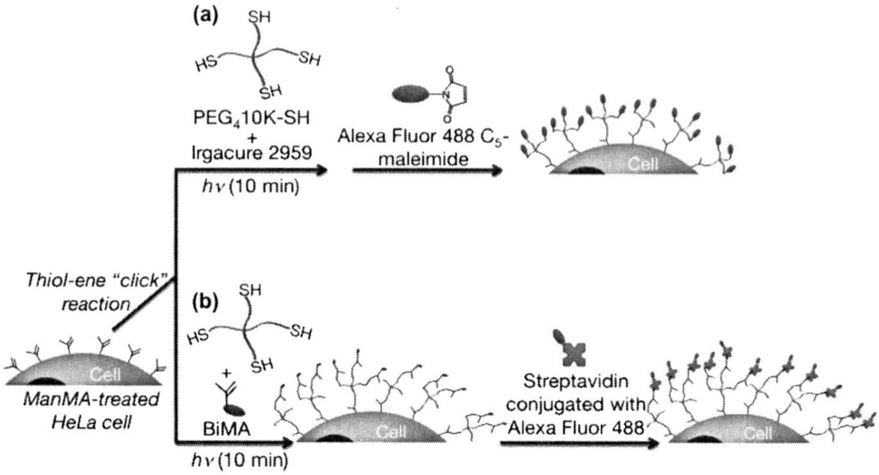


**Figure 12.16** Schematic representation of cell surface modification *via* methacryloyl groups delivered by glycosylation.
Reprinted from ref. 95 with permission from Wiley Interscience.

In the first example, Matsuno and co-workers[95] explored the alteration of human cervical carcinoma cells (HeLa) *via* glycosylation to express methacryloyl moieties directly along the cell surface (Figure 12.16). The presence of the methacryloyl functionality on the living cell surface enabled surface modification of the cells through photo-induced thiol-ene click reactions. Exposure of the cell surface to UV irradiation for 10 min at room temperature in the presence of a culture medium containing 2.5 wt% thiol-terminated 4-arm poly(ethylene glycol) and 0.05 wt% Irgacure 2959 provided PEGylated cells, with little adverse effect on cell viability upon radical-mediated thiol-ene processes. The remaining thiol functionality on the surface was used for attachment of fluorescent tags through thiol-Michael additions.

More recently, Slavin and co-workers[96] investigated covalent modification of human hair using thiol-mediated chemistries for interest in personal care products.[96] Copolymers of oligo(ethylene glycol) methyl ether methacrylate and allyl methacrylate were synthesized by cobalt catalyzed chain transfer polymerization (CCTP). The activated vinyl end groups resulting from CCTP were attached to the cysteine-derived thiol residues of α-keratin *via* a thiol-Michael addition. The use of phosphine to catalyze the thiol-Michael addition resulted in irreversible damage of the protein surface; however, ammonium thioglycolate (a known 'perming salt') provided a non-invasive catalyst while still promoting surface grafting. The polymer pendent allyl groups, moieties that are unreactive under the initial thiol-Michael grafting conditions, were subsequently post-conjugated using radical-mediated thiol-ene addition (Figure 12.17). For demonstration purposes, coumarin disulfide was reduced *in situ* to generate free thiol available for reaction with the allyl groups upon UV irradiation. Successful conjugation was observed by fluorescent microscopy.[96]

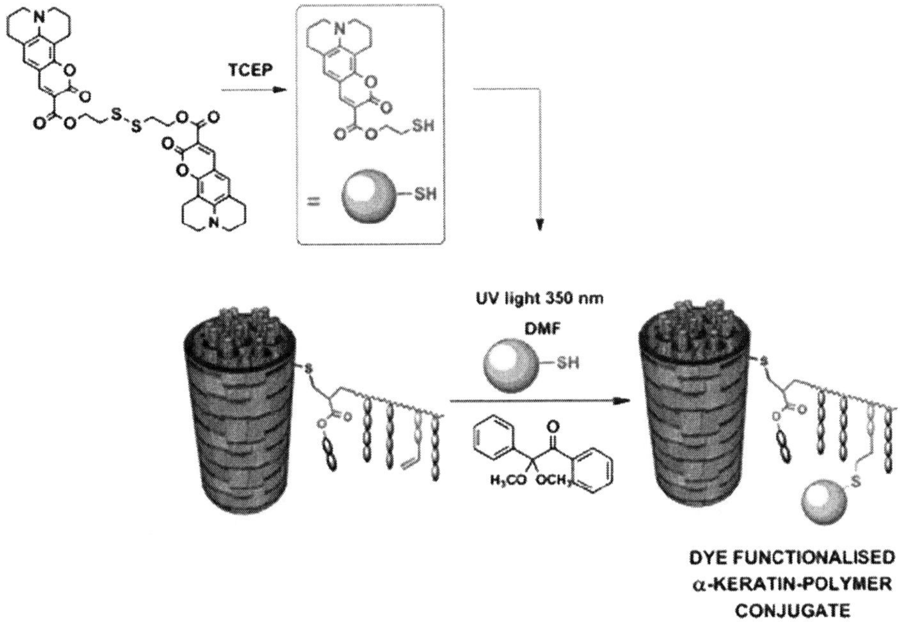


**Figure 12.17** Schematic for post-conjugation functionalization with fluorescent dye. Reproduced from ref. 96 with permission from the Royal Society of Chemistry.

These examples demonstrate not only the compatibility of thiol-click reactions under benign atmospheres and in buffer environments, but also show the potential and versatility of employing thiol-click modifications for complex biological materials.

## 12.7 Summary

In summary, thiol-click chemistry represents a powerful and versatile approach to engineer functional surfaces, as demonstrated by the broad scope of examples described in this chapter. The defining characteristics of thiol-click reactions – high efficiency, rapid kinetics, insensitivity to oxygen/water, and little to no byproduct formation – have propelled these reactions to the forefront of surface modification strategies. It is envisioned that future research on this topic will continue to focus on optimizing existing thiol-click chemistries for application to a broader scope of substrate surfaces, as well as expand opportunities to create complex, multifunctional surfaces by combining thiol-click chemistry with numerous other click chemistries in an orthogonal fashion.

## References

1. S. A. Meguid, *Surface engineering*, Elsevier Applied Science, London; New York, 1990.

2. R. Manetsch, A. Krasinski, Z. Radic, J. Raushel, P. Taylor, K. B. Sharpless and H. C. Kolb, *J. Am. Chem. Soc.*, 2004, **126**, 12809–12818.
3. H. C. Kolb, M. G. Finn and B. Sharpless, *Angew. Chem., Int. Ed.*, 2001, **40**, 2004–2021.
4. L. Nebhani and C. Barner-Kowollik, *Adv. Mater.*, 2009, **21**, 3442–3468.
5. J. M. Baskin, J. A. Prescher, S. T. Laughlin, N. J. Agard, P. V. Chang, I. A. Miller, A. Lo, J. A. Codelli and C. R. Bertozzi, *Proc. Natl. Acad. Sci. U. S. A.*, 2007, **104**, 16793–16797.
6. J. C. Jewett, E. M. Sletten and C. R. Bertozzi, *J. Am. Chem. Soc.*, 2010, **132**, 3688–3690.
7. R. Manova, T. A. van Beek and H. Zuilhof, *Angew. Chem., Int. Ed.*, 2011, **50**, 5428–5430.
8. A. Kuzmin, A. Poloukhtine, M. A. Wolfert and V. V. Popik, *Bioconjugate Chem.*, 2010, **21**, 2076–2085.
9. S. V. Orski, A. A. Poloukhtine, S. Arumugam, L. Mao, V. V. Popik and J. Locklin, *J. Am. Chem. Soc.*, 2010, **132**, 11024–11026.
10. P. T. Dirlam, G. A. Strange, J. A. Orlicki, E. D. Wetzel and P. J. Costanzo, *Langmuir*, 2009, **26**, 3942–3948.
11. X.-L. Sun, C. L. Stabler, C. S. Cazalis and E. L. Chaikof, *Bioconjugate Chem.*, 2005, **17**, 52–57.
12. T. Pauloehrl, G. Delaittre, V. Winkler, A. Welle, M. Bruns, H. G. Börner, A. M. Greiner, M. Bastmeyer and C. Barner-Kowollik, *Angew. Chem., Int. Ed.*, 2012, **51**, 1071–1074.
13. S. Arumugam, S. V. Orski, J. Locklin and V. V. Popik, *J. Am. Chem. Soc.*, 2011, **134**, 179–182.
14. S. Arumugam and V. V. Popik, *J. Am. Chem. Soc.*, 2011, **133**, 15730–15736.
15. C. Hoyle, T. Lee and T. Roper, *J. Polym. Sci., Part A: Polym. Chem.*, 2004, **42**, 5301–5338.
16. C. E. Hoyle and C. N. Bowman, *Angew. Chem., Int. Ed.*, 2010, **49**, 1540–1573.
17. C. E. Hoyle, A. B. Lowe and C. N. Bowman, *Chem. Soc. Rev.*, 2010, **39**, 1355–1387.
18. A. B. Lowe, C. E. Hoyle and C. N. Bowman, *J. Mater. Chem.*, 2010, **20**, 4745–4750.
19. J. W. Chan, C. E. Hoyle and A. B. Lowe, *J. Am. Chem. Soc.*, 2009, **131**, 5751–5753.
20. R. K. Iha, K. L. Wooley, A. M. Nystroìim, D. J. Burke, M. J. Kade and C. J. Hawker, *Chem. Rev.*, 2009, **109**, 5620–5686.
21. D. K. Schwartz, *Annu. Rev. Phys. Chem.*, 2001, **52**, 107–137.
22. B. T. Houseman, E. S. Gawalt and M. Mrksich, *Langmuir*, 2002, **19**, 1522–1531.
23. C.-Y. Lee, P.-C. T. Nguyen, D. W. Grainger, L. J. Gamble and D. G. Castner, *Anal. Chem.*, 2007, **79**, 4390–4400.
24. J. Wettero, T. Hellerstedt, P. Nygren, K. Broo, D. Aili, B. Liedberg and K.-E. Magnusson, *Langmuir*, 2008, **24**, 6803–6811.

25. E. Besson, A.-M. Gue, J. Sudor, H. Korri-Youssoufi, N. Jaffrezic and J. Tardy, *Langmuir*, 2006, **22**, 8346–8352.
26. U. Bexell, M. Olsson, P. E. Sundell, M. Johansson, P. Carlsson and M. Hellsing, *Appl. Surf. Sci.*, 2004, **231–232**, 362–365.
27. M. A. C. Campos, J. M. J. Paulusse and H. Zuilhof, *Chem. Commun.*, 2010, **46**, 5512–5514.
28. J. Mehlich and B. J. Ravoo, *Org. Biomol. Chem.*, 2011, **9**, 4108–4115.
29. T. Knieling, W. Lang and W. Benecke, *Sens. Actuators, B*, 2007, **126**, 13–17.
30. G. MacBeath, A. N. Koehler and S. L. Schreiber, *J. Am. Chem. Soc.*, 1999, **121**, 7967–7968.
31. E. A. Smith, M. J. Wanat, Y. Cheng, S. r. V. P. Barreira, A. G. Frutos and R. M. Corn, *Langmuir*, 2001, **17**, 2502–2507.
32. U. Bexell, R. Berger, M. Olsson, T. M. Grehk, P. E. Sundell and M. Johansson, *Thin Solid Films*, 2006, **515**, 838–841.
33. C. Wendeln, S. Rinnen, C. Schulz, H. F. Arlinghaus and B. J. Ravoo, *Langmuir*, 2010, **26**, 15966–15971.
34. P. Jonkheijm, D. Weinrich, M. Köhn, H. Engelkamp, P. C. M. Christianen, J. Kuhlmann, J. C. Maan, D. Nüsse, H. Schroeder, R. Wacker, R. Breinbauer, C. M. Niemeyer and H. Waldmann, *Angew. Chem., Int. Ed.*, 2008, **47**, 4421–4424.
35. D. Weinrich, M. Köhn, P. Jonkheijm, U. Westerlind, L. Dehmelt, H. Engelkamp, P. C. M. Christianen, J. Kuhlmann, J. C. Maan, D. Nüsse, H. Schröder, R. Wacker, E. Voges, R. Breinbauer, H. Kunz, C. M. Niemeyer and H. Waldmann, *ChemBioChem*, 2010, **11**, 235–247.
36. D. Weinrich, P. Lin, P. Jonkheijm, U. Nguyen, H. Schroder, C. Niemeyer, K. Alexandrov, R. Goody and H. Waldmann, *Angew. Chem., Int. Ed.*, 2010, **49**, 1252–1257.
37. V. S. Miguel, C. G. Bochet and A. del Campo, *J. Am. Chem. Soc.*, 2011, **133**, 5380–5388.
38. A. P. Pelliccioli and J. Wirz, *Photochem. Photobio. Sci.*, 2002, **1**, 441–458.
39. S. Chen and L. M. Smith, *Langmuir*, 2009, **25**, 12275–12282.
40. O. Prucker and J. Ruhe, *Langmuir*, 1998, **14**, 6893–6898.
41. S. T. Milner, *Science*, 1991, **251**, 905–914.
42. V. Khire, T. Lee and C. Bowman, *Macromolecules*, 2008, **41**, 7440–7447.
43. A. W. Harant, V. S. Khire, M. S. Thibodaux and C. N. Bowman, *Macromolecules*, 2006, **39**, 1461–1466.
44. V. S. Khire, T. Y. Lee and C. N. Bowman, *Macromolecules*, 2007, **40**, 5669–5677.
45. V. S. Khire, A. W. Harant, A. W. Watkins, K. S. Anseth and C. N. Bowman, *Macromolecules*, 2006, **39**, 5081–5086.
46. V. S. Khire, D. S. W. Benoit, K. S. Anseth and C. N. Bowman, *J. Polym. Sci., Part A: Polym. Chem.*, 2006, **44**, 7027–7039.
47. W. Senaratne, L. Andruzzi and C. K. Ober, *Biomacromolecules*, 2005, **6**, 2427–2448.
48. H. Murata, O. Prucker and J. Ruhe, *Macromolecules*, 2007, **40**, 5497–5503.

49. R. M. Hensarling, V. A. Doughty, J. W. Chan and D. L. Patton, *J. Am. Chem. Soc.*, 2009, **131**, 14673–14675.
50. R. M. Hensarling, S. B. Rahane, A. P. LeBlanc, B. J. Sparks, E. M. White, J. Locklin and D. L. Patton, *Polym. Chem.*, 2011, **2**, 88–90.
51. S. B. Rahane, R. M. Hensarling, B. J. Sparks, C. M. Stafford and D. L. Patton, *J. Mater. Chem.*, 2012, **22**, 932–943.
52. R. M. Hensarling, S. B. Rahane, A. P. Leblanc, B. J. Sparks, E. M. White, J. Locklin and D. L. Patton, *Polym. Chem.*, 2011, **2**, 88–90.
53. G. Cadenhead, *Structure and properties of cell membranes*, CRC Press, Boca Raton, FL, 1985.
54. T. Kaufmann and B. Ravoo, *Polym. Chem.*, 2010, **1**, 371–387.
55. R. Barbey, L. Lavanant, D. Paripovic, N. Schuwer, C. Sugnaux, S. Tugulu and H. A. Klok, *Chem. Rev.*, 2009, **109**, 5437–5527.
56. S. A. Curtin and T. J. Deming, *J. Am. Chem. Soc.*, 1999, **121**, 7427–7428.
57. J. Wang, M. I. Gibson, R. Barbey, S.-J. Xiao and H.-A. Klok, *Macromol. Rapid Comm.*, 2009, **30**, 845–850.
58. J. D. Lunn and D. F. Shantz, *Chem. Mater.*, 2009, **21**, 3638–3648.
59. R. J. Mart, R. D. Osborne, M. M. Stevens and R. V. Ulijn, *Soft Matter*, 2006, **2**, 822–835.
60. Y. Ito, Y. Ochiai, Y. S. Park and Y. Imanishi, *J. Am. Chem. Soc.*, 1997, **119**, 1619–1623.
61. B. J. Sparks, J. G. Ray, D. A. Savin, C. M. Stafford and D. L. Patton, *Chem. Commun.*, 2011, **47**, 6245–6247.
62. M. E. Alf, A. Asatekin, M. C. Barr, S. H. Baxamusa, H. Chelawat, G. Ozaydin-Ince, C. D. Petruczok, R. Sreenivasan, W. E. Tenhaeff, N. J. Trujillo, S. Vaddiraju, J. Xu and K. K. Gleason, *Adv. Mater.*, 2010, **22**, 1993–2027.
63. J. Lahann, *Polym. Int.*, 2006, **55**, 1361.
64. J. Lahann, *Chem. Eng. Commun.*, 2006, **193**, 1457.
65. J. Lahann and R. Langer, *Macromolecules*, 2002, **35**, 4380.
66. J.-T. Wu, C.-H. Huang, W.-C. Liang, Y.-L. Wu, J. Yu and H.-Y. Chen, *Macromol. Rapid Comm.*, 2012, **33**, 922–927.
67. T. Cai, R. Wang, K. G. Neoh and E. T. Kang, *Polym. Chem.*, 2011, **2**, 1849–1858.
68. T. Cai, K. G. Neoh and E. T. Kang, *Macromolecules*, 2011, **44**, 4258–4268.
69. C. Wang, P.-F. Ren, X.-J. Huang, J. Wu and Z.-K. Xu, *Chem. Commun.*, 2011, **47**, 3930–3932.
70. C. Wang, J. Wu and Z.-K. Xu, *Macromol. Rapid Comm.*, 2010, **31**, 1078–1082.
71. C.-S. Lee, S. E. Baker, M. S. Marcus, W. Yang, M. A. Eriksson and R. J. Hamers, *Nano Lett.*, 2004, **4**, 1713–1716.
72. W.-B. Zhang, Y. Li, X. Li, X. Dong, X. Yu, C.-L. Wang, C. Wesdemiotis, R. P. Quirk and S. Z. D. Cheng, *Macromolecules*, 2011, **44**, 2589–2596.
73. H. Kakwere and S. B. Perrier, *J. Am. Chem. Soc.*, 2009, **131**, 1889–1895.
74. G. L. Li, D. Wan, K. G. Neoh and E. T. Kang, *Macromolecules*, 2010, **43**, 10275–10282.

75. X.-J. Song, J. Hu and C.-C. Wang, *Colloids Surf., A*, 2011, **380**, 250–256.
76. G. L. Li, L. Q. Xu, X. Tang, K. G. Neoh and E. T. Kang, *Macromolecules*, 2010, **43**, 5797–5803.
77. D. Kim, E. Kim, J. Lee, S. Hong, W. Sung, N. Lim, C. G. Park and K. Kim, *J. Am. Chem. Soc.*, 2010, **132**, 9908–9919.
78. K. Hayashi, K. Ono, H. Suzuki, M. Sawada, M. Moriya, W. Sakamoto and T. Yogo, *Chem. Mater.*, 2010, **22**, 3768–3772.
79. R. D. Rutledge, C. L. Warner, J. W. Pittman, R. S. Addleman, M. Engelhard, W. Chouyyok and M. G. Warner, *Langmuir*, 2010, **26**, 12285–12292.
80. L. Wang, K. G. Neoh, E.-T. Kang and B. Shuter, *Biomaterials*, 2011, **32**, 2166–2173.
81. D. E. Achatz, F. J. Heiligtag, X. Li, M. Link and O. S. Wolfbeis, *Sens. Actuators, B*, 2010, **150**, 211–219.
82. A. Pfaff, A. Schallon, T. M. Ruhland, A. P. Majewski, H. Schmalz, R. Freitag and A. H. E. Müller, *Biomacromolecules*, 2011, **12**, 3805–3811.
83. C. Liang, L. Jing, X. Shi, Y. Zhang and Y. Xian, *Electrochim. Acta*, 2012, **69**, 167–173.
84. G. L. Li, L. Q. Xu, K. G. Neoh and E. T. Kang, *Macromolecules*, 2011, **44**, 2365–2370.
85. A. S. Goldmann, A. Walther, L. Nebhani, R. Joso, D. Ernst, K. Loos, C. Barner-Kowollik, L. Barner and A. H. E. Muller, *Macromolecules*, 2009, **42**, 3707–3714.
86. M. T. Gokmen, J. Brassinne, R. A. Prasath and F. E. Du Prez, *Chem. Commun.*, 2011, **47**, 4652–4654.
87. T. Kaufmann, M. T. Gokmen, S. Rinnen, H. F. Arlinghaus, F. Du Prez and B. J. Ravoo, *J. Mater. Chem.*, 2012, **22**, 6190–6199.
88. E. Gillitzer, D. Willits, M. Young and T. Douglas, *Chem. Commun.*, 2002, 2390–2391.
89. M. L. Flenniken, D. A. Willits, S. Brumfield, M. J. Young and T. Douglas, *Nano Lett.*, 2003, **3**, 1573–1576.
90. M. L. Flenniken, L. O. Liepold, B. E. Crowley, D. A. Willits, M. J. Young and T. Douglas, *Chem. Commun.*, 2005, 447–449.
91. S. Kang, L. M. Oltrogge, C. C. Broomell, L. O. Liepold, P. E. Prevelige, M. Young and T. Douglas, *J. Am. Chem. Soc.*, 2008, **130**, 16527–16529.
92. Q. Wang, K. S. Raja, K. D. Janda, T. Lin and M. G. Finn, *Bioconjugate Chem.*, 2002, **14**, 38–43.
93. E. Strable, J. E. Johnson and M. G. Finn, *Nano Lett.*, 2004, **4**, 1385–1389.
94. C. M. Soto, A. S. Blum, G. J. Vora, N. Lebedev, C. E. Meador, A. P. Won, A. Chatterji, J. E. Johnson and B. R. Ratna, *J. Am. Chem. Soc.*, 2006, **128**, 5184–5189.
95. Y. Iwasaki and H. Matsuno, *Macromol. Biosci.*, 2011, **11**, 1478–1483.
96. S. Slavin, E. Khoshdel and D. M. Haddleton, *Polym. Chem.*, 2012, **3**, 1461–1466.

CHAPTER 13

# *The Application of Thiol-Ene/Yne Radical Click Chemistry in Surface Modification and Functionalization*

XINMIAO LIANG,* AIJIN SHEN AND ZHIMOU GUO

Key Lab of Separation Science for Analytical Chemistry, Dalian Institute of Chemical Physics, Chinese Academy of Sciences, 116023, Dalian, P. R. China
*Email: liangxm@dicp.ac.cn

## 13.1 Introduction

Click chemistry,[1] which was introduced by Sharpless in 2001, possesses the classical characteristics of modularity, high efficiency, results in high yields with no or inoffensive byproducts, proceeds under mild reaction conditions with readily available materials, regiospecificity and stereospecificity, and orthogonality with other common organic synthesis reactions. As a classical click reaction, copper (I)-catalyzed Huisgen 1,3-dipolar cycloaddition (CuAAC) has received extensive attention. The implementation of the CuAAC click reaction in the material sciences,[1–3] surface modification[4] and the synthesis of chromatographic stationary phases,[5–8] *etc.* has achieved great success. Nonetheless, the cytotoxicity of the copper ion which is used as catalyst greatly limits

RSC Polymer Chemistry Series No. 6
Thiol-X Chemistries in Polymer and Materials Science
Edited by Andrew B. Lowe and Christopher N. Bowman

Published by the Royal Society of Chemistry, www.rsc.org

the utilization of the CuAAC reaction, especially when the products are designed to be used for biological application. Consequently, it is of great interest to introduce metal-free click reactions to avoid the use of toxic metal catalysts.

In 2009, Schubert and coworkers discussed the potential click reactions which could be performed without the presence of metal catalyst, including the azide-alkyne cycloadditions with more reactive alkynes, Michael addition of thiols with maleimide, nucleophilic substitution of the para-fluoro substituent of pentafluorophenyl groups, Diels–Alder and retro-Diels–Alder reactions as well as the radical addition between thiols and alkenes. However, most of the discussed click reactions involved complicated reactive groups and large sizes of the coupling units, which was disadvantageous for broad applications. In contrast, the simple and efficient thiol-ene radical addition was considered to be a good candidate for click reactions.[9]

Thiol-ene radical addition reaction, which involves a radical-initiated coupling of a thiol group to a terminal alkene group was regarded as a click reaction in 2007.[10] It has all the classical features of click chemistry, being highly efficient, proceeding rapidly to high yield and requiring essentially no clean up, being insensitive to ambient oxygen or water, yielding regioselective products and having a variety of readily available reagents.[11–13] The radical reaction can be induced by photo initiators or thermal initiators. The photo-initiated process can be easily controlled spatially and temporally. On the other hand, the thermally initiated thiol-ene reaction is not uncommon and has been prominent in the functionalization of biological materials. For large-scale syntheses, the need to uniformly functionalize large volumes of material limits to some extent the viability of photoinitiation, where uniform extended exposure of large volumes is limited both by reactor design and light attenuation. Therefore, to achieve uniform bulk radical generation, elevated temperatures and thermal initiators have regularly been used.[11] In addition, the thermal initiated process is of great value in the synthesis of silica-based chromatographic stationary phases. For standard porous silica used in the synthesis of chromatographic materials, the surface area (silanol groups) is exclusively within the pores, while the outer surface of the spheres is almost negligible. It is difficult to achieve light exposure within the pores where the bonding proceeds, whereas the thermally initiated thiol-ene reaction can be successfully realized. Additionally, the radical-mediated thiol-yne reaction is basically similar to the thiol-ene radical click reaction except that it involves the consecutive addition of two thiol functional groups onto the alkyne functional group. The thiol-yne radical-initiated reaction has the classical characteristics of click chemistry and is also becoming an indispensable approach in material sciences.[14–16]

In this work, we will focus on the introduction of surface modification and functionalization *via* the radical-mediated thiol-ene/yne click reaction. The representative applications of thiol radical click chemistry in the immobilization of macromolecules (proteins), surface engineering and patterning, and the modification of solid surfaces with alternative properties *etc*. in the last five years are briefly introduced. In addition, the utilization of thiol radical click chemistry in the preparation of typical chromatographic stationary phases is also introduced, such as for chiral (ion-exchange) materials, reversed-phase stationary phases, mixed-mode stationary phases and hydrophilic interaction liquid chromatography (HILIC) stationary phases.

## 13.2 Modification and Functionalization of Solid Surfaces with Tailored Properties *via* Thiol Radical Click Chemistry

Click chemistry plays an important part in surface modification and functionalization owing to its high efficiency, excellent selectivity and high yields with no or inoffensive byproducts with no chromatographic separation required. Different from the classical Huisgen azide-yne cycloaddition reaction, the thiol-ene/yne reactions can be performed without toxic transition metals and is especially beneficial for biological applications. Thiol-ene/yne click chemistry possesses several characteristic features[11,12,14] that are attractive with regard to surface modification. The introduction of thiol-functionalized molecules to terminal alkene/yne can be performed under very mild conditions and is insensitive to water and oxygen. Meanwhile, the reaction is compatible with a wide range of functional groups, *i.e.* exhibits good orthogonality with other common organic synthesis reactions. And various thiol-functionalized (bio)molecules are ready available. As a result, radial-mediated thiol-ene reaction has emerged as a versatile approach for tailoring the solid surface with specific properties, such as the immobilization of macromolecules (proteins), surface engineering and patterning *etc.*

### 13.2.1 Surface Modification Based on Thiol-Ene Click Reaction

#### *13.2.1.1 Surface Immobilization of Proteins*

Protein biochips play an important role in biotechnology and provide a unique analysis platform for advanced protein-related studies. Therefore, site-specific immobilization of proteins on solid surfaces with retention of their structure and activity is of great interest.[17–19] However, owing to the delicate and sensitive nature of proteins, mild reaction conditions are necessary to prevent protein denaturation or degradation in protein biochip preparation. Additionally, bioorthogonal chemical reactions are required in order to suppress undesired side reactions. Taking advantage of the thiol-ene click reaction, Waldmann *et al.* developed a photochemical surface patterning method for oriented protein conjugation on solid surfaces which generated homogeneous surface coverage and accessibility to the active site of the protein. The thiol-ene reaction was initiated by irradiation in the 365–405 nm range with high sensitivity and specificity under aqueous conditions. The controllable light source employed allowed fabrication of variable size and shape protein

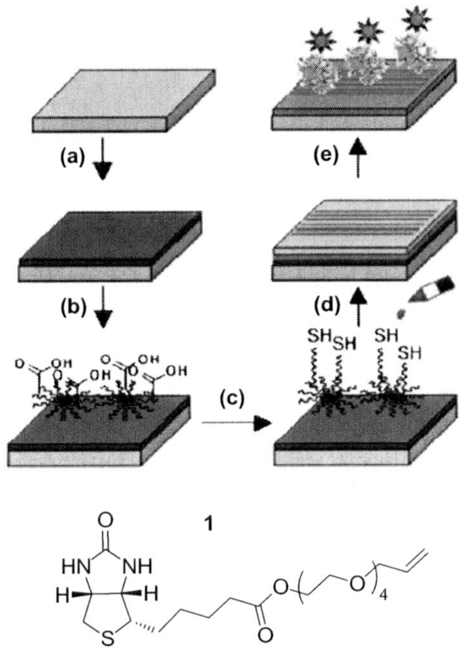


**Scheme 13.1** Surface modification: (a) activation (plasma-enhanced chemical vapor deposition, silanization); (b) optimization (aminocaproic acid linker, dendrimers); (c) functionalization with thiols; (d) drop casting of **1** (10 μL, 10 mM, ethyleneglycol) and immediate coverage with a photomask to prevent drying of the liquid film prior to irradiation (10 min, 365–405 nm); (e) removal of the mask.

biochips with a photomask, which offered great promise for the controlled fabrication of structured assemblies of proteins on surfaces.[17] Carboxylic acid-terminated polyamidoamine (PAMAM) dendrimers were first attached covalently to silicon oxide surfaces and cystamine was subsequently coupled to the surfaces. Subsequently, the reduction of the disulfide was carried out to yield thiol terminated surfaces with dendrimers. Then photochemical surface patterning was realized by coupling a terminal-olefin-functionalized biotin derivative to thiol-functionalized surfaces (Scheme 13.1). Both the irradiation time and the concentration of the biotin derivative influenced the amount of immobilized molecules. The patterned biotin surfaces could be used as templates to generate biomolecular architectures. In addition, the surface patterning method exhibited great potential in the fabrication of biofunctionalized devices used for studying protein–protein interactions. In 2010, the application of a photochemical thiol-ene reaction was successfully realized for the fast, oriented, and covalent immobilization of farnesylated, positively charged K-Ras C terminus (KRT)-bearing proteins directly from expression lysates under mild conditions (Scheme 13.2) without additional isolation, purification or chemical derivatization steps.[19]

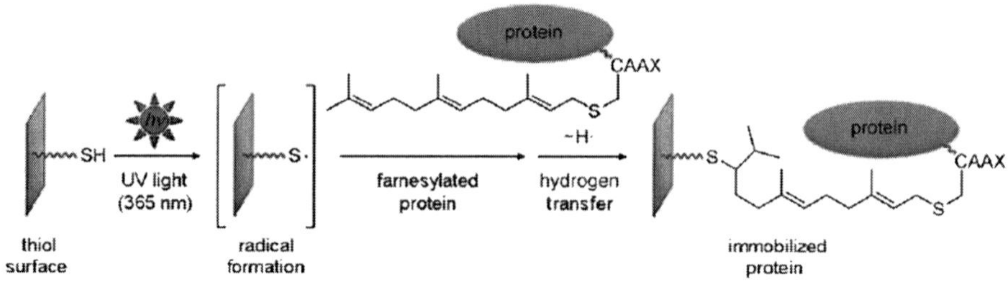


**Scheme 13.2** Thiol-ene photoimmobilization of farnesylated proteins.

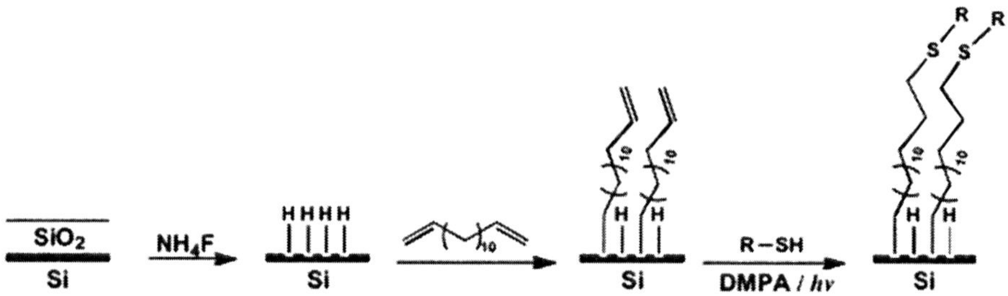


**Scheme 13.3** Functionalization of silicon surfaces using thiol-ene chemistry.

### *13.2.1.2 Surface Patterning*

Functional organic monolayers are essential to the development of sensing devices for the selective binding and detection of target biomolecules. Zuilhof *et al.* utilized a thiol-ene click strategy for the attachment of a functional monolayer onto oxide-free silicon surfaces under mild conditions.[20] H-terminated Si(111) was first reacted with neat 1,13-tetradecadiene. Then the photochemical coupling reaction between alkene-terminated monolayers and thiols was performed in the presence of 2,2-dimethoxy-2-phenylacetophenone (DMPA) as a photoinitiator (365 nm light) (Scheme 13.3). The thiol-ene click reaction was carried out at room temperature under ambient atmosphere and high surface coverages were achieved for a wide variety of functional thiols without $SiO_2$ formation. The efficient and versatile strategy allowed for successful light-induced micropatterning and provides a novel route towards biofunctional electronics. On the other hand, molecular layer deposition (MLD) of organic thin films (with controlled chemical structure) on solid surfaces was developed by Buriak and coworkers using the UV-induced thiol-ene reaction.[21] The controlled assembly of multilayer structures on solid surfaces is important for practical application in organic electronics, chemical-resistant films and biocompatible materials *etc*. As described in the article, aliphatic dienes and dithiols of various chain lengths (also aromatic dithiols) were alternatively photografted onto Si, Ge and Si/$SiO_x$ surfaces one layer at a time in the vertical direction (Scheme 13.4). Linear increases in film thickness with respect to layer number were obtained for short chain molecules ($C \leq 8$). For long chain molecules and aromatic molecules, both ends of the molecule

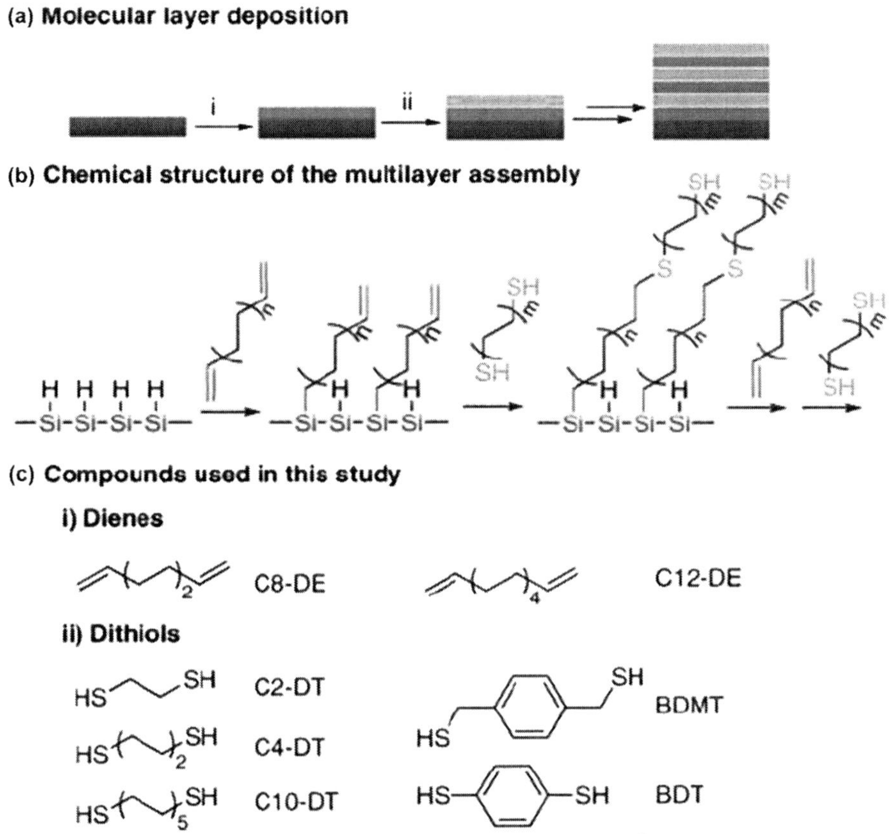


**Scheme 13.4** (a) Schematic diagram of covalent MLD assembly of a thiol-ene multilayer. (b) Chemical structure of the multilayer assembly. (c) Dienes and dithiols used in this work.

reacted with the surface, arresting film growth after several layers owing to the decrease of available functional groups.

Bowman and coworkers also reported the attachment of ultrathin polymer films (<10 nm) on surfaces with thickness and spatial control *via* the combination of thiol-ene photopolymerization and self-assembly techniques.[22] A difunctional ene and a difunctional thiol were photopolymerized on a thiol terminated self-assembled monolayers (SAMs) to form linear polymers, called "polymer brushes". The thickness of the brushes was controlled by changing the thiol-to-ene ratio or the addition of a small amount of a monofunctional ene. Brush growth was spatially controlled using an optical mask alignment system and the patterned surface was formed in regions exposed to light. Higher functionality thiols or enes could be used to graft cross-linked polymer films on the surface. Additionally, Linford *et al.* reported the modification of hydrogen-terminated silicon (111) by the direct attachment of 1,2-polybutadiene (PBd) or partially functionalized PBd, which was pre-functionalized with a variety of

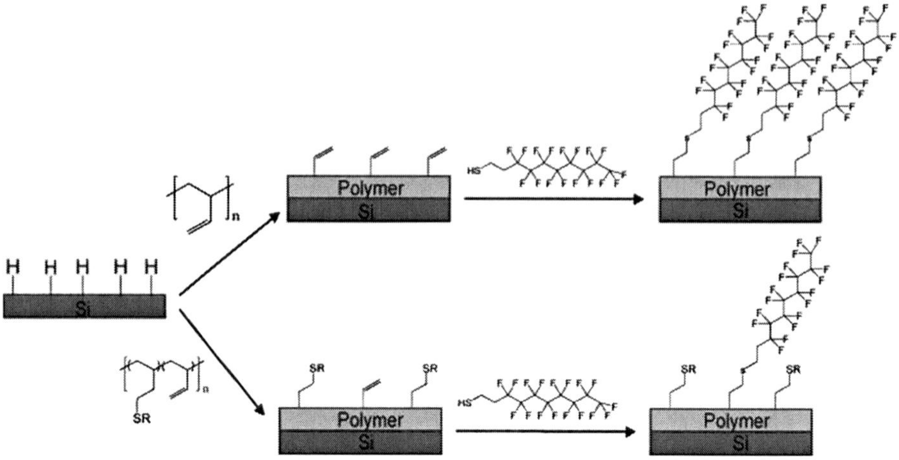


**Scheme 13.5** Light-induced surface attachment of unmodified and partially functionalized 1,2-polybutadiene.

thiols based on photoinduced thiol-ene chemistry. Further functionalization of the attached polymers through unreacted carbon–carbon double bonds was also realized using the thiol-ene click reaction (Scheme 13.5).[23]

On the other hand, engineering and patterning of surfaces is also important in biotechnology and it is of great importance that the functionalization and immobilization process occurs efficiently in a mild biological environment. The absence of transition metal catalysts, the very mild reaction conditions and the ready availability of thiol-functionalized biomolecules with regard to the thiol-ene click chemistry are desirable attributes for biochemical applications. In 2009, the applicability of this coupling chemistry for the direct modification of glass slides with polymeric and bioorganic coatings was realized by Schlaad *et al.* Surface-bound thiyl radicals were used to initiate the polymerization of methacrylic acid and the addition of α-olefins, including allyl-α-D-glucopyranoside, 1*H*,1*H*,2*H*-perfluoro-1-decene and 1,2-polybutadiene, to sulfhydrylated glass surface was successfully performed. The biological recognition test demonstrated that the immobilized glucose units retained their function in the recognition of *Concanavalin A* (Scheme 13.6).[24]

### *13.2.1.3 Modification of Surface with Alternative Properties*

Thiol-ene click reactions can be used as an efficient approach for surface functionalization with tailored properties. Müller *et al.* reported the modification of cross-linked poly(divinylbenzene) microspheres with SH-functionalized poly(*N*-isopropylacrylamide) (pNIPPAm).[25] The pNIPPAm-SH polymers were grafted to the vinyl groups on the surface of poly(divinylbenzene) particles *via* a thiol-ene approach (Scheme 13.7). The pNIPPAm-grafted particles exhibited improved hydrophilicity, *i.e.* the suspension properties of the particles in aqueous environments was enhanced.

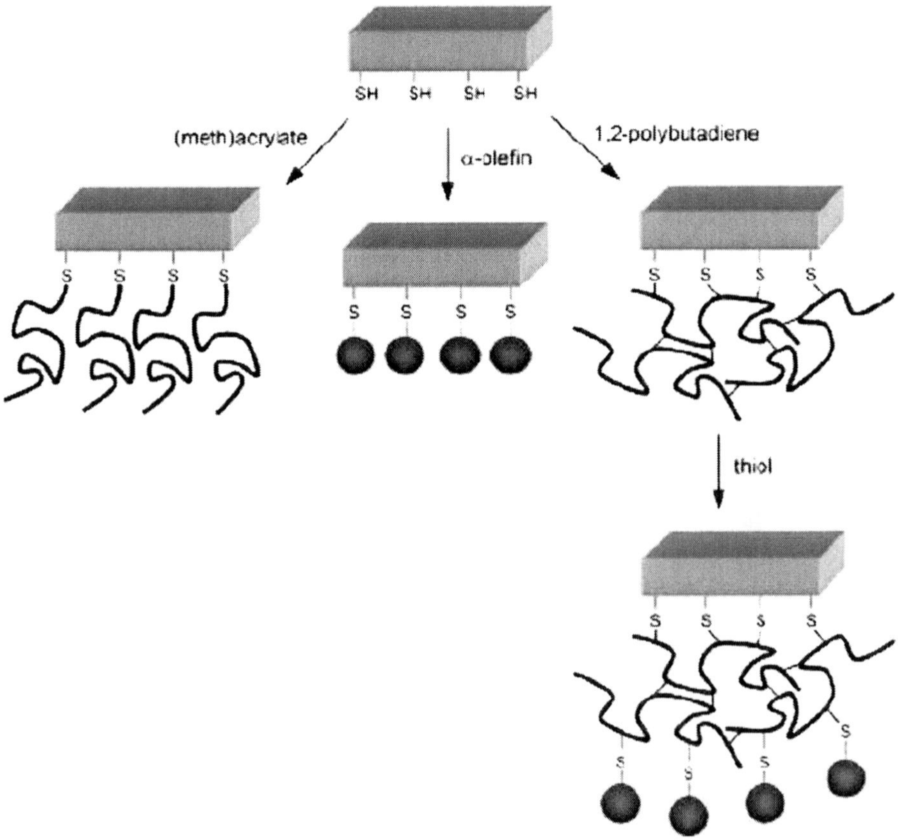


**Scheme 13.6** Functionalization of a sulfhydrylated glass slide *via* thiol-ene photochemistry.

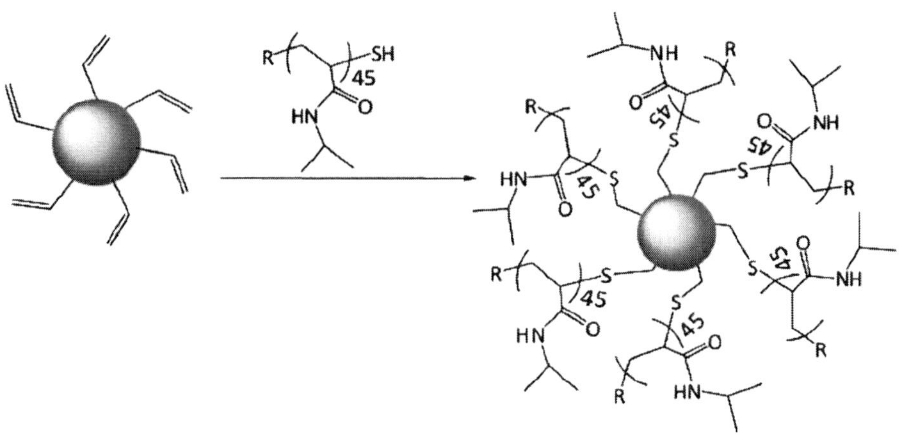

**Scheme 13.7** Thiol-ene modification of pDVB80 microspheres with $pNIPAAm_{45}$ in a one-step approach.

Additionally, photoinitiated thiol-ene click reactions were used to functionalize trialkoxysilanes, which are widely applied to modify the surface properties of materials and devices. Traditional methods for producing functionalized silanes often suffer from limited functional group tolerance, low-to-moderate product yields and tedious product purification, while the thiol-ene click reaction can overcome many of these drawbacks. Garrell *et al.* reported the synthesis of functional trialkoxysilanes by reacting either alkenes with 3-mercaptopropyltrialkoxysilane or thiols with allyltrialkoxysilanes in the presence of a photoinitiator. The click reaction achieved quantitative to near-quantitative yields with high purity. Additionally, the reaction could be scaled-up easily without sacrificing product yield or purity. The tailored silanes were also covalently bound to iron oxide superparamagnetic nanoparticles (SPNs), demonstrating the great potential of the functional trialkoxysilanes as active surface coating agents.[26]

What's more, the thiol-ene click chemistry was demonstrated to be a facile, efficient, and versatile approach for the biofunctionalization of nanoparticles. In 2011, Zuilhof and coworkers presented the preparation of alkene-terminated fluorescent silicon nanoparticles (SiNPs) and their subsequent functionalization *via* thiol-ene click chemistry.[27] Bromine-terminated silicon nanoparticles (SiNPs) were first synthesized by the oxidation of $Mg_2Si$ with bromine. Capping the resulting SiNPs with 3-butenylmagnesium bromide yielded butylene-terminated SiNPs with core size of 2.4 ($\pm$0.5) nm. The thiol-ene click reaction was performed for the modification of the SiNPs to obtain a variety of functional SiNPs, such as those terminated with thioacetic acid, mercaptoethanol, ethylene glycol, and carboxylic acid. The functionalization did not change the optical properties of the SiNPs. Furthermore, the functionalization of carboxylic acid-terminated SiNPs with amine-functional DNA strands was successfully achieved.

### 13.2.2 Surface Modification Based on Thiol-Yne Click Reaction

The thiol-yne radical click reaction, which is similar to thiol-ene radical initiated reaction, involves the consecutive addition of two thiol functional groups onto the alkyne functional group. Surface patterning based on the thiol-yne coupling procedure has also been successfully realized. Patton *et al.* utilized thiol-yne chemistry as a modular platform for surface engineering and fabricated highly functional, patterned, and multicomponent polymer brush surfaces under ambient conditions from a single substrate precursor (Scheme 13.8).[28] Silicon substrates were first functionalized with a chlorosilane derivative of 2-hydroxy-4′-(2-hydroxyethoxy)-2-methylpropiophenone photoinitiator. Then the UV-initiated polymerization of trimethylsilaneprotected propargyl methacrylate (PgMA-$SiMe_3$) was carried out and the deprotection of the p(PgMA-$SiMe_3$) was performed to obtain the yne-functionalized polymer brush as a precursor. Using the photopatterning technique, sequential and area-selective thiol-yne brush modifications were easily realized to afford micropatterned, multicomponent surfaces. What's more, the practicality of

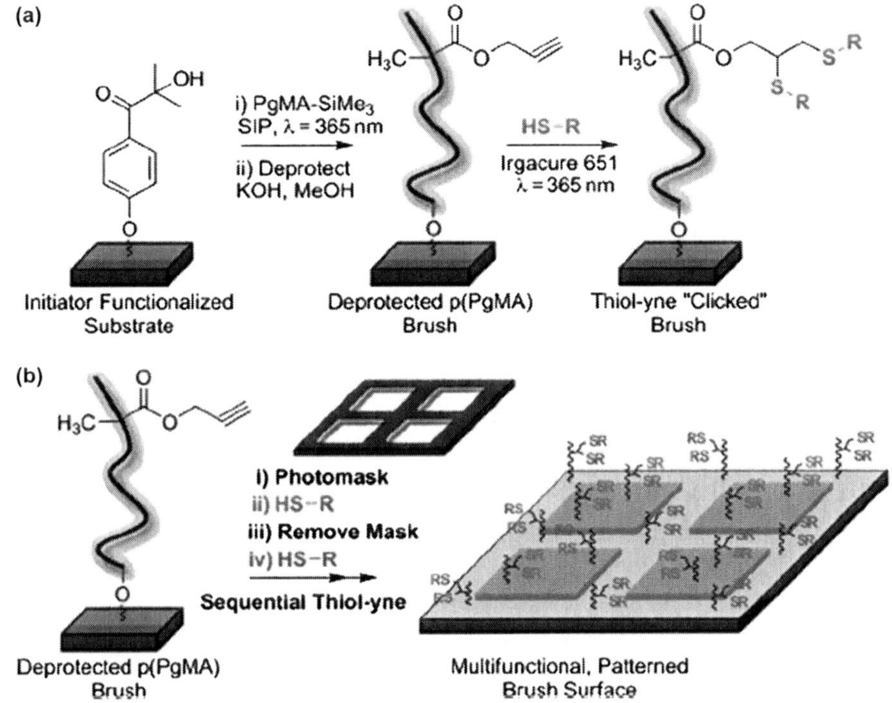


**Scheme 13.8** (a) Schematic procedure of surface-initiated photopolymerization of TMS-protected propargyl methacrylate, deprotection, and subsequent thiol-yne functionalization. (b) Schematic procedure for photopatterning "yne"-containing polymer brush surfaces with sequential thiol-yne reactions.

thiol-yne surface modification using outdoor, ambient-air reactions with sunlight as a radical source was also demonstrated.

In addition, the thiol-yne click reaction was also used in the modification of biomaterials. In 2011, Xu and coworkers first reported the glycosylation of polymer membranes for specific affinity adsorption of lectin receptors (Scheme 13.9).[29] The microporous polypropylene membrane (MPPM) was first functionalized with acrylic acid (AA). Then the polyAA-grafted membrane surface was activated with EDC/NHS and reacted with propargylamine to achieve the alkyne-modified MPPM. Subsequently, the photo-induced thiol-yne click reaction between the yne-modification membrane and 2,3,4,6-tetra-*O*-acetyl-β-D-glucopyranosyl thiol was carried out and the acetyl groups of glucose pentaacetate were deprotected to construct the glycosylated surface. The glycosylated polymer membrane displayed excellent hydrophilicity. In contrast to the traditional glycosylated modification based on the azide-yne click reaction, which showed a non-specific adsorption of proteins owing to the presence of triazoles groups, the surface glycosylation based on thiol-yne click chemistry did not exhibit specific interaction between the thiolether groups and proteins. What's more, the metal-free characteristics of the thiol-ene click reaction have great advantage for application in the fields of biomaterials and biomedicines.

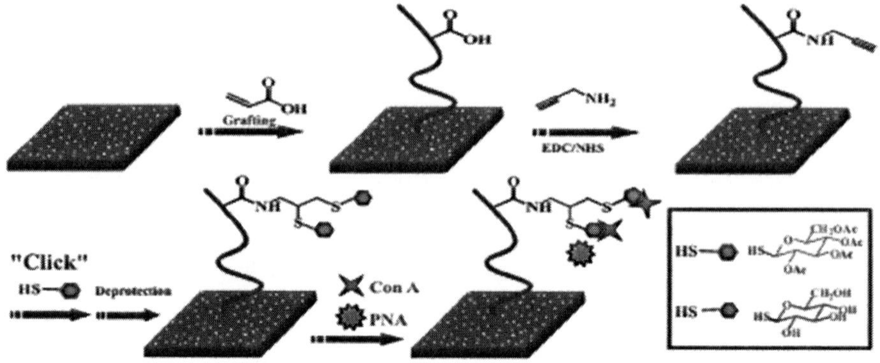


**Scheme 13.9** Schematic illustration for the construction of glycosylated surface by thiol-yne click chemistry for lectin recognition.

## 13.3 Synthesis of Chromatographic Stationary Phases Based on Radical-Mediated Thiol-Ene/Yne Click Reaction

Silica-based chromatographic stationary phases have been extensively utilized to take advantage of their high efficiency and mechanical strength. The synthesis of silica-based stationary phases with different selectivity is performed by the covalent bonding of a variety of functional groups onto the surface of silica supports. Traditional conjugation methods that have been widely used involve the nucleophilic and electrophilic reactions between the reactive groups (such as $-NH_2$, –CNO, –COCl, –COOH) on the supports and the functionalized molecules.[30–33] However, these reactions have low selectivity and efficiency. The poor selectivity of the immobilization method may cause some side reactions and result in reducing or even losing the function of the stationary phase. In addition, the reactive groups on the silica supports may not be fully converted due to the inefficient immobilization process, which will decrease the bonding density of the functional groups. The remaining reactive groups on the silica support will exhibit different retention mechanisms from the stationary phases and affect the separation selectivity. Thus, the development of a robust, reliable immobilization method with high efficiency and selectivity under mild conditions for the preparation of functionalized HPLC packings is valuable.

Fortunately, the application of "click chemistry" in the synthesis of chromatographic separation materials has been demonstrated to be a facile and efficient strategy. The implementation of classical CuAAC click reaction for the immobilization of functional molecules was first introduced in 2006.[5] Since then, various hydrophilic interaction liquid chromatography (HILIC) stationary phases have been successfully developed based on the CuAAC click reaction.[6,34,35] Nonetheless, the metal ions (metal catalyst used in the click reaction) remaining in the stationary phases is detrimental to the chromatographic separation, in particular for compounds which can form chelates with metal ions. Consequently, the development of metal-free click reactions has received great attention for this application. Among the introduced metal-free click chemistries, thiol-ene/yne radical click reactions have proved to be good candidates as a modification approach. The preparation of a variety of chromatographic stationary phases for specific applications, including chiral (ionic exchange) stationary phases, reversed-phase stationary phases, mixed-mode stationary phases and HILIC materials has been efficiently and successfully realized using thiol-ene/yne radical initiated addition.

## 13.3.1 Preparation of Chiral (Ionic Exchange) Stationary Phases

The design and synthesis of chiral stationary phases for the separation of chiral (ionic) compounds is significant, especially in drug development as individual enantiomers of chiral drugs exhibit different biological activities and pharmacological profiles.[36] Quinine and quinidine (cinchona alkaloids) have been used as a cheap natural source for stereodiscriminating auxiliaries. The arrangement of the five stereogenic centers within the molecules provides a basis for effective chiral recognition. Salvadori *et al.* developed a series of cinchona alkaloids-based chiral stationary phases (as shown in Figure 13.1) which were covalently bonded to a silica support using a thiol-ene reaction in the presence of (α,α′-azobisisobutyronitrile) AIBN as radical initiator.[37–39] In 1985, the immobilization of quinine *via* the olefinic group onto mercaptopropyl silanized silica was performed for the first time. Subsequently, both quinidine and *N*-methyl quinium iodide were similarly bonded onto thiol-modified silica to obtain the chiral stationary phases. The resulting chiral columns were utilized in the chromatographic resolution of racemic arylalkylcarbinols, binaphtol derivatives, amides and other substances of pharmaceutical interest under normal phase conditions.

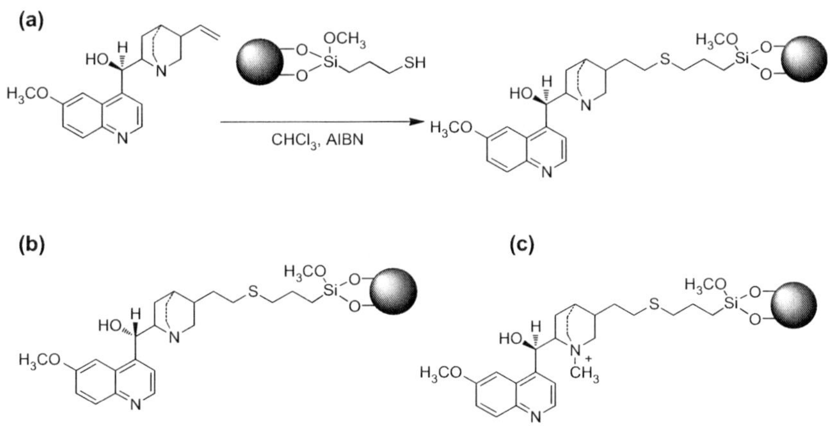


**Figure 13.1** (a) The preparation of quinine-based chiral stationary phase; (b) structure of quinidine-based chiral stationary phase; (c) structure of *N*-methyl quinium-based chiral stationary phase.

Lindner *et al.* reported a series of chiral ion exchange selectors from the immobilization of quinine and quinidine carbamates onto porous silica *via* thiol-ene radical reaction as weak chiral anion exchangers (Figure 13.2).[40] The resulting stationary phases were successfully utilized in the separation of the enantiomers of acidic analytes, such as chiral aryl-, aryloxy-, arylthiocarboxylic acids, N-derivative amino acids *etc*. The presence of long range electrostatic interactions is considered to be suitable to guide the analyte in its orientation towards the binding site of the chiral selector.[36] In addition, *tert*-butyl carbamoylated quinine and *N*-methyl *tert*-butyl carbamoylated quinine-based chiral anion exchangers (Figure 13.3) were also covalently bonded onto thiol-modified silica *via* radical addition,[41] similar to the bonding of unmodified native quinine onto mercaptopropyl-modified silica reported by Salvadori and coworkers.[38] The enantioseparation of acid chiral compounds was compared to the chiral selectors under buffered hydro-organic mobile phase conditions. The results indicated that the introduction of the carbamoyl functionality at the

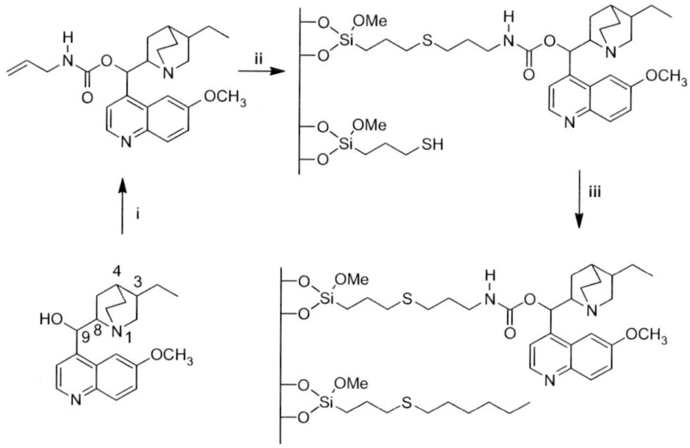


**Figure 13.2** The synthesis of the quinine and quinidine carbamates stationary phases (CSPs): (i) allyl isocyanate, dibutyl tin dilaurate, toluene; (ii) 3-mercaptopropyl silanized silica, AIBN, chloroform; (iii) 1-hexene, AIBN, chloroform.

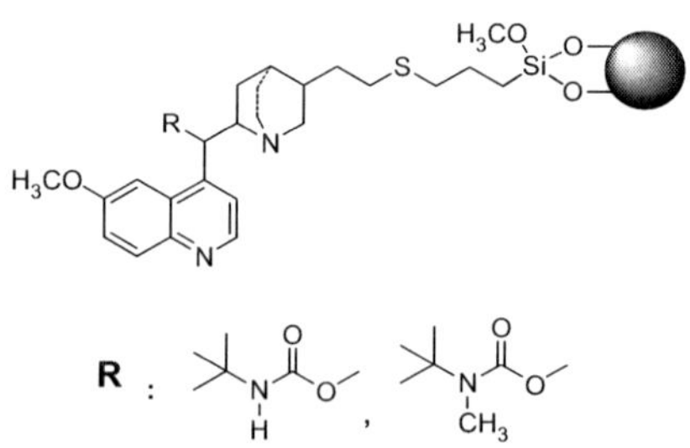


**Figure 13.3** Structures of chiral stationary phases with *tert*-butyl carbamoylated quinine or *N*-methyl-*tert*-butyl carbamoylated quinine as selectors.

secondary hydroxyl group at $C_9$ of quinine provided favorable intermolecular interactions (hydrogen bonding) with chiral analytes and greatly improved the stereorecognition capability and enantioselectivity, especially for amide, carbamate and sulfonamide derivatives of amino acids.

Lindner *et al.* also designed a variety of chiral cation exchangers. In 2002, two chiral cation-exchange stationary phases[36] were prepared by the immobilization of *N*-(3,5-dichlorobenzoyl)-*O*-allyl-tyrosine (weak chiral cation exchanger) and *N*-(4-allyloxy-3,5-dichlorobenzoyl)-1-amino-3-methylbutane phosphonic acid (strong chiral cation exchanger) on mercaptopropyl silica *via* thiol-ene reaction (Figure 13.4). The chiral selectors were developed by the combination of the "reciprocity principle of chiral recognition" and nonaqueous ion-pair capillary electrophoresis (CE). The resulting materials were packed into capillary columns and applied to the separation of chiral bases by nonaqueous capillary electrochromatography (CEC). The strong chiral cation exchanger displayed better enantioselectivity than the weak chiral cation exchanger. A broad spectrum of different chiral bases could be successfully resolved on the strong cation exchanger. In 2007, Hoffmann *et al.* further designed and developed novel brush-type chiral strong cation-exchange materials (Figure 13.5) *via* thiol-ene radical reaction.[42] The chiral stationary

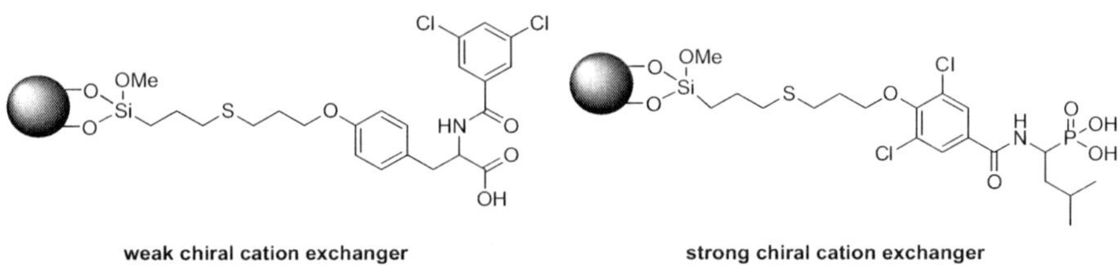


**Figure 13.4** Structures of cation-exchange type chiral stationary phases.

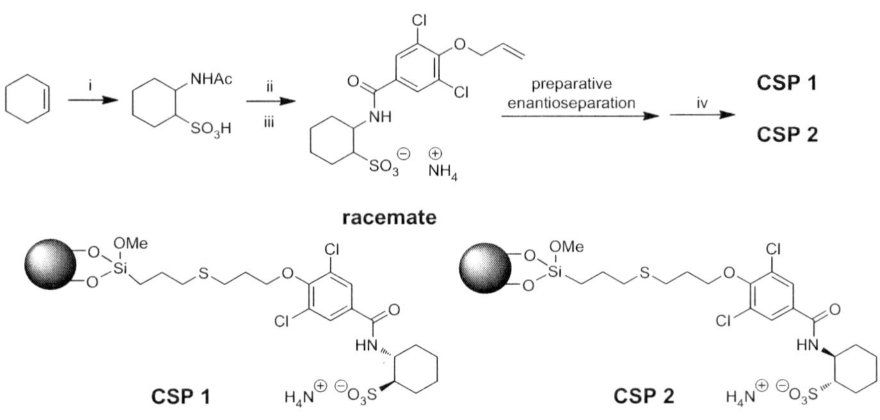


**Figure 13.5** Synthesis of chiral strong cation-exchange materials CSP **1** and **2**. (i) $SO_3$-DMF complex, $CH_3CN$, then $H_2O$ (ii) 10% aqueous HCl (iii) 4-allyloxy-3,5-dichlorobenzoic acid, DIEA, EEDQ, DMF (iv) thiol-modified silica gel, AIBN, MeOH.

phases (CSPs) were based on synthetic low molecular mass selectors (SOs) containing a β-aminocyclohexanesulfonic acid moiety. The SOs were prepared by non-enantioselective synthesis followed by preparative enantioseparation. The resulting SOs were respectively covalently bonded onto thiol-modified silica (AIBN as radical initiator) yielding CSPs with opposite configurations, CSPs **1** and **2**. The enantioselective separation of chiral amines (mainly amino-alcohol type drug molecules) was carried out under polar organic mobile phase conditions. All the analytes were successfully baseline separated with excellent peak efficiencies.

Considering the limited applicability of the cinchona alkaloid-based chiral weak anion exchangers and aminosulfonic acid-based chiral strong cation exchangers, which were restricted to oppositely charged solutes, Hoffmann and coworkers further designed and developed novel low molecular mass chiral zwitterionic selectors (Figure 13.6) based on the combinatorial synthesis approach.[43] Both anion-exchange and cation-exchange entities derived from weak anion-exchange (WAX) and strong cation-exchange (SCX) SOs were combined in the zwitterionic SOs and bonded onto thiol-modified silica *via* thiol-ene radical initiated addition. The resulting zwitterionic selectors exhibited similar enantioseparation capabilities toward chiral acids and chiral bases as the parent WAX and SCX chiral stationary phases. The zwitterionic CSPs provided excellent separation ability for the chiral amphoteric compounds including chiral amino acids and dipeptides under polar organic mobile-phase conditions. The results indicated that the individual charge centers of the new zwitterionic ion exchangers acted in a synergistic way through double ion-pairing (ion-exchange)-mediated processes for the resolution of chiral zwitterionic analytes. Additionally, a variety of quinine-based zwitterionic chiral stationary phases[44] were designed and prepared according to the same method by Hoffmann *et al.* The systematic investigation of the enantiomer recognition and retention mechanism was carried out in anion-, cation- and zwitterionic-exchange modes.

In fact, the thiol radical click reaction was applied not only in the synthesis of chiral ionic exchange stationary phases but also the preparation of the chiral stationary phases. In 2010, Li *et al.* reported the synthesis of uniform 0.66 and 0.81 μm submicron mesoporous silica spheres, which were modified with

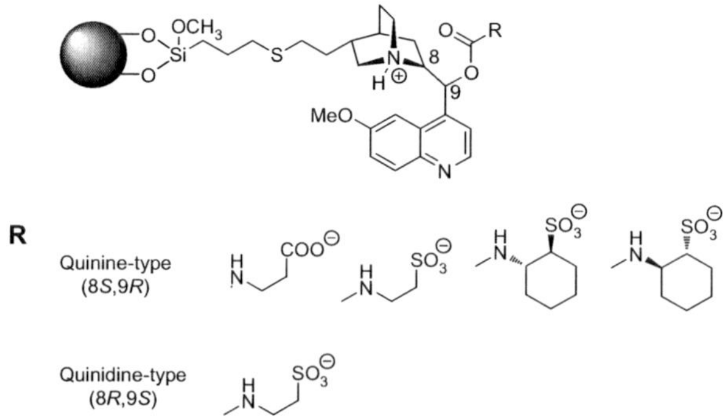


**Figure 13.6** Structures of zwitterionic chiral stationary phases.

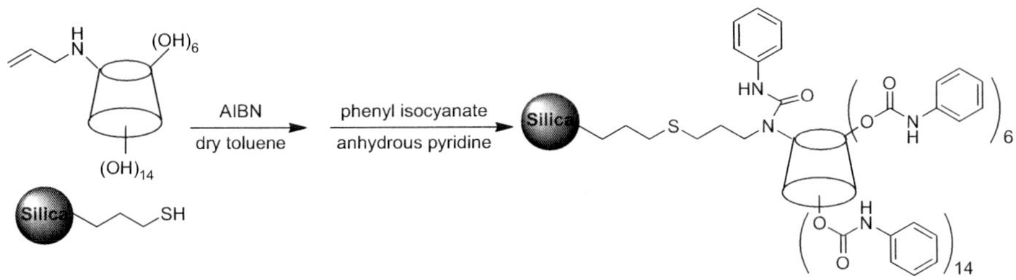


**Figure 13.7** Synthesis of phenylcarbamoylated-β-CD modified chiral stationary phase.

mono-6-phenylcarbamoylated β-CD *via* thiol-ene radical addition. The submicron chiral stationary phases were packed into capillary columns. The enantioseparation of halogen aryl alcohols and β-blockers propranolol and pindolol were efficiently realized in packed bed CEC.[45] Similarly, Ai *et al.* further developed the modification of mercaptopropyl-based sub-1-micron size (0.6–0.9 μm) mesoporous silica particles with mono-($6^A$-*N*-allyamino-$6^A$-deoxy)-β-cyclodextrin (CD) under the initiation of AIBN.[46] The CD-based chiral stationary phase (Figure 13.7) exhibited good permeability and good enantioselectivity for six neutral and basic enantiomers.

## 13.3.2 Preparation of Reversed-phase and Mixed-mode Stationary Phases

Reversed-phase liquid chromatography (RPLC) has been widely used due to its universality, high separation efficiency and good reproducibility. In 2004, Lindner *et al.* introduced the novel sulfide and thiol-group incorporated RP phases (S-RP), which were synthesized based on the radical induced thiol-ene reaction.[47] The developed hybrid RP phases could provide both hydrophobic interaction and sulfur–aromatic interaction. Comparing with conventional monomerical *n*-alkyl type RP phases, the S-RP phases not only enhanced the retention but also provided pronounced planar recognition ability especially for solutes with extended π-electron systems. Three naphthalimide-based ligands, *N*-(10-undecenyl)-1,8-naphthalene dicarboxylic imide, *N*-(2-propenyl)-1,8-naphthalene dicarboxylic imide and *N*,*N'*-bis (2-propenyl)-1,4,5,8-naphthalenetetracarboxylic diimide, were also introduced onto thiol-modified silica *via* a thermally-induced thiol-ene reaction as hybrid reversed-phase materials.[48] The electron donor/acceptor moieties within the novel stationary phases which provided π–π interactions efficiently improved the shape selectivity of the hybrid π-RP phases for aromatic analytes.

Although RPLC has extensive applicability, the separation selectivity is still limited, especially for practical applications. Mixed-mode chromatography has been developed as an alternative approach for complex samples. The application of thiol-ene click reactions in the preparation of mixed-mode stationary phases has also been achieved.[49–51] A mixed-mode reversed-phase/weak anion-exchange (RP/WAX) type stationary phase (Figure 13.8) was

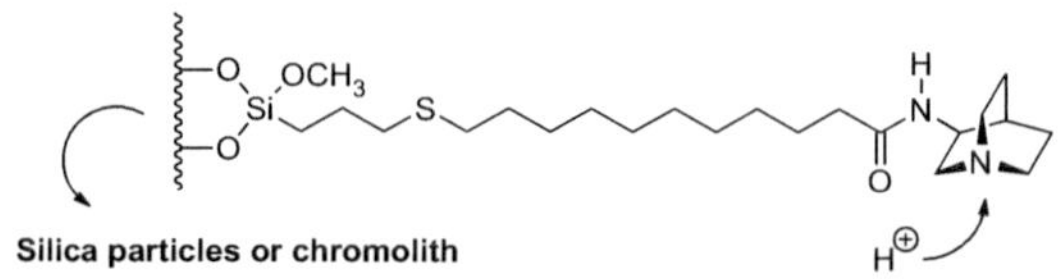


**Figure 13.8** Structures of RP/WAX mixed-mode stationary phases.

developed by the immobilization of *N*-(10-undecenoyl)-3-aminoquinuclidine selector onto thiol-modified silica particles. The RP/WAX stationary phase, which contains a lipophilic alkyl chain and a weak anion-exchange site, incorporated both reversed-phase and ion-exchange retention mechanisms. It was successfully utilized for synthetic peptide (N- and C-terminally protected tetrapeptide *N*-acetyl-Ile-Glu-Gly-Arg-*p*-nitroanilide) separation and purification. Comparing to the standard gradient elution RP-purification protocol, better selectivity was achieved and major impurities could all be removed by a single chromatographic step using the RP/WAX stationary phase. Additionally, the sample loading capacity was significantly improved, indicating the great potential of RP/WAX mixed-mode separation system in preparative scale application. In addition, the introduction of the RP/WAX stationary phase that combined orthogonal separation principles in liquid chromatography-tandem mass spectrometry proved to be an alternative separation strategy for the analysis of endogenous metabolites in fields like metabolomics and biomarker studies.[51–53] Nogueira and coworkers further reported the synthesis of a monolithic RP/WAX column based on the same coupling procedure in 2006. The immobilization of *N*-(10-undecenoyl)-3-aminoquinuclidine onto thiol-modified monolithic silica columns was realized *in situ* using AIBN as radical initiator using a flow-through mode. The RP/WAX silica-based monolith column (Figure 13.8) also exhibited reversed-phase and anion-exchange chromatography principles. The monolithic RP/WAX column displayed unique selectivity for peptide separation, especially for hydrophilic acid peptides. The upscaling of RP/WAX monolith column from the analytical to semipreparative dimension was also successfully performed.

### 13.3.3 Preparation of Hydrophilic Interaction Liquid Chromatography (HILIC) Stationary Phases

As an alternative to reversed-phase liquid chromatography which is an indispensable and versatile technique in chromatographic separation, hydrophilic interaction chromatography (HILIC)[54] plays an important role in the separation of highly polar and hydrophilic compounds, such as oligosaccharides, peptides, nucleosides, polar pharmaceuticals, *etc.* A substantial number of separation materials[6,55,56] have been designed specially for HILIC. Based on the copper-free thiol-ene click reaction, Lindner and coworkers reported the preparation of four nonionic polar stationary phases in 2008.[57] 2-Mercaptoethanol and 1-thioglycerol were first anchored onto vinyl-modified silica (ME and TG packings) and the on-phase oxidation of the embedded sulfide groups on ME and TG packings using hydrogen peroxide in aqueous

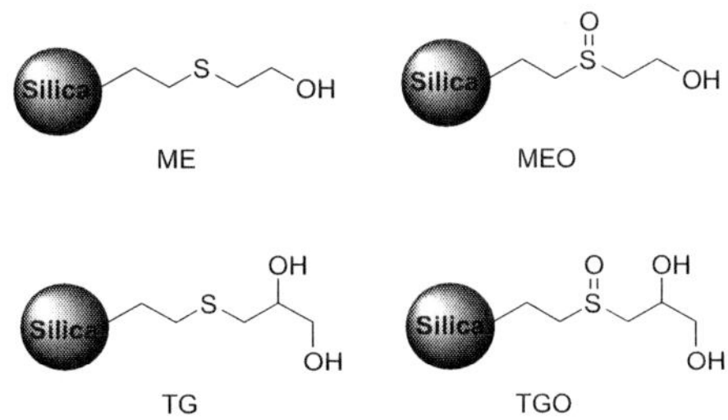


**Figure 13.9** The structures of nonionic polar stationary phases.

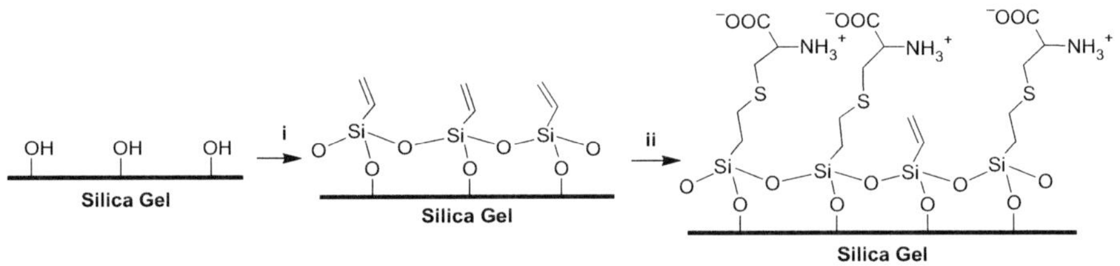


**Figure 13.10** The synthesis of Click TE-Cys stationary phases. (i) vinyl trichlorosilane, toluene, RT (ii) cysteine, AIBN, MeOH/$H_2O = 1:2$, 65 °C, nitrogen atmosphere.

medium yielded the sulfoxide group embedded stationary phases (MEO and TGO packings) (Figure 13.9). The chromatographic evaluation demonstrated that both the conversion of embedded sulfide groups to sulfoxides and/or the introduction of additional hydroxyl groups on the chromatographic ligand could improve the retention and selectivity for the separation of polar analytes. The results indicated that the retention was influenced by both partitioning and adsorption mechanisms in HILIC.

A novel zwitterionic hydrophilic interaction chromatography (HILIC) stationary phase (Figure 13.10), with the oppositely charged groups parallel to the surface of the stationary phase, was also developed by taking advantage of the thiol-ene reaction.[58,59] The utilization of thiol-ene click reactions can successfully avoid having residual metal ions in the stationary phases. The radical induced click reaction between vinyl silica and the naturally available reagent cysteine, which possesses the thiol group, was carried out using AIBN as thermal initiator. The selective reaction between the thiol group and the vinyl group was helpful for the conservation of the amino and carboxyl groups in cysteine, thus preserving the zwitterionic characteristics of the stationary phase. The developed cysteine-based HILIC stationary phase (Click TE-Cys) exhibited high hydrophilicity and excellent separation efficiency. Additionally, the Click TE-Cys stationary phase displayed great potential in the separation of peptides, oligosaccharides (Figure 13.11) and in the enrichment of glycopeptides (Figure 13.12).

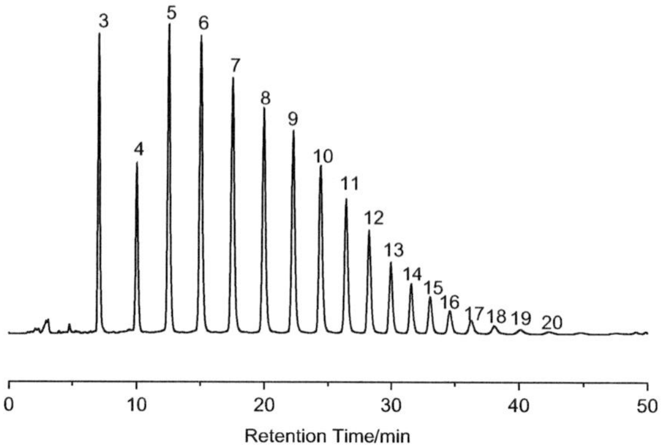


**Figure 13.11** Separation of fructo-oligosaccharides on Click TE-Cys column. The number on the peaks is the DP of oligosaccharides.

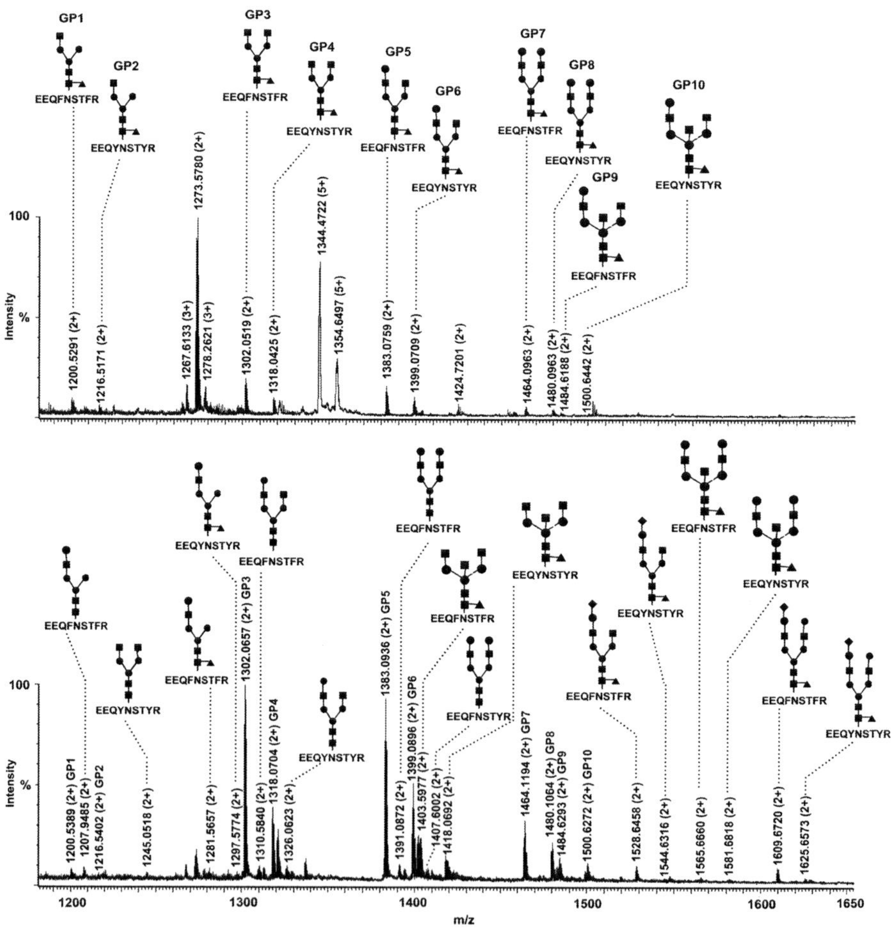


**Figure 13.12** Nano-ESI-MS of human IgG digest before and after the enrichment on Click TE-Cys material.

Although the mechanism of the thiol-yne radical click reaction is similar to the thiol-ene radical click reaction, the application of thiol-yne click chemistry for the synthesis of chromatographic stationary phases has not been reported. The implementation of thiol-yne click reaction between yne modified silica and cysteine was realized in our group (not reported), and further characterization and evaluation is under investigation.

Ihara *et al.* recently designed and synthesized a new surface-confined ionic liquid stationary phase based on the thiol-ene reaction.[60] Ionic liquid monomer pairs were copolymerized on mercaptopropyl modified silica through a surface radical chain-transfer reaction. Different properties of stationary phases for RPLC or HILIC applications could be obtained *via* the selection of appropriate monomer pairs, and the multifunctional hybrid materials exhibited excellent selectivity and stability.

## 13.4 Conclusion

The free radical addition of thiols across alkene/alkyne bonds has been emerging as a fascinating coupling approach in the modification and functionalization of surfaces with tailored properties, taking advantage of its superior characteristics of high efficiency, quantitative yields under ambient conditions, orthogonality, regioselectivity, insensitivity to oxygen and available of variety of starting compounds. There continues to be a tremendous potential for the application of radical-mediated thiol-ene/yne click chemistry in a wide variety of areas. The reaction efficiency of thiol-ene click chemistry on solid surfaces is not as high as that in solution and still needs further optimization. Additionally, the combination of vinyl functional small molecules onto thiol-modified silica suffers from the self-polymerization of the vinyl groups when the reaction is performed in solution. Thus, further investigation and implementation of radical-mediated thiol/yne click reaction will be of particular value.

## References

1. H. C. Kolb, M. G. Finn and K. B. Sharpless, *Angew. Chem., Int. Ed. Engl.*, 2001, **40**, 2004–2021.
2. M. van Dijk, D. T. S. Rijkers, R. M. J. Liskamp, C. F. van Nostrum and W. E. Hennink, *Bioconjugate Chem.*, 2009, **20**, 2001–2016.
3. F. Santoyo-Gonzalez and F. Hernandez-Mateo, *Chem. Soc. Rev.*, 2009, **38**, 3449–3462.
4. T. Lummerstorfer and H. Hoffmann, *J. Phys. Chem. B*, 2004, **108**, 3963–3966.
5. Z. M. Guo, A. W. Lei, X. M. Liang and Q. Xu, *Chem. Commun.*, 2006, 4512–4514.
6. Z. M. Guo, A. W. Lei, Y. P. Zhang, Q. Xu, X. Y. Xue, F. F. Zhang and X. M. Liang, *Chem. Commun.*, 2007, 2491–2493.
7. Y. Wang, Y. Xiao, T. Thatt and S.-C. Ng, *Tetrahedron Lett.*, 2008, **49**, 5190–5191.

8. L. Moni, A. Ciogli, I. D'Acquarica, A. Dondoni, F. Gasparrini and A. Marra, *Chem.–Eur. J.*, 2010, **16**, 5712–5722.
9. C. R. Becer, R. Hoogenboom and U. S. Schubert, *Angew. Chem., Int. Ed.*, 2009, **48**, 4900–4908.
10. A. Gress, A. Volkel and H. Schlaad, *Macromolecules*, 2007, **40**, 7928–7933.
11. C. E. Hoyle and C. N. Bowman, *Angew. Chem., Int. Ed.*, 2010, **49**, 1540–1573.
12. C. E. Hoyle, A. B. Lowe and C. N. Bowman, *Chem. Soc. Rev.*, 2010, **39**, 1355–1387.
13. A. B. Lowe, *Polym. Chem.*, 2010, **1**, 17–36.
14. R. Hoogenboom, *Angew. Chem., Int. Ed.*, 2010, **49**, 3415–3417.
15. A. B. Lowe, C. E. Hoyle and C. N. Bowman, *J. Mater. Chem.*, 2010, **20**, 4745–4750.
16. B. D. Fairbanks, T. F. Scott, C. J. Kloxin, K. S. Anseth and C. N. Bowman, *Macromolecules*, 2009, **42**, 211–217.
17. P. Jonkheijm, D. Weinrich, M. Koehn, H. Engelkamp, P. C. M. Christianen, J. Kuhlmann, J. C. Maan, D. Nuesse, H. Schroeder, R. Wacker, R. Breinbauer, C. M. Niemeyer and H. Waldmann, *Angew. Chem., Int. Ed.*, 2008, **47**, 4421–4424.
18. P. C. Lin, D. Weinrich and H. Waldmann, *Macromol. Chem. Phys.*, 2010, **211**, 136–144.
19. D. Weinrich, P. C. Lin, P. Jonkheijm, U. T. T. Nguyen, H. Schroder, C. M. Niemeyer, K. Alexandrov, R. Goody and H. Waldmann, *Angew. Chem., Int. Ed.*, 2010, **49**, 1252–1257.
20. M. A. C. Campos, J. M. J. Paulusse and H. Zuilhof, *Chem. Commun.*, 2010, **46**, 5512–5514.
21. Y. H. Li, D. Wang and J. M. Buriak, *Langmuir*, 2010, **26**, 1232–1238.
22. V. S. Khire, A. W. Harant, A. W. Watkins, K. S. Anseth and C. N. Bowman, *Macromolecules*, 2006, **39**, 5081–5086.
23. T. D. Wickard, E. Nelsen, N. Madaan, N. ten Brummelhuis, C. Diehl, H. Schlaad, R. C. Davis and M. R. Linford, *Langmuir*, 2010, **26**, 1923–1928.
24. A. Bertin and H. Schlaad, *Chem. Mater.*, 2009, **21**, 5698–5700.
25. A. S. Goldmann, A. Walther, L. Nebhani, R. Joso, D. Ernst, K. Loos, C. Barner-Kowollik, L. Barner and A. H. E. Muller, *Macromolecules*, 2009, **42**, 3707–3714.
26. A. K. Tucker-Schwartz, R. A. Farrell and R. L. Garrell, *J. Am. Chem. Soc.*, 2011, **133**, 11026–11029.
27. L. Ruizendaal, S. P. Pujari, V. Gevaerts, J. M. J. Paulusse and H. Zuilhof, *Chem.–Asian J.*, 2011, **6**, 2776–2786.
28. R. M. Hensarling, V. A. Doughty, J. W. Chan and D. L. Patton, *J. Am. Chem. Soc.*, 2009, **131**, 14673–14675.
29. C. Wang, P. F. Ren, X. J. Huang, J. A. Wu and Z. K. Xu, *Chem. Commun.*, 2011, **47**, 3930–3932.
30. L. F. Zhang, Y. C. Wong, L. Chen, C. B. Ching and S. C. Ng, *Tetrahedron Lett.*, 1999, **40**, 1815–1818.

31. S. C. Ng, L. Chen, L. F. Zhang and C. B. Ching, *Tetrahedron Lett.*, 2002, **43**, 677–681.
32. X. M. Chen, Y. Q. Liu, F. Qin, L. Kong and H. F. Zou, *J. Chromatogr., A*, 2003, **1010**, 185–194.
33. L. Chen, L. F. Zhang, C. B. Ching and S. C. Ng, *J. Chromatogr., A*, 2002, **950**, 65–74.
34. Z. M. Guo, Y. Jin, T. Liang, Y. F. Liu, Q. Xu, X. M. Liang and A. W. Lei, *J. Chromatogr., A*, 2009, **1216**, 257–263.
35. Z. M. Guo, Y. F. Liu, J. Y. Xu, Q. Xu, X. Y. Xue, F. F. Zhang, Y. X. Ke, X. M. Liang and A. W. Lei, *J. Chromatogr., A*, 2008, **1191**, 78–82.
36. E. Tobler, M. Lammerhofer, F. Wuggenig, F. Hammerschmidt and W. Lindner, *Electrophoresis*, 2002, **23**, 462–476.
37. C. Rosini, P. Altemura, D. Pini, C. Bertucci, G. Zullino and P. Salvadori, *J. Chromatogr.*, 1985, **348**, 79–87.
38. C. Rosini, C. Bertucci, D. Pini, P. Altemura and P. Salvadori, *Tetrahedron Lett.*, 1985, **26**, 3361–3364.
39. C. Rosini, C. Bertucci, D. Pini, P. Altemura and P. Salvadori, *Chromatographia*, 1987, **24**, 671–676.
40. M. Lammerhofer and W. Lindner, *J. Chromatogr., A*, 1996, **741**, 33–48.
41. A. Mandl, L. Nicoletti, M. Lammerhofer and W. Lindner, *J. Chromatogr., A*, 1999, **858**, 1–11.
42. C. V. Hoffmann, M. Laemmerhofer and W. Lindner, *J. Chromatogr., A*, 2007, **1161**, 242–251.
43. C. V. Hoffmann, R. Pell, M. Lammerhofer and W. Lindner, *Anal. Chem.*, 2008, **80**, 8780–8789.
44. C. V. Hoffmann, R. Reischl, N. M. Maier, M. Lammerhofer and W. Lindner, *J. Chromatogr., A*, 2009, **1216**, 1147–1156.
45. L. S. Li, Y. Wang, D. J. Young, S. C. Ng and T. T. Y. Tan, *Electrophoresis*, 2010, **31**, 378–387.
46. F. Ai, L. S. Li, S. C. Ng and T. T. Y. Tan, *J. Chromatogr., A*, 2010, **1217**, 7502–7506.
47. J. Horak and W. Lindner, *J. Chromatogr., A*, 2004, **1043**, 177–194.
48. J. Horak, N. M. Maier and W. Lindner, *J. Chromatogr., A*, 2004, **1045**, 43–58.
49. R. Nogueira, M. Lammerhofer and W. Lindner, *J. Chromatogr., A*, 2005, **1089**, 158–169.
50. R. Nogueira, D. Lubda, A. Leitner, W. Bicker, N. M. Maier, M. Lammerhofer and W. Lindner, *J. Sep. Sci.*, 2006, **29**, 966–978.
51. W. Bicker, M. Lammerhofer and W. Lindner, *Anal Bioanal Chem*, 2008, **390**, 263–266.
52. W. Bicker, M. Lammerhofer and W. Lindner, *J. Chromatogr., B*, 2005, **822**, 160–169.
53. W. Bicker, M. Lammerhofer, T. Keller, R. Schuhmacher, R. Krska and W. Lindner, *Anal. Chem.*, 2006, **78**, 5884–5892.
54. A. J. Alpert, *J. Chromatogr.*, 1990, **499**, 177–196.
55. P. Hemström and K. Irgum, *J. Sep. Sci.*, 2006, **29**, 1784–1821.

56. T. Ikegami, K. Tomomatsu, H. Takubo, K. Horie and N. Tanaka, *J. Chromatogr., A*, 2008, **1184**, 474–503.
57. J. Wu, W. Bicker and W. Lindner, *J. Sep. Sci.*, 2008, **31**, 1492–1503.
58. A. J. Shen, Z. M. Guo, L. Yu, L. W. Cao and X. M. Liang, *Chem. Commun.*, 2011, **47**, 4550–4552.
59. A. Shen, Z. Guo, X. Cai, X. Xue and X. Liang, *J. Chromatogr., A*, 2012, **1228**, 175–182.
60. H. D. Qiu, A. K. Mallik, T. Sawada, M. Takafuji and H. Ihara, *Chem. Commun.*, 2012, **48**, 1299–1301.

# Subject Index